L. F. Titov

WIND-DRIVEN WAVES

(Vetrovye volny)

Gidrometeorologicheskoe Izdatel'stvo
Leningrad 1969

Translated from Russian

Israel Program for Scientific Translations
Jerusalem 1971

Published Pursuant to an Agreement with
THE NATIONAL OCEANIC AND ATMOSPHERIC ADMINISTRATION,
U.S. DEPARTMENT OF COMMERCE
and
THE NATIONAL SCIENCE FOUNDATION, WASHINGTON, D.C.

Translated by D. Lederman
Edited by P. Greenberg

Printed in Jerusalem by Keter Press
Binding: Wiener Bindery Ltd., Jerusalem

Available from the
U.S. DEPARTMENT OF COMMERCE
National Technical Information Service
Springfield, Va. 22151

X/6/3

Contents

List of Russian Abbreviations v

Preface vii

INTRODUCTION 1

   1. Historical note 1

   2. Basic definitions. Wind-wave elements. State of the sea.
      Wave strength 6

Chapter 1. FUNDAMENTALS OF WAVE THEORY 19

   1. Small-amplitude potential waves 19

   2. Small-amplitude waves in fluids of finite depth 23

   3. Small-amplitude waves in fluids of infinite depth 31

   4. Long and short waves 33

   5. Some results of the theory of finite-amplitude waves 36

   6. Trochoidal wave theory 38

   7. Energy of a trochoidal wave 54

   8. Energy transport by wave motion. Group velocity of waves 57

Chapter 2. FUNDAMENTALS OF THE THEORY OF WIND WAVES 65

   1. Generation of wind waves 65

   2. Transfer of wind energy to waves and dissipation of this energy . . . 72

   3. Statistical regularities of wind waves 87

   4. Wave-generating factors 103

   5. Dependence of wind-wave elements on the wave-generating
      factors 112

   6. Variety of wind-wave elements and the variability of wave-
      generating factors. Regime-climatic distribution functions of
      wave elements 122

   7. The energy balance equation for wind waves 134

   8. Solutions of the energy balance equation for wind waves 141

   9. Decay of wind waves. Swell 152

10. The spectral method of wind-wave study 157

11. Wind waves in shallow water and at the coast. Wave
    refraction 164

Chapter 3. FUNDAMENTALS OF TECHNIQUES OF COMPUTING
           WIND-WAVE ELEMENTS 182

1. Purposes and substance of computations of wind-wave elements.
   Reliability considerations 182

2. Determination of wave-generating factors 183

3. Computation of wind-wave elements for deep-water seas 187

4. Computation of wind-wave elements in shallow water and
   shore shoals 221

5. Computation of climatic-regime characteristics of wind waves . . 225

Bibliography 239

LIST OF RUSSIAN ABBREVIATIONS

Abbreviation	Full name (transliterated)	Translation
AN SSSR	Akademiya Nauk SSSR	Academy of Sciences of the USSR
DAN SSSR	Doklady Akademii Nauk SSSR	Reports of the Academy of Sciences of the USSR
GGI	Gosudarstvennyi Gidrologicheskii Institut	State Hydrological Institute
GMI	Gidrometeorologicheskii Institut	Hydrometeorological Institute
GOIN	Gosudarstvennyi Okeanograficheskii Institut	State Oceanographic Institute
GUGMS	Glavnoe Upravlenie Gidrometeorologicheskoi Sluzhby SSSR	Main Administration of the Hydrometeorological Service of the USSR
IOAN SSSR	Institut Okeanologii Akademii Nauk SSSR	Institute of Oceanology of the Academy of Sciences of the USSR
LGMI	Leningradskii Gidrometeorologicheskii Institut	Leningrad Hydrometeorological Institute
MGI	Moskovskii Geodezicheskii Institut	Moscow Geodetic Institute
OGIZ	Ob"edinenie Gosudarstvennykh Izdatel'stv	State United Publishing Houses
ONTI	Otdel Nauchno-Tekhnicheskoi Informatsii	Department of Scientific and Technical Information
VNII	Vsesoyuznyi Neftegazovyi Nauchno-Issledovatel'skii Institut	All-Union Petroleum and Gas Scientific Research Institute

This book is a survey of theory and computation of wind waves in the sea. The theory of wind waves is presented, existing techniques for calculating wind-wave elements are discussed, and attention is given to their practical utilization.

The book is intended for specialists in oceanography, and may also serve as a practical textbook for shipbuilders, sailors, hydraulic engineers, hydrometeorologists and others whose work involves the effect of wind waves.

The book may also serve as a textbook for students of hydrometeorological institutes and universities attending courses in Physics of the Ocean, Sea Forecasting, and General Oceanography.

To the fond memory of
Vsevolod Vsevolodovich Timonov

The author

PREFACE

The aim of this book is to provide insight into the main problems and directions in the study of wind waves (wind-driven waves). It is based on lectures given by the author to students of oceanography at the Leningrad Hydro-meteorological Institute, as well as on results of the author's research.

Extensive use has recently been made of calculations of wind-wave elements. Therefore, in addition to fundamentals of wind-wave theory the book also gives attention to the simplest techniques for their computation.

Not all problems treated in the book are examined to the same extent, due somewhat to the book's limited scope. Only general concepts of many problems are presented, for example, of modern studies of wind-wave spectra. Methods of observing wind waves and data-processing procedures are not considered.

The bibliography contains only the most basic sources which are readily accessible and which themselves contain more extensive bibliographies.

The rough draft of the book was examined by Professor D. D. Lappo, who made a number of important modifications in its content and structure. Some chapters of the book were discussed with M. M. Zubova.

The author expresses his heartfelt thanks to the aforementioned, as well as to N. I. Egorov, who undertook the difficult task of editing the book.

L. F. Titov

May 1969

INTRODUCTION

1. Historical note

Leonardo da Vinci, famous naturalist and scientist of the fifteenth century, made a number of contributions to the understanding of wave motion, some of which he obtained from Arabic sources. One of the first compilations of wind-wave measurements was made by Marsigli (1725) and pertains to the waters of the Mediterranean. More systematic observations of wind waves in various regions of the World Ocean were carried out during round-the-world expeditions at the end of the 18th and beginning of the 19th centuries, some by Russian sailors.

Kotsebu, lieutenant in the Russian Navy (sailed round the world on the brig "Ryurik" in 1815—1818), remarking on the necessity of wave observations, stated: "The theory of this motion is as yet highly imperfect and the object of measurement is so transient and unsuitable for comprehension, that even ordinary determination of the length, height, width and velocity of these water masses, which acquire a large variety of shapes, may contribute to the communication of information in mathematical physics, which is also important for naval navigation" (Kotsebu, O. E. Puteshestviya vokrug sveta (Travels Around the World), 2nd edition. 1948).

However, the lack of suitable instruments at that time precluded any estimates of wave dimensions, save by sight. At the same time, the complexity of sea-wave processes and their high variability required the accumulation of a large variety of data. In addition, observed data at this time were not sufficiently objective. As a result, theoretical studies of wind-wave processes were based on formal postulates of hydrodynamics and were carried out for a long time without regard to observed data. This approach should be termed hydrodynamic.

The initial results of theoretical studies were contradictory, which did not contribute to their reliability. Newton (1687) was the first to apply mathematical analysis to the solution of the wave problem. A century later, in 1776, Laplace found that wave oscillations in water are formed by water particles moving along elliptical orbits which flatten out toward the bottom, while at the very bottom they degenerate into straight lines. At the same time Lagrange concluded that wave velocity depends only on depth, and not on the ratio of wave length to the depth, as postulated by Laplace.

Gerstner (1802) developed trochoidal wave theory, according to which particles in water of infinite depth move in circular orbits and wave speed depends only on the length of the wave.

Thus two theories appeared at the beginning of the 19th century. According to Laplace water particles moved in elliptical orbits and according to Gerstner, in circular orbits.

The following stage in the development of wave theory is the work by
Poisson and Cauchy (1815), both of whom concluded that wave speed does
not depend on wave length, and that waves propagate with increasing speed.

Subsequently these contradictions were eliminated, when it became
possible to establish the limits of applicability of these theories and to
determine their interrelationship. This was done in later studies by Green
(1839), Stokes (1847), Rayleigh (1878), Reynolds (1878) and other prominent
mathematicians and physicists of the 19th century.

Their studies constitute the completion of investigations into waves of
stable shape and of infinitesimal amplitude, so-called linear wave theory.
Stokes studied wave motions on the assumption of finite wave amplitude
(nonlinear theory). He proved a number of important properties of wave
oscillations, including the existence of water-mass transport. Stokes'
solutions were approximate. A rigorous proof of the existence of waves
studied by Stokes was given by Nekrasov (1921) and somewhat later by
Levi-Civita (1925). These studies are of major importance, since lack of
a rigorous solution of the Stokes problem would cast doubt on the suitability
of approximate solutions for describing the wind motion under study.

It may at present be assumed that the establishment of nonlinear wave
theory for deep water is completed. On the other hand, only individual, not
sufficiently interrelated elements of the hydrodynamic theory of waves were
developed for shallow water; these include the linear theory of progressive
waves, nonlinear theory of waves of steady shape at constant depth, nonlinear
theory of long waves of unstable form. Hence a great deal of work is done
on developing a hydrodynamic theory of waves which would unite all existing
particular solutions.

From the physical point of view the generation of wind waves in the
primary and most general form was studied by Kelvin (1871) and somewhat
later by Helmholtz (1895). Thirty years later Jeffreys (1924, 1935) developed
a more refined theory of generation and development of wind waves for
conditions actually observed at sea.

As long as the contradiction in theoretical results was not resolved,
major contributions to our knowledge of wind waves were made by experi-
menters. Work by Scott Russel (1837—1840) and Airy (1843) resulted in
the discovery of a number of very important phenomena, associated primarily
with propagation of long waves, which exerted a major effect on the future
development of studies into the wave problem as a whole. Observations of
the actual dimensions and forms of wind waves were for a long while insufficient for checking theoretical results. Determination of the dimensions
of sea waves was limited only to measurement by sight of their height, which
in a number of cases yielded contradictory results.

At the same time we should note one event of this period which is of great
importance in the accumulation of data on sea waves. This is the International Meteorological Congress which took place in London (1874) and which
established a "sea-state scale." This information started to be recorded
in ship logs together with data on wind force and direction.

However, all the observed data accumulated at this time could not
satisfy the ever-increasing requirements, nor were they sufficient for
verifying theoretical results. Subsequent investigators continued the
refinement of observational methods with the purpose of attaining the highest
possible objectivity of results.

Abercrombie (1888) combined visual observation of waves in the South
Pacific with the use of a sensitive aneroid for determining wave height,
following a suggestion by von Neumayer. This approach was followed also
by other investigators. The aneroid was subsequently used by Shuleikin
(1926) in a microbarograph, which was used to obtain interesting results
when sailing from Odessa to Vladivostok. Methods of visual observations
were perfected to the highest degree by Cornish (1934), who obtained very
objective data using these observational methods.

Froude (1865), making use of the rapid damping of wave oscillations
with depth, used a specially designed floating staff for assessing wave height.
Similar and quite thorough observations were subsequently carried out by
Stevenson (1851) at the shores of England and by Gaillard (1904) at American
lakes and at the Atlantic shores of the U.S.A.

Berezkin (1929, 1930), Bogdanovich (1930) and Loidis (1936) greatly
improved methods of shore observations of waves in USSR seas. Observational methods were subsequently improved by Shuleikin (1934), who
developed "optical wavemeters" and various other instruments and techniques
for observing waves at shores of USSR seas.

To determine the true form of waves in the sea and to accumulate more
objective data on their dimensions, a number of investigators used stereo-
photogrammetry, i.e., exact photographic surveying of waves. Such studies
of waves were carried out in the Indian Ocean in 1906 by Coolshutter and
in the Atlantic Ocean in 1929 by Schumacher.

The first stereophotographs of waves in Russia were obtained by Krylov
in 1907 in the Black Sea. This work was continued in 1931 by the State
Hydrological Institute (Titov) in the Black Sea and other seas of the USSR
(1932—1940), as well as by ships in the open sea and at shores. First
Soviet phototheodolites for sea-wave surveying were constructed in 1937
(Titov, Korshak).

Methods of wave stereophotometry were developed in 1946—1950 by
Pugin, Rekhtzamer, Kudritskii and others. Subsequently stereophotography
of waves has been carried out in the USSR by aircraft, and has been used
even more extensively in connection with work by Cherkasskii, Sharikov and
others. An original photographic wave recorder was designed by Ivanov
(1953).

The State Hydrological Institute was the first (in 1935) to carry out
synchronous observations of waves (Titov, Bruns) involving 35 shore
hydrometeorological stations and more than 30 ships.

The desire to coordinate wave studies in individual countries caused the
14th International Maritime Conference in 1926 to establish an International
Committee on the Study of the Effect of Waves on Structures. The first
session of this Committee was held in 1928.

At this time, wave recorders capable of continuous recording of wave
height and period were introduced. Instruments of this kind were designed
by Bruns (1934), Telyaev and Morozov (1948), Vilenskii and Glukhovskii
(1951) and others. Extensive modifications were made in wave-pressure
recorders. The construction of automatic instruments such as wave-height
and wave-pressure recorders allowed very important characteristics of
wind waves to be obtained. Special wave-gauging stations were set up for
obtaining data on waves in shore waters, of particular interest to hydraulic
engineers. These stations are located in Dieppe (France), Genoa (Italy),
and Leningrad (1932), Katsiveli (1935) and Sochi (1948) (USSR).

Observations under natural conditions at USSR seas and lakes were carried out by Andreyanov (1939), Ivanov (1938), Sysoev (1936), Titov (1931—1940), Shishov (1938—1948) and others.

At the same time additional techniques were developed for laboratory studies of wave motion of fluids. These are modeled in special open and closed channels and tanks, some of which are quite large. The waves are produced by wave generators, and the air flow [above the water] is also simulated. One of the largest, ingeniously constructed wind-wave tanks is the "storm" tank, constructed by Shuleikin (1948) at Katsiveli (Crimea). As a result of experiments carried out with this tank Shuleikin developed in 1954 a new theory of the origin and development of wind waves, which replaces the Jeffreys theory.

At the beginning of the fifties the number of studies of wind waves increased together with the overall development of oceanographic studies.

The use of various instruments for observing wind waves at natural and water-storage lakes, seas and oceans, the perfection of methods and scales of laboratory experiments, all supply new information on this natural phenomenon. The desire to gain insight into waves has resulted in various hypotheses presented in the form of physicomathematical theories, which require further checking and refinement.

At this time it became obvious, on the basis of instrumental measurements under natural conditions, that consideration must be given to the large variety of sizes and forms of wind waves. This variety was first expressed by certain phenomenological statistical regularities (Selyuk, 1950; Morozov, 1953; Vilenskii and Glukhovskii, 1955), and subsequently from the standpoint of probability theory (Krylov, 1956; Brovikov, 1954).

The variety of wind waves and the laws governing them not only markedly affected the methods of observing wave elements, but also showed that wind-wave processes must be expressed simultaneously by both hydro-dynamic and statistical relationships. Presently the best studied statistical features of wind waves are their statistical characteristics in deep water. Information on the statistics of wind waves in shallow water and at the shore, as well as on swell waves, is much more scarce.

A great deal of attention has been paid lately to statistical characteristics of wind waves over extended periods (month, season, year). These studies resulted in wind-wave atlases, giving the distribution of wind waves of different sizes for the majority of Soviet seas and for individual oceans. The construction of practical aids is preceded by studies of regularities in the distribution of wind-wave elements over extended time periods, namely, climate-regime distribution functions. The development of methods for calculating these functions forms a large proportion of the work involved in atlas compilation. Work in this field was carried out by Zubova (1956), Davidan (1961, 1965) and others.

An important step in the study of the origin and development of wind waves was the use of the wind-wave energy equation, first obtained by Makkaveev (1937). This equation was used in many recent studies, both in the Soviet Union and elsewhere. Another feature of these "energy" studies is that they employ various methods of investigating wind-wave processes, i.e., alongside with theoretical methods they use results of laboratory experiments and correlations of natural observations. The most complete and rational combination of these methods was developed in fundamental studies of Shuleikin (1956) on the kinematics, dynamics and computation of wind waves.

An important prerequisite for the development of these studies was the undertaking of many experimental observations of wind-wave elements. It was mentioned previously that stereophotography of waves as well as wave recorders are being extensively used. Numerous Soviet oceanographic expeditions during recent years on the "Ob," "M. V. Lomonosov," "Slava" and other vessels measured wind-wave elements in oceans. Similar observations have also been performed in various seas. This information, together with previously obtained data, made it possible to find empirical relationships between wave elements and wind characteristics, to determine certain constants, and to gain insight into given features in the generation of wind waves. Results obtained by Ivanov (1953), Labzovskii (1952), Titov (1951) and others were found suitable, to one degree or another, for use in solving equations of wind-wave energy balance. Results of laboratory experiments by Bonchkovskii (1954), Dobroklonskii (1947), Kondrat'ev (1953), Kononkova (1953) and others are extensively utilized. Similar studies, based on the wave-energy balance equation, were carried out in the West (Sverdrup and Munk (1947), Neumann (1953), Bretschneider (1959) and others).

The aforementioned Soviet studies on the development of wind waves yielded results suitable for computing wave elements under specific weather conditions. Titov (1951), and Krylov and Sorkina (1960) wrote handbooks on calculating wave elements. Similar handbooks have also been published lately in the U. S. A., France and England.

In spite of significant achievements in the study of the origin and development of wind waves, based on solving equations of their energy balance, these studies are as yet incomplete. Substantial contradictions exist in estimating and interpreting the processes under study. In particular, it is apparently necessary to use direct observations to determine, under natural conditions, the mechanism of wind-energy transfer to waves and of dissipation of this energy. It is necessary to refine a number of constants, for example, those pertaining to energy dissipation in wave motions. The damping of waves and generation and propagation of ripples have not been sufficiently studied.

Establishing the probability nature of wind waves served as a point of departure for a new approach to the study of wind-wave processes—the spectral method of studying this phenomenon. Underlying this approach is the description of visible wind waves as an ensemble of sinusoids. The energy distribution among these elementary waves is characterized by the energy spectrum. Study of the energy spectrum of sea waves enables one to approach the analysis of the physical substance of wind-wave processes. The random nature of the combination of these individual elementary waves allows one to treat this phenomenon as a probabilistic process and to apply the theory of random processes.

This trend was started by American scientists (Pierson, 1955 and Longuet-Higgins, 1957). The formation of the wind-wave spectrum is discussed by Phillips (1957) and Miles (1960). Similar studies were carried out in the Soviet Union by Firsov (1958), Rakhmanin (1958), Krylov (1967) and others.

It should be noted in surveying studies of wind waves during recent years, that the main problem (gaining insight into the generation of waves by wind) has not been completely solved. No qualitative expression was obtained for the mechanism of forming wind waves of different size and form. This

variety apparently results from the complex structure of wind flow. Henc[e]
direct observations are needed of the interaction of processes in the surfa[ce]
layer of the atmosphere and wave oscillations in the sea. This problem
should be solved by statistical hydrodynamics, in order to efficiently com-
bine wave statistics and energetics with the hydrodynamics of wave motion[.]
Techniques for calculating wind-wave elements, particularly those used
for water bodies with complex underwater topography and complex shore
outlines, are still insufficiently refined. Information on the dimensions of
wind waves in many parts of the World Ocean is as yet incomplete. Techniques for compiling various practical aids, atlases, etc., require refinemen[t]

2. Basic definitions. Wind-wave elements. State of the sea. Wave strength

The high mobility of water is responsible for the generation of oscillatory motions in the sea. The propagation of the latter in space is termed
w a v e s. Waves appear in the sea for a number of reasons. W i n d w a v e[s]
are generated by winds; a n e m o b a r i c w a v e s (s e i c h e s) are generat[ed]
by relatively rapid variations in atmospheric pressure over the sea;
s e i s m i c w a v e s (t s u n a m i s) are due to seismic activity of the sea
bottom; t i d a l w a v e s are generated by astronomical forces.

In the study of sea waves, irrespective of their origin, consideration
must be given to two principal forces: gravity and the deflecting force
produced by the earth's rotation. Only when considering waves several
centimeters long (capillary waves) is it necessary to make allowance for
surface tension, but such waves have no practical significance in the sea.
Sea waves are gravity waves. The deflecting force produced by the earth'[s]
rotation should be taken into account primarily in the study of very long
waves, for example, tidal waves.

Sea waves generated by various mechanisms differ from one another in
form, and consequently their elements also differ. This difference can be
seen most graphically by comparing wave periods (Figure 2.1).

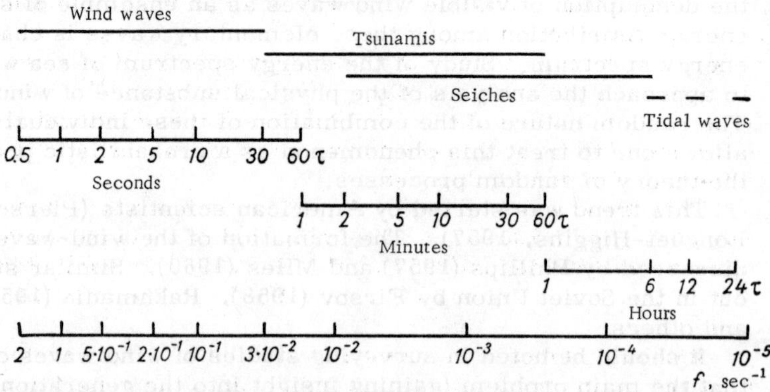

FIGURE 2.1. Spectrum of wind periods encountered in seas.

The lowermost scale shows wave frequency $\left(f = \frac{1}{\tau}\right)$.

Thus, ripples have a period shorter than 1 sec. The period of wind waves does not exceed 30 secs. Seiches and seismic waves have periods from minutes to several tens of minutes. Periods of tidal waves are measured in hours. Wind waves can be classified as short-period waves and tidal waves as long-period waves.

Subsequently we shall consider only wind waves, which play a significant role in a number of ocean processes. In the open ocean on the average the fifty-meter water layer is mixed by wind waves. In addition to redistribution of heat, wave mixing adjusts the distribution of oxygen and other gases in sea water, as well as the salinity, involving also the density of water masses. Wind waves have a marked effect on evaporation of water from oceans, primarily because strong wind waves lift minute water droplets into the air. Turbulent air motion lifts them further into layers with lower water vapor content, where they evaporate. At the same time salt crystals are carried into the atmosphere; due to their hygroscopic nature they become, together with negative ions of the air, excellent condensation nuclei, which are needed for the formation of fog and clouds. At the same time this process determines the chemical interaction between ocean and atmosphere. Wind waves increase the sea surface by up to approximately 5 % (depending on wave intensity) as compared with the area during calm, thus increasing the total amount of evaporated moisture.

The complex, often irregular topography of the sea surface produced by wind waves, due to fluctuations of wind flux above the sea surface, in its turn affects the wind flux, producing in it additional, higher-order oscilla-tions. This interaction of horizontal fluxes of air masses and of oscillatory processes at the sea surface is apparently the cause of many, as yet not entirely clear, interactions at the interface between the two moving shells of our planet. Atmospheric and sea disquiet are two interrelated processes.

Wind waves affect the propagation of light and sound in water and reflect radio waves; they are also responsible for the generation of microseismic disturbances of the earth's mantle.

In addition to heat conversion, the enormous mechanical energy of wave motion is expended primarily on degrading the sea bottom in relatively shallow coastal regions and on shore erosion. In spite of its slow rate, this process occurs on a grandiose scale. The products of shore erosion (bottom sediments) are moved by waves over hundreds of miles, and determine the shape of shores and the topography of shallow shore regions. To this de-structive and, at the same time, creative activity of wind waves one should add their contribution to the destruction of ice cover. In all ice-covered seas, the destruction of ice cover by wind waves is a common spring and winter occurrence.

The above comments pertain to the role and significance of wind waves in processes occurring in oceans.

As to the practical significance of wind waves, man encountered this phenomenon when his activity spread to the seas. At present there is virtually no branch of engineering related to the sea in which allowance for wind waves is not a condition for successful solution of given engineering problems, for example, hydraulic structures, maritime transportation (shipbuilding and operation), mining of minerals from the sea bottom.

Wind waves are subdivided into forced, free and mixed. Forced waves are waves generated by wind and continuously sustained by it. Free wind waves or swell is the term applied to waves which have

escaped the influence of the generating wind upon a strong reduction in its
strength or complete cessation of wind, or upon passage of forced wind
waves from the region in which they originated into another, wind-free
region. Mixed waves are wind waves which form as a result of the
simultaneous existence of swell and forced wind waves.

Wind waves can also be subdivided into regular and irregular.
The former, unlike the latter, do not change dimension and form at the
same point in the sea over time intervals equal to the wave period.

If the orbits of water particles involved in wave motion are directed
parallel to some vertical plane, coinciding with the direction of wave
motion, but do not depend on the distance from it, then these waves are
termed two-dimensional or plane waves, as distinguished from
three-dimensional or spatial waves, in which the water particle
orbits do not lie in parallel vertical planes.

The superimposition of two or more waves is termed wave interference, and the resulting waves are called interference waves.
The superposition of identical two-dimensional waves traveling toward
one another results in the formation of two-dimensional standing
waves. The interaction of waves traveling in several directions produces
a complex, disordered wind motion.

Tidal waves are traveling waves which continuously or periodically
convey breakers (foaming vortices), as a result of partial breakup due to
small depths in the tidal strip of seas or in shallow waters. Wave motion
is termed developing, developed (or established) and decaying,
depending on whether the averaged value of elements of not less than 100
waves following one another through a given point of the body of water
increases, remains unchanged or decreases.

The wave forms and dimensions are described by the wave elements. The
principal geometric elements of two-dimensional (plane) waves (Figure 2.2)
are wave height and length. Using thse two elements we may define wave
steepness (δ) as the ratio of its height (h) to its length (λ):

$$\delta = \frac{h}{\lambda}. \tag{2.1}$$

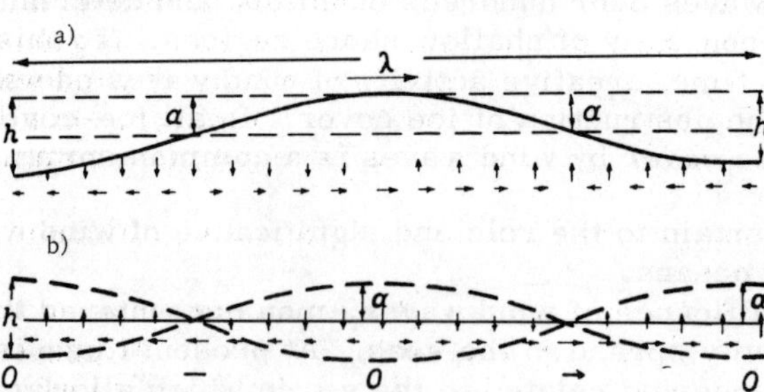

FIGURE 2.2. Progressive [traveling] (a) and standing (b) waves.

The arrows show the direction of motion of fluid particles. α is
the wave amplitude.

The kinematic elements of a wave are its period (τ) and speed (c).

Below we define elements of two-dimensional progressive waves.

W a v e   h e i g h t (h) is twice the amplitude of oscillations, and is deter-mined by the distance from the bottom of the wave trough to the tip of the crest.

W a v e   l e n g t h (λ) is the horizontal distance, in the direction of wave motion, between two particles in the same oscillatory phase. It is deter-mined by the horizontal distance between two successive crests or troughs.

D i r e c t i o n   o f   w a v e   a d v a n c e , or w a v e   r a y   , is defined by a line directed perpendicular to the wave front in the direction from which the wave moves. It is measured in compass points with respect to that point from which the waves travel.

W a v e   p e r i o d (τ) is the time during which the water particles trace out their orbit. . The wave period is determined by the period of time needed for passage of two successive wave crests through the same point on the sea surface.

W a v e   v e l o c i t y (c) is the distance traveled by the wave profile per unit time. It is determined by the distance traveled per second by the wave crest in the direction of wave advance.

The height, length and period of a wave, which define its principal elements, are termed wave parameters. As is known,

$$\lambda = c\tau. \tag{2.2}$$

W a v e   p r o f i l e is the vertical cross section of the wave surface in the direction of wave travel. W a v e   c r e s t is that part of the wave lying above the neutral level, i.e., above the water level in a calm sea. W a v e   r i d g e is the term applied to the uppermost part of the wave crest, while the t r o u g h is the lowest point of the wave hollow. The latter is that part of the wave which lies between two crests below the neutral level. *

The sea surface under the effect of wind waves has a complex topography. This is seen from photographs and is particularly noticeable in sea-surface plans, constructed by stereophotography (Figure 2.3). The main feature of this topography is that wave crests and troughs frequently do not alternate in orderly fashion and are of limited length. Hence, strictly speaking, there is no such thing as a regular two-dimensional wind wave. The limited length of wind-wave crests made it necessary to introduce a new character-istic length, the c r e s t   l e n g t h (l_{cr}).

Figure 2.4 shows a small segment of sea surface (approximately 10,000 m^2) with wind waves of strength V on the wave scale (in Table 2.2) and wave-surface profiles obtained by stereophotography. The profiles are drawn to an arbitrary scale at 10 m intervals in the general direction of wave motion (from left to right in Figure 2.4). The profiles show eleva-tions of wave surface in meters for the highest and lowest points of each profile. These elevations are measured from an arbitrarily assumed zero surface. The profile generatrix corresponds to an elevation of 0.75 m, this surface being taken as the neutral sea level. Its location is usually

---

* [Usually no distinction is made in English-language literature between wave crest and wave ridge, or between wave trough and wave hollow. Hence subsequently only the terms crest and trough will be used, as is standard practice.]

defined as the difference between the highest and lowest elevations of the
wave surface. The areas of wave profiles below and above this line are
approximately equal to one another. That part of the wave which projects
above the neutral sea level is termed the wave crest. The crest contours
in Figure 2.4 are outlined by thin lines coinciding, for each profile, with
the neutral level. The term w a v e c r e s t l i n e or w a v e f r o n t is
applied to the horizontal line through the peaks of crests and coinciding,
at its initial points, with the neutral-level plane. The length of this line is
the wave f r o n t l e n g t h.

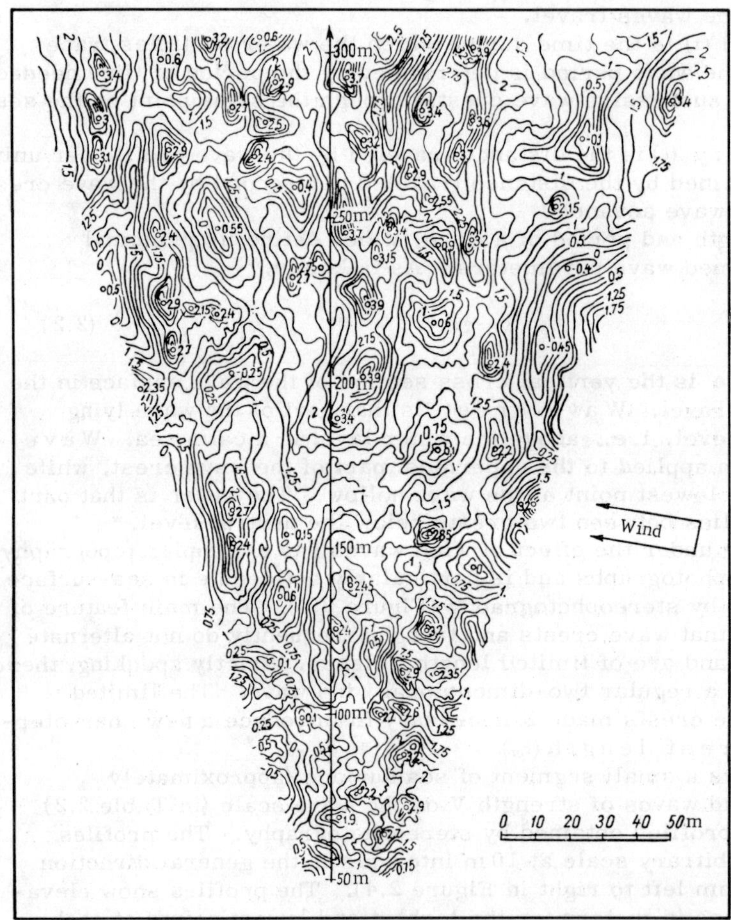

FIGURE 2.3. Wind waves on the sea surface (stereophotograph). Strength V on the
wave scale [in Table 2.2].

Horizontals were drawn at 0.25 m intervals. Figures at points denote elevations, in
m, above an arbitrary zero level (positive values pertain to points above this level,
and negative values to points below this level).

Wave fronts in Figure 2.4 are delineated by heavy lines. The w a v e
t r o u g h is that part of the wave between crests and lying below the calm
water level. The dashed line in Figure 2.4 shows the deepest part of the
trough, i.e., approximately the location of its bottom. It is easy to see
from this figure that the wave crests are of different shape in the plan and
profile, differing in their dimensions. For example, the wave crest at the
central part of the figure is clearly delineated; it passes out of the figure.
Its length is apparently in excess of 100 m. Wave crests to the left of it
are much shorter, correspondingly ~ 25 and ~ 30 m. The wave crest to the
right at the bottom of the figure is round and of length $15 - 20$ m.

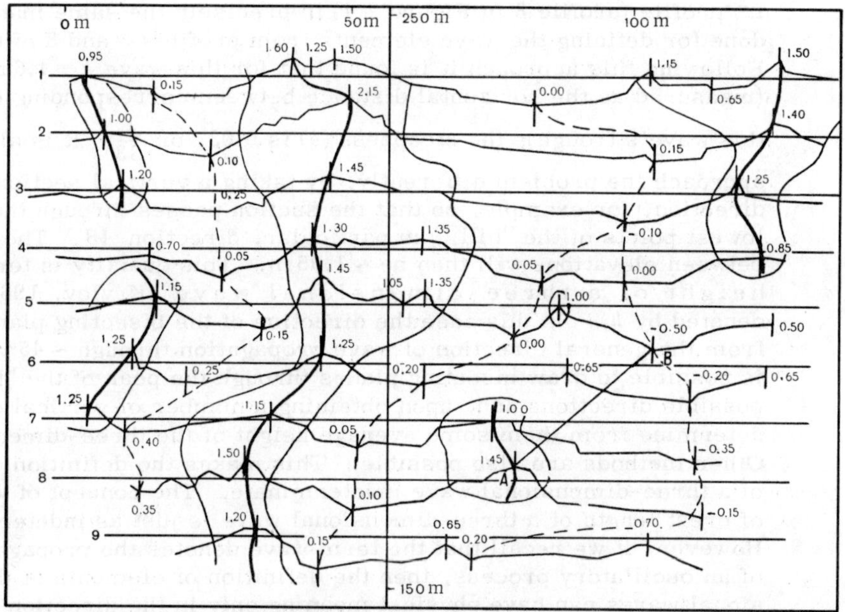

FIGURE 2.4. Wind waves at the sea surface (stereophotograph). Strength V on the wave scale [in
Table 2.2].

The figure is explained in the text.

Together with its surrounding trough this wave is of a characteristic
"hill" shape (Krylov, 1966). Such a "hill" can be treated as a three-
dimensional wave. However, the waves in the upper part of Figure 2.4
are closer by their outlines in the plan and profiles to two-dimensional
waves (for example, profile 2). The definitions of wave elements given
for two-dimensional waves can be extended to these waves, although obviously involving some approximation.

The wave height represented by profile 2, which passes through the topmost part of the crest (2.15 m) and through the lowest part of the trough of
this very wave (0.10 m), is 2.05 m ($h = 2.05$ m). The wave length, measured
as the horizontal distance between neighboring wave crests (Figure 2.4),

is $\lambda = 38$ m, while the wave steepness $\delta = 0.053$; or $\dfrac{h}{\lambda} = \dfrac{1}{19}$.

It is seen by examining profile 3, which adjoins profile 2, that the vertical section of the sea surface in this case did not pass through either the topmost point of the crest or the lowest part of the trough. The difference between these elevations is also termed wave height, the latter then being treated as relative to the given wave profile. The wave height for this profile is 1.20 m and the length is 35 m. Consequently, $\delta = 0.034$; or $\frac{h}{\lambda} = \frac{1}{28}$. One's attention is attracted by the marked difference in wave elements determined by neighboring profiles.

The elements of the three-dimensional wave singled out in Figure 2.4 can be defined in two ways. Assuming that it propagated along or close to the general direction of wave propagation, its elements can be defined using its profile (profile 8 in Figure 2.4) in precisely the same manner as was done for defining the wave elements from profiles 2 and 3 of this figure. Following this approach it is found that for this wave $h = 1.6$ m and $\lambda \approx 55$ m (measured as the horizontal distance between corresponding elevations of this wave's trough); the steepness (δ) is 0.03, or $\frac{1}{33}$. It is also possible to approach the problem differently, by taking a vertical section in any other direction, for example, so that the section passes through the highest and lowest points of the "hill," in particular, direction AB. The difference between elevations will then be ~ 1.95 m. This quantity is termed the height of a three-dimensional wave (Krylov, 1966) and is denoted by h_{3D} In this case the direction of the bisecting plane will be shifted from the general direction of wave propagation through $\sim 45°$. Finally, it is possible to draw bisecting planes through the peak of the "hill" in all possible directions and, upon obtaining a number of vertical distances, to determine from them some average height of the three-dimensional wave. Other methods are also possible. This makes the definition of the height of a three-dimensional wave indeterminate. The concept of wave length and of crest length of a three-dimensional wave is just as indeterminate. However, if we recall that the term wave denotes the propagation in space of an oscillatory process, then the definition of elements of three-dimensional waves can have physical meaning only in the direction of propagation of such waves. This direction for individual waves does not differ by more than 25° from the general direction of motion of the remaining waves. In all other cases, when defining the difference in "hill" elevations it is possible to refer to differences in level fluctuations.

The extension of the concept of a three-dimensional wave to wind waves depends on the extent to which these waves are typical of wind-wave motion. This question is controversial. The validity of always identifying real wind waves with three- or two-dimensional waves is open to discussion.

It is the common experience of seamen and numerous investigators that the form of wind waves depends, on the one hand, on the effect exerted on them by the wind (i.e., on the latter's velocity) and on the wave development stage, and on the other hand, on morphometric conditions, i.e., on sea depth, shore outline and such phenomena as, for example, strong currents.

Since all these conditions exist in infinite variety, the form of real wind waves is also variable. However, it can be noted that the principal factors in the open sea are the velocity of the prevailing wind and the stability of its direction, while under shore conditions an additional factor is the effect of sea depth.

The state of the sea, which is expressed by a scale from 0 to 9, has for a long time now served seamen in assessing wind strength. At present this estimate has lost its meaning, but it is useful to present it in conjunction with the question of variation in forms of wave crests and of the waves proper when subjected to the wind's force (Table 2.1).

TABLE 2.1. Sea-state scale* (Titov, 1952)

Scale	Sea state	Sea criteria
0	Calm	Mirrorlike
1	Smooth	Ripples
2		Small wavelets, crests have glassy appearance rather than being topped by white foam
3	Moderate	Large wavelets, crests begin to break, scattered whitecaps
4		Wave shapes clearly delineated, whitecaps everywhere
5		High wave crests, their foaming ridges continuously elongating, white foam from breaking waves
6	Rough	Long storm waves begin to form and white foam from breaking waves begins to be blown in streaks along wave slopes
7		Foam blown in well-marked streaks lines the wave slopes; at places the streaks join and reach the wave troughs
8		Wide, ever-combining foam streaks line the wave slopes, the sea surface takes on a white appearance; foam-free patches observed only locally in some wave troughs
9	Exceptionally rough	Entire sea surface covered by dense foam, air filled with foam and spray, visibility sharply curtailed

* In effect for hydrometeorological observations in seas, natural lakes and large water-storage lakes as of 1953.

Examination of Table 2.1 shows that starting at approximately scale 6 the appearance of long-crested storm waves is expected. Obviously, such long-crested waves do not occur in seas below scale 6; on the other hand, it is emphasized that the waves are small and that their crests break up. Mention of clearly delineated waves is made only at scale 4. Apparently this indicates that developing waves are more frequently three- rather than two-dimensional. During strong winds, particularly when winds blow for a long time, the waves are most frequently two- rather than three-dimensional.

This point of view is verified by inspection of presently available numerous plan views of sea waves, as well as of aerial surveys and photographs. For example, Figure 2.5 reproduces the sea surface near the shore with 10 m/sec winds, when the wind waves are at the initial development stage. Depths range from 15 m at the right and top parts of the figure, to 5 m and less at its left side, at the pier, which is represented by the barbed line. The pier projects +2.7 m above mean sea level. Wave crests delineated by solid lines, particularly 100 m and more away from the pier, show three-dimensional formation. These waves are unaffected

here by sea depth, unlike waves near the pier, where the wave crests are
aligned along a front, three-dimensionality features having been lost.

Figure 2.6 reproduces the sea surface at the same shore region with a
25 m/sec wind. There is a glaring contrast between this figure and
Figure 2.5. The waves have a clearly delineated two-dimensional form.
Several profiles of the crests of these waves and profiles of the waves pr
are shown in Figure 2.7. Departure from two-dimensionality is noticeab
only close to the pier, due to the reflection of waves from it. The wave
fronts, however, are still noticeable and retain their shape and orientatic
which is perpendicular to the general direction of wave propagation.

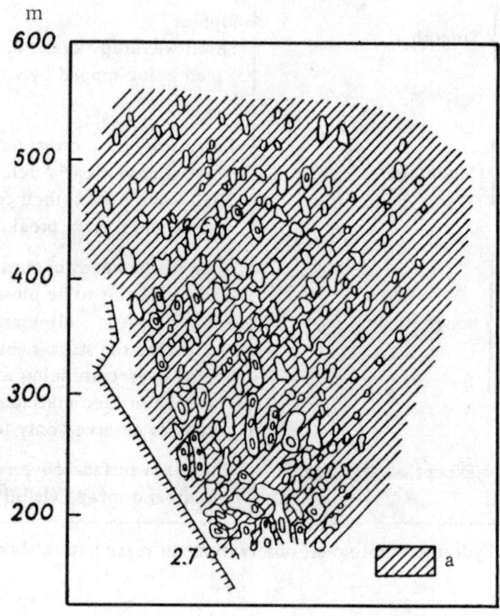

FIGURE 2.5. Sea surface with a 10 m/sec wind.
Stereophotograph of the sea near the shore:

a denotes that part of the sea surface located
below the neutral oscillation level, corresponding
in this figure to the 0.0 m elevation, i.e., to mean
sea level.

Figure 2.8 shows scale VIII wind waves near Cape Horn. The two-
dimensionality of waves is quite clearly expressed. Figure 2.9 is an aeri
photograph of a sea surface, in which it is impossible to regard all the
waves as three-dimensional.

It can apparently be concluded from the above examples that, when wa
are generated in the open sea as a result of wind of sufficiently long durat
their form in the overwhelming majority of cases is closer to two- rather
than three-dimensional. This is especially so when wind waves propagate
near shores. Hence the identification of wind-wave profiles with those of
two-dimensional waves is apparently quite valid in many cases.

As a measure of three-dimensionality of waves one may use a three-
dimensionality factor (Krylov, 1966), defined as the ratio of wave length to
the length of its crest:

$$k_{3D} = \frac{\lambda}{l_{cr}} \, . \tag{2.3}$$

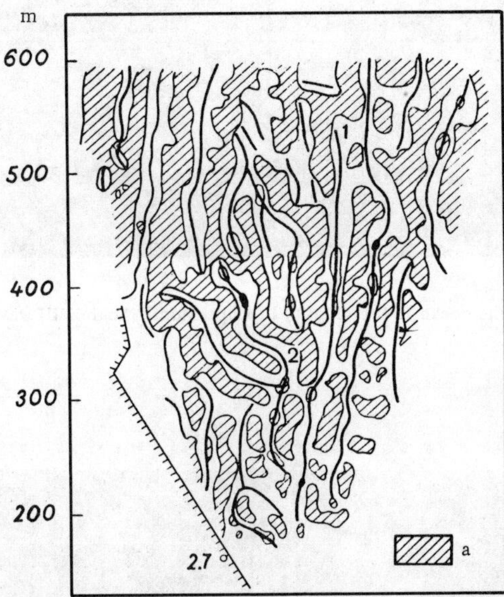

FIGURE 2.6. Sea surface with a 25 m/sec wind. Stereophotograph of the region shown in Figure 2.5:

a denotes that part of the sea surface located below the
neutral oscillation level, corresponding in this figure to
the + 0.5 m elevation above mean sea level. Heavy lines
delineate the highest wave crest elevations.

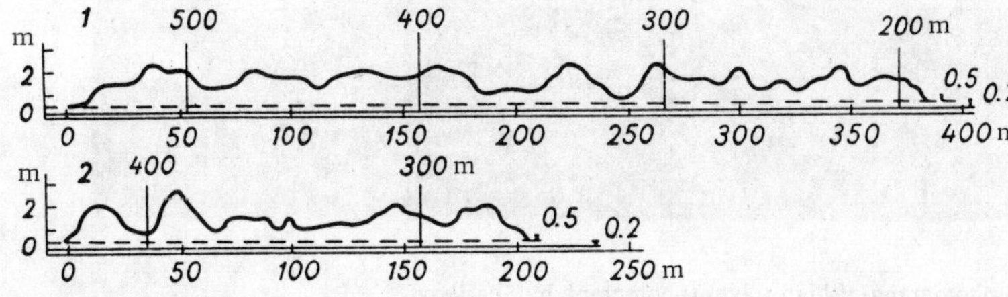

FIGURE 2.7. Profiles of wave crests shown in Figure 2.6 (1 and 2).

Vertical lines marked 500, 400 m, etc., correspond to distances along the direction of
the photograph in Figure 2.6. The mean oscillation level is 0.5 m and the [mean] sea
level is 0.2 m.

FIGURE 2.8. Wind waves in the ocean in the vicinity of Cape Horn. Scale VIII waves.

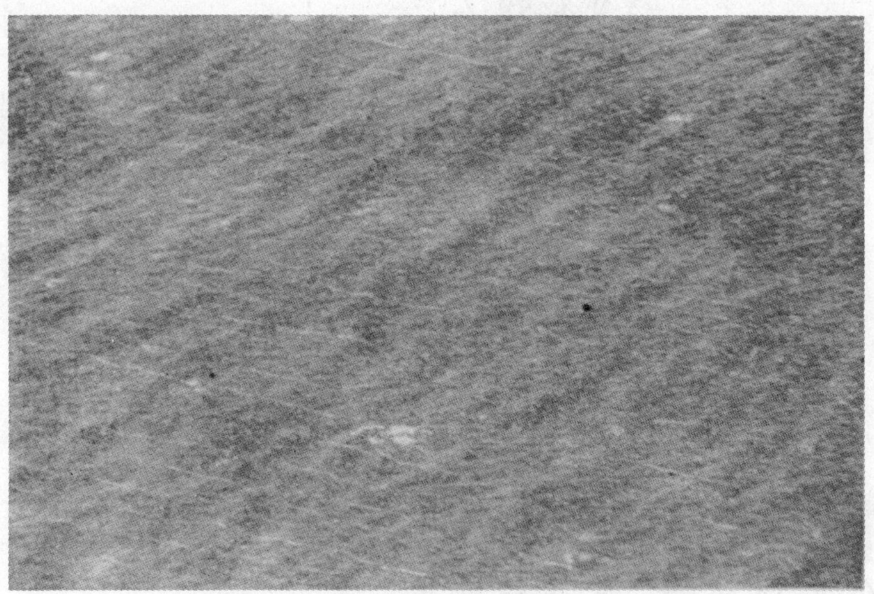

FIGURE 2.9. Aerial photograph of high waves (photograph by Sharikov).

For two-dimensional waves, subject to the rigorous definition of this term,

$$l_{cr} \to \infty,$$

in which case

$$k_{3D} \to 0.$$

For virtually all cases of maximum three-dimensionality of wind waves k_{3D} on the average does not exceed unity.

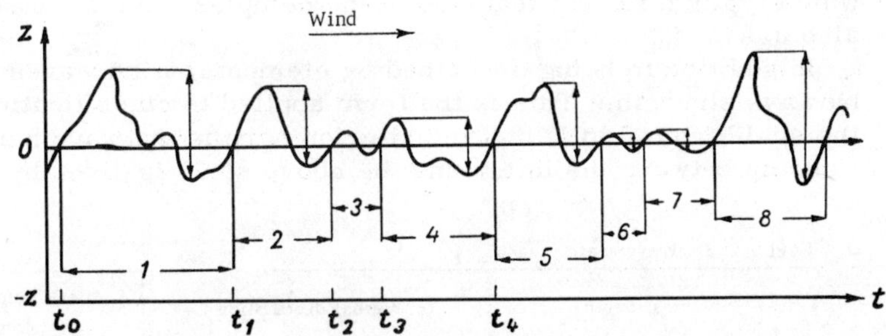

FIGURE 2.10. Recording wave oscillations at a point using a wave recorder (after Glukhovskii).

Returning to the previously detected difference in the values of wind-wave elements measured by wave profiles (Figure 2.3), allowance must be made for both the variety of their magnitudes and their arbitrariness and indeterminacy. The latter is not reduced even when the wave elements are estimated by means of automatic wave recorders, which record variation in sea level at a given point of the sea surface (such as wave-height staffs, gauges, and so on). The form of recording is shown in Figure 2.10. The level variation is expressed by a complex curve with individual maxima (peaks) and minima (valleys) relative to the mean wave line. Its location is defined as the difference between the highest and lowest curve heights. The term individual wave is usually applied to oscillations such as those located between two neighboring intersections of the mean wave line of the leeward (or windward) slope. Horizontal arrows in Figure 2.10 represent individual waves, numbered 1, 2, etc. The vertical distances, representing peak-to-peak level oscillations, are termed wave height. The wave period is defined as the difference between the recordings of the time recorder (bottom horizontal line in Figure 2.10), corresponding to points of successive intersection of mean wave lines with wave slopes.

The measured wave heights and periods are significant if, as mentioned above, sea-level oscillations are recorded in the direction of wave propagation. This is not always possible. In addition, both when determining wave height from wave profiles of Figure 2.4 and in the case considered here, there is no assurance that the given point on the sea surface will always be traversed by the uppermost part of the crest or by the lowest part of the trough. The estimate of wave period is just as indeterminate.

The advent of spectral methods in the study (Chapter 2, Section 10) resulted in new definitions of wave elements (Longuet-Higgins, 1962). For example, instead of heights of three-dimensional waves, discussed above, one considers vertical deviations of the maxima and minima of the undular surface from the horizontal plane. Wave length is defined as twice the distance of successive intersections between the wave profile and the mean horizontal plane. These points are the zeros. The direction in which this average length is minimum is termed the general direction of wave propagation. The speed of three-dimensional waves is identified with the rate of displacement of the zeros of the wave profile in the general direction. Twice the time interval between two successive traverses of the level through its mean position is taken as the wave period. Other definitions, which typify different features of the complex wave-field structure, are also used.

In addition to being described by elements, wind waves can be classified by wave strength. This is the term applied to classification of waves on the scale from 0 to IX. The criterion here is the wave height. The relationship between the latter and the above scale is given by Table 2.2.

TABLE 2.2. Wave scale*

Wave strength	Verbal description	Wave height, m (with 3% probability)
0	Calm	—
I	Smooth	< 0.25
II	Slight	0.25—0.75
III	Moderate	0.75—1.25
IV		1.25—2.0
V	High	2.0—3.5
VI		3.5—6.0
VII	Very high	6.0—8.5
VIII		8.5—11.0
IX	Exceptional	≥ 11.0

* In effect for hydrometeorological observations in seas, natural lakes and large water-storage lakes as of 1953.

The experienced seaman who wishes to determine wave strength does usually make a special estimate of wave height, but gains an insight into wave strength from the dimensions of the most prominent waves. The relationships between wave height and wave strength, as expressed by a scale, are found to be quite stable within the ranges of Table 2.2 (Titov, 1952). Hence information on wave strength can serve as a quantitative estimate of the wind waves encountered in a given sea or ocean.

Wave strength classification has been in existence from the start of the 19th century and is used by the USSR Navy up to the present. On the other hand, it went out of use in other countries. Naturally, the wave strength estimate should not be confused with the assessment of the state of the sea (Table 2.1). As previously mentioned, the latter aided seamen (and is used even now in individual cases) in estimating wind strength on a scale and typifies the effect of the latter on the sea surface.

Chapter 1

FUNDAMENTALS OF WAVE THEORY

1. Small-amplitude potential waves

The wave theory under consideration does not address itself to problems of generation and development of wave motions. It is assumed that initially there exists a disturbance in the fluid, i.e., deviation of its state from equilibrium due to an instantaneous external impulse (wind gust, sharp change in atmospheric pressure, etc.). The operation of these forces, although of short duration, sets into motion the water particles on the free surface. This motion to some extent propagates to lower-lying water layers. The removal of the disturbing force in the presence of the gravity force (the intrinsic force of the water particles) results in reversing the direction of particle motion, which tends to return them to the position occupied before application of the force. However, the inertia forces developed by the induced motion tend to carry the water particles past their rest position. As a result the fluid is set into oscillatory motion. It is assumed that such wave motion occurs in a homogeneous, frictionless, incompressible fluid. All the fluid particles move parallel to the same vertical plane xoz (Figure 1.1.1), which is taken as the coordinate plane. Here the velocity and pressure are independent of the y coordinate, i.e., the motion in every plane parallel to the xoz plane is identical. It is also assumed that the fluid extends to infinity on both sides of the ox axis. It is limited from the bottom by an immovable wall (the sea bottom), while the top comprises a free surface at which the waves are visible.

Of great importance in the theory of waves are solutions involving the potential motion of fluids. It is known that there exists a function φ_1, the partial derivatives of which in any direction equal the velocity projection in this direction:

$$\frac{\partial \varphi_1}{\partial x} = u; \qquad \frac{\partial \varphi_1}{\partial z} = w. \qquad (1.1.1)$$

Fluid motion satisfying this condition is termed motion with a velocity potential, or potential motion. Irrotational or potential motion lends itself easier to study than vortex motion, because once function φ_1 is known the velocity at any point of space is obtainable by differentiation.

The continuity equation for an incompressible fluid is

$$\frac{\partial u}{\partial x} + \frac{\partial w}{\partial z} = 0.$$

With the aid of (1.1.1) this equation may be expressed as a Laplace equation:

$$\frac{\partial^2 \varphi_1}{\partial x^2} + \frac{\partial^2 \varphi_1}{\partial z^2} = 0. \tag{1.1.2}$$

A function satisfying equation (1.1.2) is called a harmonic function, and only such functions can define potential flow in an incompressible fluid.

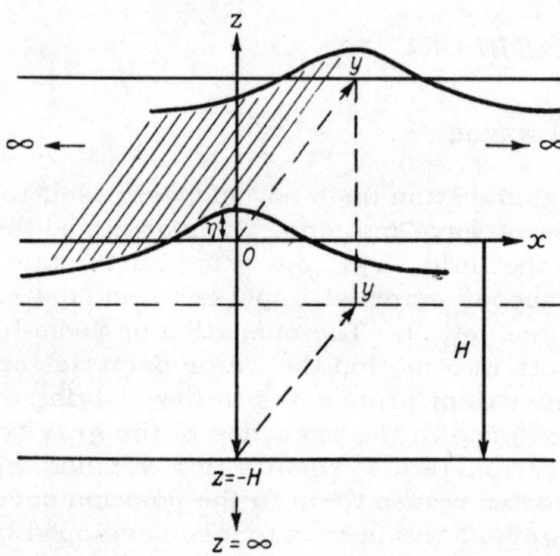

FIGURE 1.1.1.

It is known from hydromechanics that the potential motion of an ideal fluid can be expressed in terms of the Lagrange-Cauchy integral, which may be written in the form

$$\frac{\partial \varphi_1}{\partial t} + \frac{v^2}{2} - V + \frac{p}{\rho} = f(t), \tag{1.1.3}$$

where V is potential of external forces (gravity forces) acting on unit mass; $v^2 = u^2 + w^2$; ρ is fluid density; p is pressure; $f(t)$ is an arbitrary function of time.

In terms of the assumed coordinate system (Figure 1.1.1), potential V can be written as

$$V = -gz. \tag{1.1.4}$$

To eliminate arbitrary function $f(t)$ from equation (1.1.3), we replace φ_1 by the function

$$\varphi = \varphi_1 - \int_0^t f(t)\, dt + \frac{p_0}{\rho} t, \tag{1.1.5}$$

where p_0 is atmospheric pressure at the free surface. Function φ also satisfies Laplace equation (1.1.2). From equation (1.1.5) we have

$$\frac{\partial \varphi_1}{\partial t} = \frac{\partial \varphi}{\partial t} + f(t) - \frac{p_0}{\rho}.$$

Substituting the above expression into equation (1.1.3) and making use of equation (1.1.4),

$$\frac{\partial \varphi}{\partial t} + \frac{v^2}{2} + gz + \frac{p - p_0}{\rho} = 0. \qquad (1.1.6)$$

If we consider only slow motions, i. e., $\frac{v^2}{2}$ is approximately zero, then equation (1.1.6) can be replaced by

$$\frac{\partial \varphi}{\partial t} + gz + \frac{p - p_0}{\rho} = 0. \qquad (1.1.7)$$

Equation (1.1.7) can be used to determine the pressure at any point of the fluid, provided φ is known. However, equation (1.1.2), from which φ should be determined, has a number of solutions, the required one being chosen in conformance to boundary and initial conditions.

Boundary conditions are defined at the free surface and at the bottom. It is assumed that fluid pressure p is constant along the free surface and equal to atmospheric pressure at any time t. Consequently,

$$p_0 = p. \qquad (1.1.8)$$

Let the equation of the free surface at some time t have the form

$$z = \eta(x,\, t).$$

It then follows from equations (1.1.7) and (1.1.8) that

$$g\eta + \frac{\partial \varphi}{\partial t}\bigg|_{z=\eta} = 0.$$

We now make the major assumption that waves generated at the fluid surface are so small that the ordinates of the points on the surface in the presence of waves differ little from the ordinates of the free surface with the fluid at rest. This is equivalent to assuming infinitesimal wave amplitude. As a result

$$\frac{\partial \varphi}{\partial t}\bigg|_{z=\eta} \cong \frac{\partial \varphi}{\partial t}\bigg|_{z=0}$$

and η may be expressed in the form

$$\eta = -\frac{1}{g}\frac{\partial \varphi(x,\, 0,\, t)}{\partial t}. \qquad (1.1.9)$$

As is known from hydromechanics, function η and potential φ are inter-related by the relationship

$$\left(\frac{\partial \varphi}{\partial z}\right)_{z=0} \cong \left(\frac{\partial \eta}{\partial t}\right)_{z=0}. \tag{1.1.10}$$

This expresses the fact that in this type of motion a fluid particle at the surface cannot move with time into the fluid, but remains forever at the surface.

Differentiation of equation (1.1.9) with respect to t, making use in the process of equation (1.1.10), yields

$$\left.\frac{\partial \varphi}{\partial z}\right|_{z=0} + \frac{1}{g}\left.\frac{\partial^2 \varphi}{\partial t^2}\right|_{z=0} = 0. \tag{1.1.11}$$

Equations (1.1.9) and (1.1.11) are the boundary conditions at the surface.

The bottom is a solid immovable horizontal surface, and so the normal velocity component at the bottom vanishes:

$$v_z = \left.\frac{\partial \varphi}{\partial z}\right|_{z=-H} = 0. \tag{1.1.12}$$

When the fluid depth is infinite

$$v_z = \left.\frac{\partial \varphi}{\partial z}\right|_{z=-\infty} = 0. \tag{1.1.13}$$

The initial conditions are determined on the assumption that the initial pressure was applied at a free surface, i.e., at $z=\eta$ or, for small-amplitude waves, approximately at $z=0$. Hence the velocity potential at the free surface should initially satisfy the condition

$$\varphi(x,\ 0,\ 0) = f_1(x). \tag{1.1.14}$$

The initial pressure will produce some initial disturbance, which corresponds to some initial position of the free surface. The level rise η defining this state of the free surface is found at $t=0$ from the expression

$$z = \eta(x,\ 0) = f(x).$$

Equation (1.1.9) at $t=0$ is

$$\eta = -\frac{1}{g}\frac{\partial \varphi(x,\ 0,\ 0)}{\partial t}.$$

When $gf(x)$ is replaced by $f_2(x)$, the latter two equalities yield the second initial condition in the form

$$\left.\frac{\partial \varphi}{\partial t}\right|_{z=0,\ t=0} = f_2(x). \tag{1.1.15}$$

It should be noted that the problem of wave motion is definable only when functions $f_1(x)$ and $f_2(x)$ are known a priori.

The study of infinitesimal-amplitude surface waves in heavy incompressible fluids thus reduces to determining, from continuity equation (1.1.2), the velocity potential φ, satisfying boundary conditions (1.1.12) or (1.1.13) at the bottom, conditions (1.1.9) or (1.1.11) at the free surface, and obeying initial conditions (1.1.14) and (1.1.15). Here $f_1(x)$ and $f_2(x)$ are assumed known. After function φ is determined, the free surface of the fluid is defined by equation (1.1.9).

However, the problem of wave motion of fluids can be solved only approximately. The main simplifying assumptions here are smallness of wave amplitude (as compared with wave length) and disregarding the squares of velocities in the Lagrange integral as quantities of a much lower order of magnitude as compared with other terms. The solution also excludes regard for water viscosity.

Determination of potential φ as such is a difficult task, the main difficulty being that the wave profile $z = \eta(x)$ is unknown and is subject to determination in the course of the solution, while equation (1.1.9) must be satisfied for $z = \eta(x)$.

2. Small-amplitude waves in fluids of finite depth

The nature of wind motion, as shown by experience, is a periodic function of x and t and a nonperiodic function of z. On this basis the velocity potential can be expressed as a product of some arbitrary function $f(z)$ and the elementary function $\sin \theta$, where

$$\theta = kx - \sigma t, \qquad (1.2.1)$$

i.e., in the form of a simple sinusoidal wave with amplitude expressed as a function of z. In equation (1.2.1)

$$k = \frac{2\pi}{\lambda} \quad \text{and} \quad \sigma = \frac{2\pi}{\tau}.$$

Hence,

$$\varphi(x,\, z,\, t) = f(z) \sin \theta, \qquad (1.2.2)$$

from which follows

$$\frac{\partial^2 \varphi}{\partial x^2} = -k^2 f(z) \sin \theta,$$

$$\frac{\partial^2 \varphi}{\partial z^2} = f''(z) \sin \theta.$$

Substitution of these expressions into continuity equation (1.1.2) yields

$$[-k^2 f(z) + f''(z)] \sin \theta = 0.$$

In general $\sin \theta$ is not zero; therefore

$$f''(z) - k^2 f(z) = 0.$$

The general solution of this equation has the form

$$f(z) = C_1 e^{kz} + C_2 e^{-kz},\tag{1.2.3}$$

where C_1 and C_2 are constants of integration.

Substitution of function (1.2.3) into equation (1.2.2) yields

$$\varphi = (C_1 e^{kz} + C_2 e^{-kz}) \sin \theta.\tag{1.2.4}$$

Hence boundary condition (1.1.12) assumes the form

$$\left. \frac{\partial \varphi}{\partial z} \right|_{z=-H} = k (C_1 e^{-kH} - C_2 e^{kH}) \sin \theta = 0.$$

Since $k \sin \theta \neq 0$, the expression in parentheses must be zero. Consequently

$$C_2 = C_1 e^{-2kH}.\tag{1.2.5}$$

Function (1.2.4) therefore becomes

$$\varphi = C_1 e^{-kH} \left[e^{k(z+H)} + e^{-k(z+H)} \right] \sin \theta.$$

Replacing the expression in brackets by twice the hyperbolic cosine, and introducing the notation

$$2C_1 e^{-kH} = C,$$

the expression for the velocity potential reduces to

$$\varphi = C \operatorname{ch} k(z+H) \sin \theta.\tag{1.2.6}$$

The form of the free surface is assessed from boundary condition (1.1.9), into which the value of the first derivative of equation (1.2.6) with respect to t at $z = 0$ is substituted:

$$\eta = \frac{\sigma C}{g} \operatorname{ch} kH \cos (kx - \sigma t).\tag{1.2.7}$$

Introducing further the notation

$$\frac{\sigma C}{g} \operatorname{ch} kH = \frac{h}{2} = a,\tag{1.2.8}$$

(1.2.7) finally reduces to

$$\eta = a \cos (kx - \sigma t).\tag{1.2.9}$$

Equation (1.2.9) is the equation of an undulated free surface. It follows that the profile of this surface is described by a cosine wave, while the motion proper is a harmonic oscillation. The cosine sign in equation (1.2.9) pertains not only to the coordinate x, but also to t; consequently the wave-surface profile moves in the positive direction. These waves are therefore termed progressive waves. From (1.2.8) we have that

$$C = \frac{ag}{\sigma \operatorname{ch} kH}.$$

and so (1.2.6) may be replaced by

$$\varphi = \frac{gh}{2\sigma} \frac{\operatorname{ch}[k(z+H)]}{\operatorname{ch} kH} \sin(kx - \sigma t). \qquad (1.2.10)$$

This is the final form of the velocity potential for progressive waves propagating over the surface of a water body of depth H.

The relationship between the elements of waves under study is assessed from boundary condition (1.1.11), in which function φ is substituted from equation (1.2.10). If function (1.2.6) is differentiated with respect to z and t:

$$\frac{\partial \varphi}{\partial z} = kC \operatorname{sh} k(z+H) \sin \theta,$$

$$\frac{\partial^2 \varphi}{\partial t^2} = -\sigma^2 C \operatorname{ch} k(z+H) \sin \theta,$$

and these values are substituted into equation (1.1.11) at $z = 0$, we obtain

$$k \operatorname{sh} kH - \frac{\sigma^2}{g} \operatorname{ch} kH = 0,$$

or

$$\sigma^2 = gk \operatorname{th} kH, \qquad (1.2.11)$$

whence

$$c^2 = \frac{\lambda^2}{\tau^2} = \frac{g}{k} \operatorname{th} kH. \qquad (1.2.12)$$

This is the expression for wave speed. It can be used to determine how sea depth affects rate of wave propagation.

If we assume that H is very large (or infinite), i.e., $H \to \infty$, then

$$\operatorname{th} 2\pi \frac{H}{\lambda} \approx 1$$

and consequently,

$$c^2 = \frac{g\lambda}{2\pi}. \qquad (1.2.13)$$

If, conversely, it is assumed that H is very small, then

$$\operatorname{th} 2\pi \frac{H}{\lambda} \approx 2\pi \frac{H}{\lambda}$$

and consequently,

$$c^2 = gH. \tag{1.2.14}$$

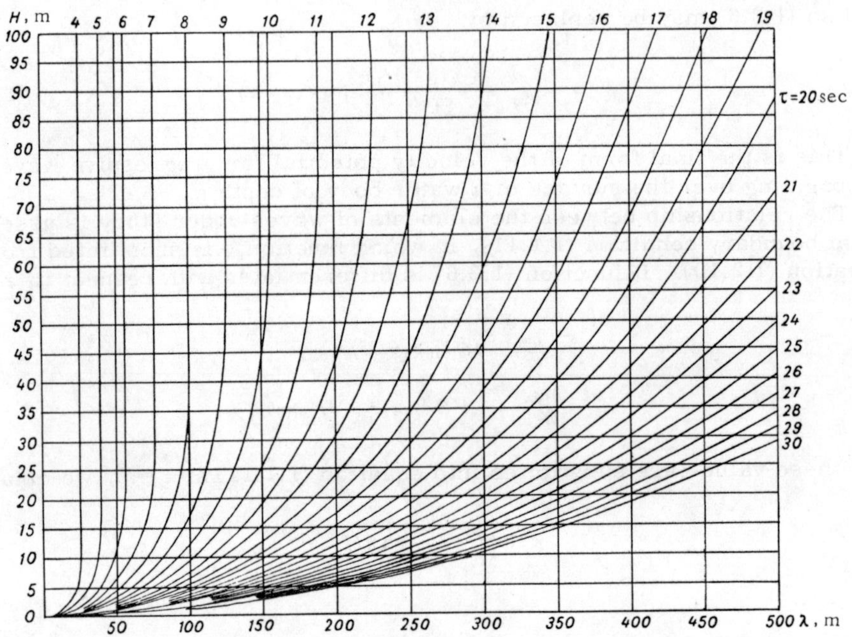

FIGURE 1.2.1.   Nomogram for determining wave length in shallow water (after Lappo).

In the last equation, known as the Lagrange formula, wave speed depends only on depth and not on wave length.   The wave period for finite-depth conditions is found from expression $\lambda = c\tau$.   Substitution of $c^2 = \dfrac{\lambda^2}{\tau^2}$ in equation (1.2.12) yields

$$\tau^2 = \frac{2\pi\lambda}{g} \operatorname{cth} 2\pi \frac{H}{\lambda}. \tag{1.2.15}$$

For constant τ the wave length here varies as a function of H.   By calculating curves of $\lambda = f(H)$ one may obtain graphs of λ as a function of H for different values of τ, shown in Figure 1.2.1 (Lappo, 1956).

Projections of the speeds and orbits of fluid particles may be derived from expression (1.2.10):

$$\frac{\partial \varphi}{\partial x} = \frac{hgk}{2\sigma} \frac{\operatorname{ch} k(z+H)}{\operatorname{ch} kH} \cos(kx - \sigma t), \qquad (1.2.16)$$

$$\frac{\partial \varphi}{\partial z} = \frac{hgk}{2\sigma} \frac{\operatorname{sh} k(z+H)}{\operatorname{ch} kH} \sin(kx - \sigma t). \qquad (1.2.17)$$

Integration with respect to t yields equations of particle motion as a function of time. First variables x and z in the right-hand sides of the above equations are replaced by constants x_0 and z_0, i.e., coordinates of the particle at rest. This assumption is based on the fact that the theory at hand pertains only to small-amplitude oscillations, enabling one to assume that the velocity projections at the crest and trough levels are identical and equal to those at rest. With reference to the above, integration of equations (1.2.16) and (1.2.17) yields

$$x = -\frac{hgk}{2\sigma^2} \frac{\operatorname{ch} k(z_0+H)}{\operatorname{ch} kH} \sin(kx_0 - \sigma t) - C_1. \qquad (1.2.18)$$

$$z = \frac{hgk}{2\sigma^2} \frac{\operatorname{sh} k(z_0+H)}{\operatorname{ch} kH} \cos(kx_0 - \sigma t) - C_2. \qquad (1.2.19)$$

Each of these equations defines harmonic oscillations of the fluid particles about some position, described by coordinates C_1 and C_2. These coordinates correspond precisely to the initial position of the particle, i.e., the state of rest, and can be replaced by x_0 and z_0. Keeping this in mind, replacing σ^2 by expression (1.2.11) and transforming, the following approximate equations of motion of individual particles may be derived:

$$x = x_0 + r_x \sin(kx_0 - \sigma t), \qquad (1.2.20)$$

$$z = z_0 + r_z \cos(kx_0 - \sigma t), \qquad (1.2.21)$$

where

$$r_x = \frac{h}{2} \frac{\operatorname{ch} k(z_0+H)}{\operatorname{sh} kH}, \qquad (1.2.22)$$

$$r_z = \frac{h}{2} \frac{\operatorname{sh} k(z_0+H)}{\operatorname{sh} kH}. \qquad (1.2.23)$$

Eliminating t, by squaring and adding equations (1.2.20) and (1.2.21), we have

$$\frac{(x - x_0)^2}{r_x^2} + \frac{(z - z_0)^2}{r_z^2} = 1, \qquad (1.2.24)$$

which is the canonical equation of an ellipse with horizontal semiaxis r_x and vertical semiaxis r_z.

Consequently, the orbits of particles comprising progressive waves for finite-depth fluids are ellipses (Figure 1.2.2), the axes of which are aligned so that the major axis is always horizontal.

The equations defining the shape of particle orbits are correct only for small-amplitude waves. In waves of finite amplitude the water particles do not travel in closed orbits (Chapter 1, Section 5).

At the surface ($z_0 = 0$),

$$r_x = \frac{h_0}{2} \operatorname{cth} kH, \qquad\qquad (1.2.25)$$

$$r_z = \frac{h_0}{2}. \qquad\qquad (1.2.26)$$

At the bottom ($z_0 = -H$),

$$r_x = \frac{h_0}{2}\,\frac{1}{\operatorname{sh} kH}, \qquad\qquad (1.2.27)$$

$$r_z = 0. \qquad\qquad (1.2.28)$$

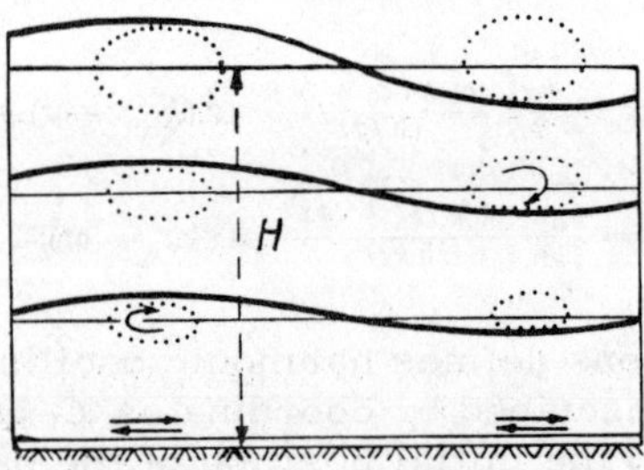

FIGURE 1.2.2. Variation in orbits of a "long" wave with downward propagation of wave oscillations.

Consequently, vertical particle motions vanish at the bottom, while horizontal motion exceeds approximately twofold the value it would have had at this depth had there been no bottom. Expressing the hyperbolic functions in terms of exponentials,

$$\frac{r_x}{\frac{h_0}{2}\exp[-kH]} = \frac{2}{1 - \exp[-2kH]}.$$

When $\dfrac{H}{\lambda} = 0.5$, the above expression yields

$$\frac{r_x}{\frac{h_0}{2}\exp[-kH]} \approx 2.09,$$

while when $\dfrac{H}{\lambda} = 0.1$ it gives ~ 3.0.

It follows from the continuity condition that vanishing of vertical oscillations at the very bottom must (in the case of finite depth under study) induce doubling of horizontal particle motion at this level.

The rotational velocity of the particles in their orbits is determined as follows. The rotational period is the same, and so as one moves deeper into the fluid the particle velocity should decrease. If f is the number of particle rotations per second (the frequency), then the orbital speed of the particles is

$$v = 2\pi r f = \frac{2\pi r}{\tau},$$ (1.2.29)

where r is the radius of the orbit, while $f = \dfrac{1}{\tau}$ (τ is the period). However, since

$$\lambda = c\tau,$$

then

$$v = krc = \frac{2\pi}{\lambda}\, rc.$$

The velocity of orbital motion may be expressed at the water surface first for the horizontal and then for the vertical directions. This is done by means of equation (1.2.29), from which it follows that $v^2 = \dfrac{4\pi^2 r^2}{\tau^2}$. Then, utilizing equations (1.2.15) and (1.2.22), for the horizontal axis at $z = 0$

$$v_x = \frac{h_0}{2}\left(\frac{2\pi g}{\lambda}\,\mathrm{cth}\,kH\right)^{\frac{1}{2}}.$$ (1.2.30)

For the vertical axis, using equations (1.2.15) and (1.2.23),

$$v_z = \frac{h_0}{2}\left(\frac{2\pi g}{\lambda}\,\mathrm{th}\,kH\right)^{\frac{1}{2}}.$$ (1.2.31)

Finally, for the bottom ($z_0 = -H$),

$$v_z = 0,$$
$$v_x = \frac{\pi h_0}{\left(\dfrac{\pi\lambda}{g}\,\mathrm{sh}\,2kH\right)^{\frac{1}{2}}}.$$ (1.2.32)

The velocity of orbital motion at intermediate levels is

$$v_x = h_0\,\mathrm{ch}\,k(z_0 + H)\left(\frac{\pi g}{\lambda\,\mathrm{sh}\,2kH}\right)^{\frac{1}{2}},$$ (1.2.33)

$$v_z = h_0\,\mathrm{sh}\,k(z_0 + H)\left(\frac{\pi g}{\lambda\,\mathrm{sh}\,2kH}\right)^{\frac{1}{2}}.$$ (1.2.34)

If the wave length is large compared with depth ($kH \leqslant 0.1$), ch kH in equation (1.2.22) can be replaced without much error by unity, while sh kH may be represented by kH. Equation (1.2.22) then assumes the form

$$r_x = \frac{h_0}{2}\,\frac{1}{kH}.$$ (1.2.35)

Using the fact that, according to equation (1.2.29),

$$v = krc$$

and substituting appropriately, we derive equation (1.2.30) in the form

$$v_x = \frac{h_0}{2} \frac{c}{H}. \tag{1.2.36}$$

Since $c^2 = gH$,

$$v_x = \frac{h_0}{2} \left(\frac{g}{H} \right)^{\frac{1}{2}}. \tag{1.2.37}$$

Consequently, the horizontal velocity of particles in very long waves is virtually identical over the entire depth, down to the bottom.

Replacing sh $k(z_0 + H)$ by $k(z_0 + H)$ and sh kH by kH in equation (1.2.23), on the basis of the previous considerations it is found that in very long waves

$$r_z = \frac{h_0}{2} \frac{z_0 + H}{H}$$

and, from (1.2.29),

$$v_z = \frac{2\pi}{\tau} \frac{h_0}{2} \frac{z_0 + H}{H}. \tag{1.2.38}$$

Thus, the vertical axis of the ellipse and the vertical velocity of particles decrease with depth, becoming zero at the bottom.

Equation (1.2.15), upon suitable substitutions, may be expressed in the following form for wave length:

$$\lambda = c^2 \frac{2\pi}{g} \operatorname{cth} kH. \tag{1.2.39}$$

Wave height is not related to the other elements and must therefore be determined directly by observation.

The wave pressure is determined from Lagrange-Cauchy integral (1.1.7). Here consideration is given only to the pressure in excess of atmospheric, and hence it is assumed that $p_0 = 0$. Equation (1.1.7) then reduces to

$$p = -\rho g z - \rho \frac{\partial \varphi}{\partial t}, \tag{1.2.40}$$

where derivative $\dfrac{\partial \varphi}{\partial t}$ is obtained by differentiating velocity potential (1.2.10) with respect to time:

$$\frac{\partial \varphi}{\partial t} = -\frac{hg}{2} \frac{\operatorname{ch} k(z_0 + H)}{\operatorname{ch} kH} \cos(kx - \sigma t). \tag{1.2.41}$$

Substituting equation (1.2.41) into equation (1.2.40) and replacing ρg by γ (specific weight),

$$p = -\gamma z + \gamma \frac{h}{2} \frac{\operatorname{ch} k\,(z_0 + H)}{\operatorname{ch} kH} \cos(kx - \sigma t). \qquad (1.2.42)$$

The first term in equation (1.2.41) expresses the hydrostatic pressure and the second, the wave pressure, whose amplitude decreases with depth, according to equation (1.2.42):

$$\frac{p_z}{p} = \frac{\operatorname{ch} k\,(z_0 + H)}{\operatorname{ch} kH}. \qquad (1.2.43)$$

If the wave length is large compared with depth, $\dfrac{p_z}{p} = 1$ and, consequently, the wave pressure in a long wave is the same over the entire depth and proportional to the wave height at the surface. If $z_0 = -H$, then

$$\frac{p_H}{p} = \frac{1}{\operatorname{ch} kH}. \qquad (1.2.44)$$

3. Small-amplitude waves in fluids of infinite depth

An expression for the velocity potential and other wave-motion characteristics at $H = \infty$ can be obtained by analyzing the appropriate formulas presented in Section 2.

Consider the expression for the velocity potential (1.2.4), and boundary condition (1.1.13):

$$\varphi = (C_1 e^{kz} + C_2 e^{-kz}) \sin \theta,$$
$$\left. \frac{\partial \varphi}{\partial z} \right|_{z = -\infty} = 0.$$

Compatibility of these equations implies $C_2 = 2$, since otherwise the second term of the velocity potential becomes infinite and boundary condition (1.1.13) is not satisfied. Hence the velocity potential at infinite depth takes the form

$$\varphi = C e^{kz} \sin(kx - \sigma t). \qquad (1.3.1)$$

Equation (1.2.11) is written as

$$\sigma^2 = gk. \qquad (1.3.2)$$

Consequently, expression (1.2.8) is replaced by

$$\frac{h}{2} = \frac{\sigma C}{g},$$

or

$$C = \frac{hg}{2\sigma}. \tag{1.3.3}$$

The equation of the free surface is, as before, (1.2.9). The velocity at which waves propagate is expressed by equation (1.2.13):

$$c^2 = \frac{g\lambda}{2\pi}, \tag{1.3.4}$$

and consequently the wave period is

$$\tau^2 = \frac{2\pi}{g}\lambda. \tag{1.3.5}$$

After substituting the value of c from equation (1.3.3), the velocity potential takes the form

$$\varphi = \frac{hg}{2\sigma} e^{kz} \sin(kx - \sigma t). \tag{1.3.6}$$

The velocity projections are

$$\frac{\partial \varphi}{\partial x} = \frac{hgk}{2\sigma} e^{kz} \cos(kx - \sigma t), \tag{1.3.7}$$

$$\frac{\partial \varphi}{\partial z} = \frac{hgk}{2\sigma} e^{kz} \sin(kx - \sigma t). \tag{1.3.8}$$

The equations of motion of individual particles have the general form of equations (1.2.20) and (1.2.21), but in the case under study, as $H \to \infty$, they are written in the form

$$r_x = r_z = r = \frac{h}{2} e^{kz_0}. \tag{1.3.9}$$

Hence

$$x = x_0 - r\sin(kx_0 - \sigma t), \tag{1.3.10}$$
$$z = z_0 + r\cos(kx_0 - \sigma t). \tag{1.3.11}$$

Consequently, equation (1.2.24) reduces to

$$(x - x_0)^2 + (z - z_0)^2 = r^2. \tag{1.3.12}$$

It may be concluded from equations (1.3.9)—(1.3.12) that for wave motion in deep water: 1) individual fluid particles describe closed orbits in the form of circles of the same radius r, decreasing with increasing depth; 2) all the water particles which are at rest in a given horizontal plane describe circles of equal radius; 3) particles at rest at the free surface describe circles of diameters equal to the wave height. At depth $z_0 = -0.5$ the circle diameters equal $0.043\,h$, i.e., the wave motion is virtually damped out at this depth. The wave pressure is given by

$$p = -\gamma z + \gamma \,\frac{h}{2}\, e^{kz} \cos(kx - \sigma t), \tag{1.3.13}$$

where $\gamma = \rho g$.

Maximum pressure is calculated from the expression

$$p = -\gamma z \pm \gamma \,\frac{h}{2}\, e^{kz_0}, \tag{1.3.14}$$

and the corresponding particle position is given by

$$z = z_0 \pm \frac{h}{2}\, e^{kz_0}. \tag{1.3.15}$$

4. Long and short waves

It follows from equations (1.2.22) and (1.2.23), defining the semiaxes of the ellipse (Kochin, Kibel' and Roze, 1948), that, since the hyperbolic cosine is always larger than the hyperbolic sine of the same argument, the elliptical orbits of water particles are flattened in vertical directions and elongated in horizontal directions. With increasing depth of submersion of the centers of elliptical orbits, i.e., with decreasing water depth, the vertical axes of ellipses will "shrink" faster than the horizontal. The ellipses whose absolute dimensions will thus decrease (Figure 1.2.2) will at the same time become elongated in the vertical direction. At the bottom the water particles will move only back and forth along straight lines.

If we assume that the sea is infinitely deep, then the ellipses become circles, i.e., in this case we have a condition corresponding to waves propagating at the surface of an infinite-depth fluid, examined in Section 3.

The limits of applicability are determined as follows. The ratio is written of the vertical semiaxis of the ellipse, equation (1.2.23), to its horizontal semiaxis at the surface, equation (1.2.22):

$$\frac{r_z}{r_x} = \operatorname{th} kH. \tag{1.4.1}$$

This ratio is tabulated in Table 1.4.1 as a function of $\frac{H}{\lambda}$. This table also lists ratios $\dfrac{c}{\left(\dfrac{g\lambda}{2\pi}\right)^{1/2}}$ and $\dfrac{c}{(gH)^{1/2}}$ as a function of $\frac{H}{\lambda}$, obtained from 1.3.4),

(1.2.12) and (1.2.14). The data of the table are illustrated by Figure 1.4.1.

For a sea depth which is only one-half the wave depth, the vertical semiaxis differs from the major axis by only 0.4%. Even when $\frac{H}{\lambda} = 0.3$ the minor axis is by approximately 4.5% shorter than the major, horizontal axis.

The ellipse starts elongating very markedly at smaller depths, for example, when $\frac{H}{\lambda} = 0.1$ the vertical semiaxis is shorter than the horizontal by as much as 43%. At the very bottom, as mentioned above, vertical oscillations vanish, while horizontal oscillations exceed approximately twice the amplitude which would have existed at the same depth had there been no bottom.

TABLE 1.4.1.

$\dfrac{H}{\lambda}$	th kH	$\dfrac{c}{(gH)^{1/2}}$	$\dfrac{c}{\left(\dfrac{g\lambda}{2\pi}\right)^{1/2}}$	$\dfrac{H}{\lambda}$	th kH	$\dfrac{c}{(gH)^{1/2}}$	$\dfrac{c}{\left(\dfrac{g\lambda}{2\pi}\right)^{1/2}}$
0.00	0.000	1.000	0.000	0.1	0.557	0.941	0.746
0.01	0.063	0.999	0.250	0.2	0.850	0.823	0.922
0.02	0.125	0.997	0.354	0.3	0.955	0.713	0.977
0.03	0.186	0.994	0.432	0.4	0.987	0.627	0.993
0.04	0.246	0.990	0.496	0.5	0.996	0.563	0.998
0.05	0.304	0.984	0.552	0.6	0.999	0.515	0.999
0.06	0.360	0.917	0.600	0.7	1.000	0.477	1.000
0.07	0.413	0.970	0.643	0.8	1.000	0.446	1.000
0.08	0.464	0.961	0.681	0.9	1.000	0.421	1.000
0.09	0.512	0.951	0.715	1.0	1.000	0.399	1.000

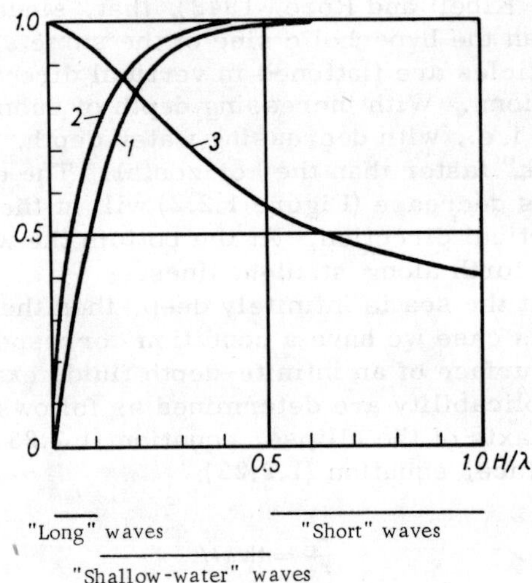

FIGURE 1.4.1. Pertaining to Table 1.4.1:

$$1 - \text{th } kH; \quad 2 - \frac{c}{\left(\frac{g\lambda}{2\pi}\right)^{0.5}}; \quad 3 - \frac{c}{(gH)^{0.5}}.$$

We now return to the manner in which ratio $\dfrac{c}{\left(\dfrac{g\lambda}{2\pi}\right)^{1/2}}$ varies as a function of $\dfrac{H}{\lambda}$, i.e., of the ratio of wave speeds in shallow and deep water with equal wave lengths. Here, as in assessing the effect of shallowness on orbit shape, it should be noted that a marked difference is observed only at very small $\frac{H}{\lambda}$. When $\frac{H}{\lambda} = 0.5$ the phase velocity according to (1.3.4) differs only by 0.2 % from that given by (1.2.12). Hence, for practical purposes one may use (1.3.4) for depths equal to or greater than $\frac{H}{\lambda} = 0.5$. When $\frac{H}{\lambda} = 0.2$ the wave speed at finite depth as given by expression (1.2.12) differs by only 8 % from the speed given by (1.3.4), and decreases sharply upon further reduction in depth. Ratio $\frac{H}{\lambda} = 0.5$ is thus a criterion for establishing the range of applicability of the theory of small-amplitude progressive waves for finite-depth conditions ($\frac{H}{\lambda} \leqslant 0.5$) and of the theory for infinite fluid depth ($\frac{H}{\lambda} \geqslant 0.5$).

TABLE 1.4.2. Characteristics of short and long waves (after Zubov)

Characteristic	Short-period waves		Long-period waves
	short	long	long
Rate of propagation	Depends on wave length, but not on water depth	Depends on wave length and water depth	Depends on water depth, but not on wave length
Particle orbits	Circles whose radii shrink rapidly with depth and become virtually zero at a depth equal to the wave length	Ellipses with horizontal major axis which shrinks rapidly with depth	Ellipses highly elongated in a horizontal direction. Horizontal particle motion virtually independent of depth
Vertical motion of particles	Rapidly decrease with depth and become virtually zero at a depth equal to the wave length	Rapidly decrease with depth and become virtually zero at a depth equal to the wave length	Decrease linearly from the sea surface to its bottom
Pressure distribution	The wave has no effect below depths equal to the wave length	The effect of the wave is the greater, the smaller the ratio of water depth to wave length	The wave affects the pressure distribution in the same manner at all depths, the degree of influence being a function only of wave height
Effect of the earth's rotation	Imperceptible	Imperceptible	If the period of the wave is close to that of the earth's rotation, this affects the rate of propagation and the particle motion
Formation of currents and inception of suspended-particle transport, etc.	Virtually none	No currents (within the proper meaning of the term) form, but transport may be appreciable	Currents and transport phenomena are sometimes quite pronounced, as, for example, during tidal motions. Maximum current velocity is proportional to wave height
Typical kinds of waves	Wind waves and deep-water swell	Wind waves and shallow-water swell	Tidal waves

It is usually assumed in practical engineering calculations that the error in calculating wave elements should not exceed $\sim 5\%$. Then, as is seen from Table 1.4.1 and Figure 1.4.1, when $\dfrac{H}{\lambda} \leqslant 0.1$ one should use expression (1.2.14):

$$c^2 = gH.$$

When $\dfrac{H}{\lambda} \geqslant 0.3$ it is permissible to use (1.3.4):

$$c^2 = \frac{g\lambda}{2\pi}.$$

Only when $0.1 < \dfrac{H}{\lambda} < 0.3$ should one use (1.2.12):

$$c^2 = \frac{g\lambda}{2\pi}\,\mathrm{th}\,kH.$$

According to these limits of applicability of formulas for the phase velocity and other wave elements, surface waves are subdivided into: a) shallow-water or "long" waves ($\dfrac{H}{\lambda} \leqslant 0.1$); b) intermediate-water waves ($0.1 < \dfrac{H}{\lambda} < 0.3 - 0.5$); and c) deep-water or "short" waves ($\dfrac{H}{\lambda} > 0.3 - 0.5$).

It follows that the characteristic distinguishing short from long waves is the ratio of sea depth to wave length. Hence, the same wave when passing from deep- to shallow-water regions may change from a short into a long wave. For example, if wind waves 800 m long with a period of 23 sec move into the shore region, where the assumed depth is, say, about 80 m, then the ratio of the depth to the length will be close to 0.1. Consequently, this wave, which was short before, has now become long although its period has not changed. For this reason short-period waves (see Introduction, Section 2) may be both short and long. The principal characteristics of short and long waves are listed in Table 1.4.2.

5. Some results of the theory of finite-amplitude waves

It was assumed in the wave theory examined in the preceding sections that wave amplitude is an infinitesimal quantity. It cannot, however, be extended to all kinds of real sea waves. For example, tidal waves, seismic waves and very long swell may be counted among waves of whose amplitude hardly anything is known as compared with their length. However, the amplitude of wind waves is quite appreciable. The ratio of their height to their length is appreciable. The large variety of waves actually encountered in the sea requires a separate discussion of results of classical relationships in their application to finite-amplitude waves.

Equations defining the motion of waves on fluid surfaces involving finite-amplitude waves, are obtained by solving equations (1.1.2) and (1.1.3), i.e., we again consider potential motion. Now, however, unlike equation (1.1.7) we retain nonlinear terms in equation (1.1.3); the wave height in (1.1.9) is treated as a finite, rather than an infinitesimal quantity.

An approximate solution of the problem as stated above is due to Stokes (1847), whose work was developed by Rayleigh (1878). Exact solutions were obtained by Nekrasov (1921) and Levi-Civita (1925).

One result of these solutions is that the phase velocity of the wave is expressible in the form (Sretenskii, 1936)

$$c^2 = \frac{g\lambda}{2\pi}\left(1 + \frac{\pi^2 h^2}{\lambda^2}\right).$$

(1.5.1)

The velocity of a progressive wave for finite-amplitude waves thus depends not only on the wave length, but also on its height (amplitude). The velocity also increases with an increase in the latter. High waves propagate faster than low waves.

Denoting $\frac{h}{\lambda}$ by δ, we can write (1.5.1) as

$$c^2 = \frac{g\lambda}{2\pi}(1 + \pi^2\delta^2).$$

(1.5.2)

It follows from this expression that the phase velocity of waves is the higher, the steeper the wave. As to the wave form, at moderate δ it is almost trochoidal but, as δ increases, the troughs become wider and flatter and the crests become steeper. It is found that the maximum steepness that a wave may have before breaking up is 0.142 ($^1/_7$) and the vertex angle is then 120° (Figure 1.5.1). When the steepness increases past the above value, the wave breaks up. The rate of propagation of such a wave is 12 % higher than the speed of a wave with the same length but infinitesimal amplitude.

FIGURE 1.5.1.

One of the most interesting results of the above solutions obtained from the theory of finite-amplitude waves is the determination of the shape of the fluid-particle orbits. Each fluid particle, in addition to its oscillatory motion about some average position, also performs translational motion in the direction of wave propagation. This translational motion is the same for all points lying at the same depth and damps out rapidly with depth. After traversing each almost-circular path the particle moves forward through some distance in the direction of wave propagation and carries with it a small mass, the so-called "wave" current. If u denotes the average velocity at which this mass is carried forward during one wave period, then

$$u = \pi^2\delta^2 c \exp\left[-\frac{4\pi z_0}{\lambda}\right],$$

(1.5.3)

or

$$u = \delta^2 g^{\frac{1}{2}}\left(\frac{2\pi}{\lambda}\right)^{\frac{3}{2}} \exp\left[-\frac{4\pi z_0}{\lambda}\right].$$

(1.5.4)

Equations (1.5.3) and (1.5.4) are applicable to deep water. For finite-depth conditions

$$u = \frac{k^2 h^2 c^2}{g}\frac{\text{ch}\,2k\,(z_0 + H)}{\text{sh}^2\,kH}.$$

(1.5.5)

Integration of (1.5.3) from 0 to H yields an expression for the water flow rate (Q) due to wave transport:

$$Q \text{ cm}^2/\text{sec} = \delta^2 g^{\frac{1}{2}} \left(\frac{\pi}{2\lambda}\right)^{\frac{1}{2}} (1 - e)^{-\frac{4\kappa z_0}{\lambda}}. \qquad (1.5.6)$$

The same water flow rate for finite-depth water is given by the expression

$$Q \text{ cm}^2/\text{sec} = \frac{h^2 kc}{g} \text{ cth } kH. \qquad (1.5.7)$$

Surface flow of fluid is an inevitable result of the velocity potential inherent to the wave motions under study (Sretenskii, 1936).

The above expressions ((1.5.1) and those following) become inapplicable when the water depth becomes much smaller than the wave length. This situation has another nonlinear solution. The latter in its turn is inapplicable for deep-water waves. Lately much work has been done on developing a more general theory of waves which would be universal and apply to all the above particular ranges.

6. Trochoidal wave theory

Trochoidal wave theory represents a particular solution of finite-amplitude waves, obtained when the fluid particles move in closed circular orbits. In addition, the fluid motion produced by these wave oscillations does not have a velocity potential. Hence this theory expresses the property of a particular case of vortex waves. At the same time, trochoidal wave theory yields exact solutions in terms of elementary functions which satisfactorily agree with wave motions under natural conditions. The latter circumstances allow extensive application of this theory, though only to situations for which the theory is applicable (sea depth large compared with wave length at the surface, steady type of wave oscillations). The latter is most typical of swell.

Consider wave motion of flat, finite-amplitude waves under the following conditions: 1) fluid-particle orbits are circles of radius r; 2) surface pressure is constant and equal to atmospheric pressure; 3) wave form and dimension do not change with time; 4) waves have finite amplitude.

The coordinate origin is taken at the still-water level (Figure 1.6.1). The ox axis is directed along the surface, while the oz axis is drawn perpendicularly downward, which is taken as the positive direction. A fluid particle, whose density is taken as constant and equal to ρ, will perform oscillations in the plane parallel to xz. The coordinates of the center of the circular orbits of this particle are denoted by a and b. Initially this particle (M) was located at the lower end of the vertical diameter of the orbit, and toward some time t it, moving counterclockwise, will be found in position M_1. The equation of motion of a particle moving in a circle of radius r can be expressed as

$$x = r \sin \theta,$$

$$z = r \cos \theta,$$

and when referred to the origin of our coordinate system

$$x - a = r \sin \theta,$$
$$z - b = r \cos \theta. \tag{1.6.1}$$

Radius r depends in some way on r_0, the radius of the particle orbit at the surface, and on distance b from the sea surface:

$$r = f(r_0, b).$$

Also involved is θ, when the wave form propagates to the right of the coordinate origin:

$$\theta = ka - \omega t, \tag{1.6.2}$$

where

$$k = \frac{2\pi}{\lambda}; \qquad \omega = \frac{2\pi}{\tau}.$$

Further analysis involves the use of equations of motion of the fluid in Lagrangian form. The external-force components are, respectively,

$$X = 0; \qquad Z = g.$$

Then the Lagrangian equations take the form

$$\frac{\partial^2 x}{\partial t^2}\frac{\partial x}{\partial a} + \left(\frac{\partial^2 z}{\partial t^2} - g\right)\frac{\partial z}{\partial a} + \frac{1}{\rho}\frac{\partial p}{\partial a} = 0,$$

$$\frac{\partial^2 x}{\partial t^2}\frac{\partial x}{\partial b} + \left(\frac{\partial^2 z}{\partial t^2} - g\right)\frac{\partial z}{\partial b} + \frac{1}{\rho}\frac{\partial p}{\partial b} = 0. \tag{1.6.3}$$

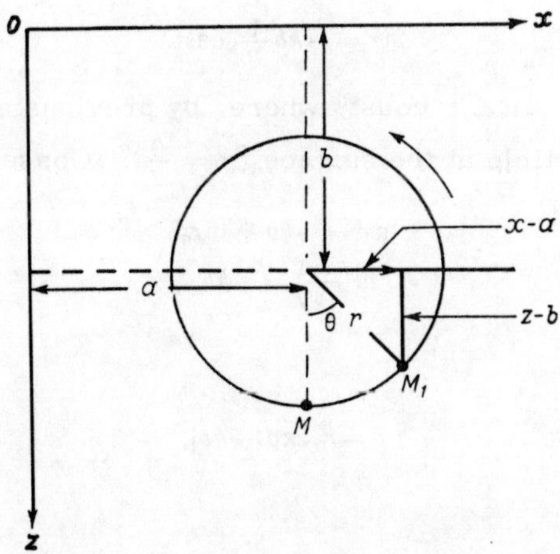

FIGURE 1.6.1.

Equations (1.6.3) are supplemented by the continuity equation in Lagrangian form

$$\frac{\partial}{\partial t}\left(\frac{\partial x}{\partial a}\frac{\partial z}{\partial b} - \frac{\partial x}{\partial b}\frac{\partial z}{\partial a}\right) = 0. \tag{1.6.4}$$

The relationships contained implicitly in (1.6.1) and (1.6.2) are defined by using equation (1.6.4), in which the partial derivatives, with the aid of (1.6.1), are given by

$$\frac{\partial x}{\partial a} = 1 + kr\cos\theta, \qquad \frac{\partial z}{\partial a} = -kr\sin\theta,$$

$$\frac{\partial x}{\partial b} = \sin\theta\,\frac{\partial z}{\partial b}, \qquad \frac{\partial z}{\partial b} = 1 + \cos\theta\,\frac{\partial z}{\partial b}. \tag{1.6.5}$$

Substitution of (1.6.5) into (1.6.4) yields

$$\frac{\partial x}{\partial a}\frac{\partial z}{\partial b} - \frac{\partial x}{\partial b}\frac{\partial z}{\partial a} = (1 + kr\cos\theta)\left(1 + \cos\theta\frac{\partial r}{\partial b}\right) +$$

$$+\, kr\sin^2\theta\,\frac{\partial r}{\partial b} = 1 + kr\,\frac{\partial r}{\partial b} + \left(kr + \frac{\partial r}{\partial b}\right)\cos\theta.$$

According to equation (1.6.4) the time derivative of this expression should be zero. However, according to equation (1.6.2), it is a priori known to be a function of t. This means that the continuity equation can be satisfied only when the coefficient of $\cos\theta$ is zero. Consequently

$$kr + \frac{\partial r}{\partial b} = 0,$$

whence

$$\frac{\partial r}{r} = -k\partial b.$$

Integration of this latter relationship yields

$$\ln r = -kb + \text{const.}$$

When $b = 0$, $r = r_0$, $\ln r_0 = \text{const}$, where, by previous assumption, r_0 is the radius of the particle at the surface ($r_0 = \frac{h}{2}$). Consequently

$$\ln r = -kb + \ln r_0,$$
$$\ln\left(\frac{r}{r_0}\right) = -kb,$$

whence

$$r = r_0\exp[-kb], \tag{1.6.6}$$

where

$$k = \frac{2\pi}{\lambda}.$$

Thus, the orbital radii (r) decrease exponentially moving deeper into the sea the more rapidly, the shorter the wave lengths. The governing quantity is the ratio $\frac{b}{\lambda}$, i. e., the ratio of the distance of the particle from the surface to the wave length. When $b = \frac{\lambda}{2}$, r equals $0.043\, r_0$. Table 1.6.1 lists values of r as a function of the wave length ($b = \lambda$).

TABLE 1.6.1.

Depth, as a fraction of wave length λ	0.05	0.1	0.2	0.3	0.4	0.5
r, as a fraction of $r_0 = 1$	0.73	0.53	0.28	0.15	0.08	0.04
E_z (E is the wave energy), as a fraction of E_0 at the surface ($E_0 = 1$)	0.53	0.28	0.08	0.02	0.007	0.002

Depth, as a fraction of wave length λ	0.6	0.7	0.8	0.9	1.0
r, as a fraction of $r_0 = 1$	0.02	0.01	0.007	0.003	0.002
E_z (E is the wave energy), as a fraction of E_0 at the surface ($E_0 = 1$)	0.0005	0.0002	0.00004	0.00004	—

To determine the surface wave form one substitutes the value of r from (1.6.6) into (1.6.1) (with consideration of (1.6.2)). Hence

$$x = a + r_0 e^{-kb} \sin(ka + \omega t),$$
$$z = b + r_0 e^{-kb} \cos(ka + \omega t). \qquad (1.6.7)$$

Differentiation of (1.6.7) with respect to time yields the horizontal and vertical components of the particle velocity:

$$u = v_x = \frac{\partial x}{\partial t} = r_0 e^{-kb} \omega \cos(ka + \omega t),$$

$$w = v_z = \frac{\partial z}{\partial t} = -r_0 e^{-kb} \omega \sin(ka + \omega t). \qquad (1.6.8)$$

It follows that the maximum horizontal velocity components will be found at the wave crest and trough, while the maximum vertical components occur at points on the profile intermediate between the crest and trough.

In order to obtain a parametric equation of the wave profile at the surface, it suffices to set in (1.6.7)

$$b = b_0 = 0,$$

whereupon expressions (1.6.7) take the form

$$x = a + r_0 \sin(ka + \omega t),$$
$$z = r_0 \cos ka. \qquad (1.6.9)$$

Setting $t = 0$, we may replace (1.6.9) by

$$x = a + r_0 \sin ka,$$
$$z = r_0 \cos ka; \qquad (1.6.10)$$

when $t = 0$

$$\theta = ka - \omega t = ka,$$

whence

$$a = \frac{\theta}{k}.$$

Now the parametric equation of the wave profile at $t = 0$ is reduced to the form

$$x = \frac{1}{k}\theta + r_0 \sin\theta,$$

$$z = r_0 \cos\theta. \tag{1.6.11}$$

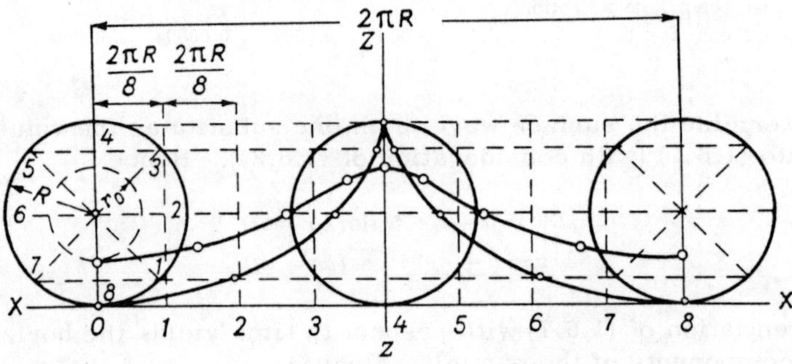

FIGURE 1.6.2.

When $r_0 > \dfrac{1}{k}$, system (1.6.11) represents a looped trochoid. This case should be excluded, since such a wave profile cannot be real. When $r_0 = \dfrac{1}{k}$ (1.6.11) represents a cycloid (Figure 1.6.2), and when $r_0 \leqslant \dfrac{1}{k}$ it represents a loopless trochoid. Consequently, $r_0 \leqslant \dfrac{1}{k} = \dfrac{\lambda}{2\pi}$, which implies that the wave height $h_0 = 2r_0$ cannot exceed $\dfrac{\lambda}{\pi} = 0.32\,\lambda$.

The required curve can be obtained by means of the following simple operation. It suffices to take a disk of radius

$$\frac{1}{k} = \frac{\lambda}{2\pi} = R$$

rolling without sliding over a horizontal line and to fasten on it (perpendicular to its plane) a stylus whose tip is located distance r_0 from the center of the disk. Such a stylus will draw on a stationary vertical screen a curve, which in parametric form is expressed by

$$x = R\theta + r_0 \sin\theta,$$

$$z = r_0 \cos\theta. \tag{1.6.12}$$

This is the equation of a trochoid. The circle of radius R is the rolling circle, and the circle of radius r_0 is the generating circle. If r_0 is very small (small-amplitude) as compared with R, then the generated trochoid has a shape very close to a sinusoid with the same wave length λ and the same height $2r_0$. The higher $\dfrac{r_0}{R}$, the more explicit the difference between the trochoidal and sinusoidal profiles, i.e., the sharper the wave crest and the flatter its trough (Figure 1.6.3).

The remaining relationships are obtained by using equations of motion (1.6.3), in which one should substitute expressions for second derivatives of x and z with respect to t and use (1.6.6):

$$\frac{\partial^2 x}{\partial t^2} = -r_0 \omega^2 \sin\theta = -r_0\omega^2 e^{-kb}\sin\theta,$$

$$\frac{\partial^2 z}{\partial t^2} = -r_0\omega^2 \cos\theta = -r_0\omega^2 e^{-kb}\cos\theta, \qquad (1.6.13)$$

as well as expressions for the first derivatives $\dfrac{\partial x}{\partial a}$, $\dfrac{\partial x}{\partial b}$, $\dfrac{\partial z}{\partial a}$, $\dfrac{\partial z}{\partial b}$ (expressions (1.6.5)). Then, with consideration of equation (1.6.6),

$$\frac{\partial x}{\partial a} = 1 + kr_0 e^{-kb}\cos\theta, \qquad \frac{\partial x}{\partial b} = -kr_0 e^{-kb}\sin\theta,$$

$$\frac{\partial z}{\partial a} = -kr_0 e^{-kb}\sin\theta, \qquad \frac{\partial z}{\partial b} = 1 - kr_0 e^{-kb}\cos\theta. \qquad (1.6.14)$$

This yields two equations

$$\frac{1}{\rho}\frac{\partial p}{\partial a} = (1 + kr_0 e^{-kb}\cos\theta)\,\omega^2 r_0 e^{-kb}\sin\theta - (\omega^2 r_0 e^{-kb}\cos\theta + g),$$

$$kr_0 e^{-kb}\sin\theta = r_0 e^{-kb}(\omega^2 - kg)\sin\theta,$$

$$\frac{1}{\rho}\frac{\partial p}{\partial b} = -kr_0 e^{-kb}\sin\theta\,\omega^2 r_0 e^{-kb}\sin\theta + (\omega^2 r_0 e^{-kb}\cos\theta + g),$$

$$(1 - kr_0 e^{-kb}\cos\theta) = g - k\omega^2 r_0^2 e^{-2kb} + r_0 e^{-kb}(\omega^2 - kg)\cos\theta,$$

the first of which is multiplied by da and the second by db. Then, upon adding left-hand sides, we obtain the total differential of the pressure, divided by the density, and adding right-hand sides we obtain the rather simple expression

$$\frac{1}{\rho}\left(\frac{\partial p}{\partial a}da + \frac{\partial p}{\partial b}db\right) = \frac{1}{\rho}dp = r_0 e^{-kb}(\omega^2 - kg)\sin\theta\,da +$$

$$+ \left[g - k\omega^2 r_0^2 e^{-2kb} + r_0 e^{-kb}(\omega^2 - kg)\cos\theta\right]db.$$

Integrating this equation gives

$$\frac{1}{\rho}p = C_1 + gb + \frac{1}{2}\omega^2 r_0^2 e^{-2kb} -$$

$$- \frac{r_0}{k}e^{-kb}(\omega^2 - kg)\cos\theta. \qquad (1.6.15)$$

In particular, when substituting $b = 0$ and $p = p_0$, the following equation is derived for the wave surfaces:

$$\frac{1}{\rho} p_0 = C_1 + \frac{1}{2} \omega^2 r_0^2 - \frac{r_0}{k} (\omega^2 - kg) \cos \theta. \qquad (1.6.16)$$

In total calm it may be assumed that pressure p_0 at all points of the undulated sea surface is constant, irrespective of phase angle θ. Consequently, satisfaction of this boundary condition requires that the factor in parentheses in the coefficient of $\cos \theta$ in (1.6.16) be zero:

$$\omega^2 - kg = 0,$$

or

$$\omega^2 = kg. \qquad (1.6.17)$$

Both parts of this equation are divided by k^2. Then

$$\left(\frac{\omega}{k} \right)^2 = \left(\frac{\lambda}{\tau} \right)^2 = c^2,$$

and

$$\frac{g}{k} = \frac{g\lambda}{2\pi}.$$

Consequently, the rate of propagation of deep-water waves (c) depends solely on the wave length:

$$c^2 = \frac{g\lambda}{2\pi}. \qquad (1.6.18)$$

This result coincides with (1.3.4), obtained for waves of infinitesimal amplitudes (Section 3). When λ is very small compared with depth H, one may introduce c into (1.6.2):

$$\theta = ka - \omega t.$$

Since

$$\omega = \frac{2\pi}{\tau}; \qquad \tau = \frac{\lambda}{c}; \qquad k = \frac{2\pi}{\lambda},$$

we have $\omega = \dfrac{2\pi c}{\lambda}$ and consequently

$$\theta = ka - ktc = k(a - tc).$$

Recalling also the expression from (1.6.6),

$$r = r_0 e^{-kb},$$

we can introduce into the general formulas (1.6.1), namely

$$x - a = r \sin \theta,$$
$$z - b = r \cos \theta,$$

values of 0 and r expressed by the above relationships. This yields

$$x - a = r_0 e^{-kb} \sin k(a - ct),$$
$$z - b = r_0 e^{-kb} \cos k(a - ct). \tag{1.6.19}$$

This is the equation of wave motion of a surface situated distance b downward from the still-water sea surface. This surface, as that of the sea bounding the air, was a plane. Now there propagate over it waves with velocity c and with amplitude decreasing exponentially with increasing distance from the surface.

The value of p at all points of this undulating surface remains constant and does not depend on t. The last term in equation (1.6.16) vanishes and therefore

$$\frac{1}{\rho} p_0 = C_1 + \frac{1}{2} \omega^2 r_c^2.$$

The equation for the constant of integration has the form

$$C_1 = \frac{1}{\rho} p_0 - \frac{1}{2} \omega^2 r_0^2.$$

This expression is substituted into equation (1.6.15), in which the last term also vanishes:

$$\frac{1}{\rho} p = C_1 + gb + \frac{1}{2} \omega^2 r_0^2 e^{-2kb}.$$

Simple transformations yield

$$p - p_0 = \rho g \left[b - \pi \frac{r_0}{\lambda} \left(1 - e^{-4\pi \frac{b}{\lambda}} \right) \right], \tag{1.6.20}$$

where p_0 is the pressure at the surface. The undulated surface is thus found to be a true isobaric surface: pressure p at every point does not depend on t, but depends only on distance b separating the corresponding plane within the water from the flat surface of the calm sea.

In practice one more often encounters the need to determine the pressure not at some oscillating isobaric surface, but the pressure distribution at a stationary point, situated at a given constant distance downward from the calm-sea level. This is found from equation (1.6.20).

When the sea surface is undulated, various isobaric surfaces will start passing through the point located distance b from the calm-sea level. In particular, twice during each period this point will be traversed by that isobaric surface which corresponds to the value of p obtained by substituting the given distance b into equation (1.6.20); this will happen when the oscillating isobaric surface is situated approximately at the midpoint of its extreme upper and extreme lower positions.

It may consequently be asserted that the pressure at a stationary point (for example, at the end of a pole fastened at the sea bottom) at depth z from the calm-water level will oscillate about $\overline{p} = \rho g b$ by an amount approximately equal to

$$\Delta p \approx \pm \rho g r_0 e^{-\frac{2\pi b}{\lambda}}.$$

It follows that every isobaric surface which is horizontal in calm sea undulates (under steady wave motion conditions) along trochoids, corresponding to the same wave length, but with a radius rapidly shrinking with depth. The distance between the isobars varies along the wave, attaining a maximum at the vertex and a minimum at the trough (Figure 1.6.3). It follows from this reduction in particle [orbit] radii with depth that, when the sea becomes undulated, particles, which under stationary conditions are located on the same vertical, line up along certain curves, termed dynamic verticals, which oscillate in the plane of wave propagation. The distance between the dynamic vertical changes constantly, attaining a maximum at the trough and a minimum at the crest. The initial distance between the water particles at the trough increases by a factor of $1 + \pi \dfrac{h}{\lambda}$, and at the crest it decreases by a factor of $1 - \pi \dfrac{h}{\lambda}$. Only two dynamic verticals remain truly vertical at any given time: the vertical of the crest and the vertical of the trough; all the others are inclined toward the nearest crest (Figure 1.6.3). Thus, water particles of each rectangle formed in calm water by two horizontals and two verticals are transformed by the wave motion into curvilinear parallelograms, bounded by two trochoids and two dynamic verticals, the area of each such parallelogram remaining at all times equal to the area of the initial rectangle. Here the vertical water layers oscillate about their calm-water position in a manner similar to elastic struts fastened by their bases at a depth where there are no waves (Figure 1.6.3). Water layers which under calm-water conditions have the same thickness become, upon being converted into trochoidal layers, thinner at the trough and thicker at the crest (Figure 1.6.3). This is an inescapable result of the transition of water particles from a state of rest to wave motion. These are the only conditions under which it is possible that a layer situated at some depth will experience the same hydrostatic pressure at each point, i. e., it will be at rest and have a horizontal surface, which is precisely what is observed at depths somewhat greater than the wave length.

This becomes understandable by examining Figure 1.6.4. The circle represents the orbit along which particle n moves at a certain depth of an undulated sea. When the particle is located at the uppermost point of the orbit, the gravity force g is directed downward along the radius, while the centrifugal force f is directed along the same radius upward. Consequently, the gravity force is reduced by the magnitude of the centrifugal force and the weight of the particle is expressed in terms of quantity g_1, which is smaller than g. At its lowermost position the weight g_3 of the particle is greater than g, since in this position the centrifugal force acts in the same direction as gravity. This explains the increasing thickness of water layers in the upper part of the wave, at the crest, and its reduction at the wave trough.

The preceding discussion allows one to assess the nature of this kind of wave motion. Any fluid particle (for which a and b are constant) moves uniformly in a circular orbit, executing a complete revolution in time $t =$ $= \dfrac{2\pi}{n}$, where n is the number of revolutions performed by the particle per second.

Particles with orbital centers on the same vertical line (for which a has the same value) are constantly in phase, the phase angle being given by

$$\theta = k(a - ct).$$

The radii of circular orbits decrease with depth in geometric progression while the depth increases in arithmetic progression:

$$r = r_0 e^{-kb}. \tag{1.6.21}$$

For all the particles lying under calm-water conditions at the same depth, the value of p is the same. Hence fluid particles situated under calm-water conditions at the same horizontal plane ($r = b$) will be situated during wave motion at the same surface. The wave form is a trochoid.

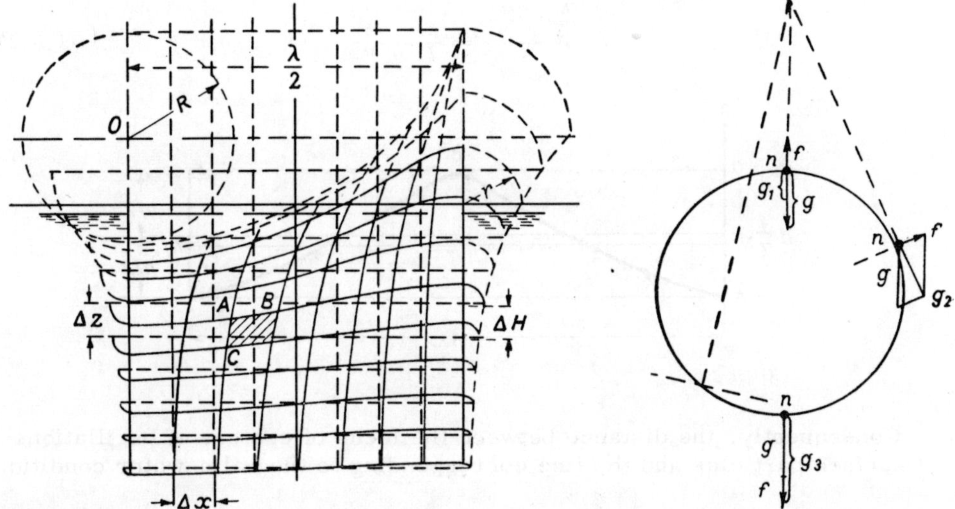

FIGURE 1.6.3. Internal structure of a trochoidal wave.   FIGURE 1.6.4.

In determining the height of the crest of a trochoidal wave over the calm-water surface, or in finding the depth of the trough from the same level, it is of interest to find the distance from the locus of the centers of oscillation of surface particles and the line corresponding to the calm-water position of the surface.

Figure 1.6.5 shows the surface ABC of a trochoidal wave. The coordinate axes have been drawn such that the x axis is tangential at points corresponding to the uppermost position of moving water particles. A horizontal tangent to the lowermost points of the trochoid is drawn parallel to the x axis at distance $2r_0 = h_0$ from it. The area contained between this horizontal (AC) and the trochoidal-wave surface (ABC) can be determined, over length λ, by the expression

$$s = \int_0^\lambda (h - z)\,dx. \tag{1.6.22}$$

Making use of equation (1.6.10) and considering the fact that

$$dx = (1 + r_0 \cos ka)\, da,$$

and

$$z = r_0 \cos ka,$$

substituting into (1.6.22) and integrating we find that

$$s = \frac{h\lambda}{2} - \frac{\pi h^2}{4}.$$

The height of the rectangle $AN = CN_1$, bounded by NN_1, the calm-water level and line AC, is

$$\frac{\dfrac{h\lambda}{2} - \dfrac{\pi h^2}{4}}{\lambda} = \frac{h}{2} - \frac{\pi h^2}{4\lambda}. \tag{1.6.23}$$

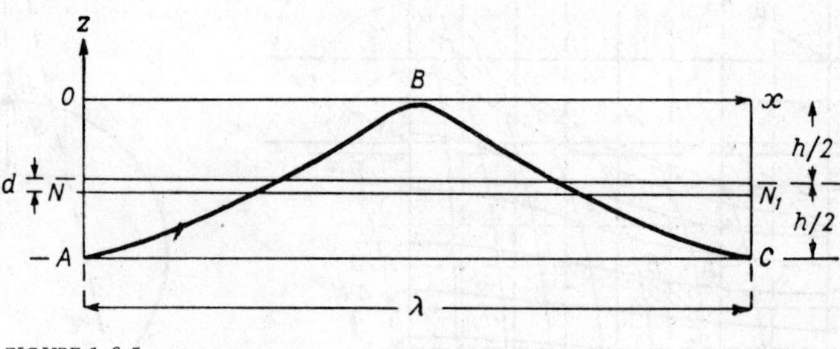

FIGURE 1.6.5.

Consequently, the distance between the locus of centers of oscillations of surface particles and the line corresponding to the calm-water condition is

$$d_0 = \frac{\pi h_0^2}{4\lambda} = \frac{\pi r_0^2}{\lambda}. \tag{1.6.24}$$

This means that the wave crests lie above the calm-water level by the amount

$$\frac{h}{2} + \frac{\pi h_0^2}{4\lambda},$$

while the wave troughs are removed from the calm-water level by the distance

$$\frac{h}{2} - \frac{\pi h_0^2}{4\lambda}.$$

It is quite obvious that (1.6.24) can be used for determining d at any level of wave oscillations with the aid of (1.6.6).

The length, period and speed of a wave are given by

$$\lambda = \frac{g\tau^2}{2\pi},$$

(1.6.25)

$$\tau = \left(\frac{2\pi\lambda}{g}\right)^{\frac{1}{2}},$$

(1.6.26)

$$c = \left(\frac{g\lambda}{2\pi}\right)^{\frac{1}{2}}.$$

(1.6.27)

The results obtained from these latter expressions are tabulated in Table 1.6.2.

TABLE 1.6.2. Tables for calculating elements of trochoidal waves

Wave length (λ) as a function of the period (τ)

τ, sec	0.0	0.1	0.2	0.3	0.4	0.5	0.6	0.7	0.8	0.9
					λ, m					
2	6	7	8	8	9	10	11	11	12	13
3	14	15	16	17	18	19	20	21	23	24
4	25	26	27	29	30	32	33	35	36	37
5	39	41	42	44	46	47	49	51	52	54
6	56	58	60	62	64	66	68	70	72	74
7	76	78	81	83	86	88	90	92	95	97
8	100	102	105	108	110	113	115	118	121	124
9	126	129	132	135	138	141	144	147	150	153
10	156	159	162	166	169	172	175	179	182	185
11	189	192	196	199	203	206	210	214	217	221
12	225	228	232	236	240	244	248	252	256	260
13	264	268	272	276	280	284	288	293	297	301
14	306	310	314	319	323	327	332	337	342	346
15	351	356	360	365	370	375	380	384	389	394
16	399	404	409	414	420	425	430	435	440	446
17	451	456	462	467	472	478	483	489	494	500
18	505	511	517	522	528	534	540	546	551	557
19	563	569	575	581	587	593	599	605	612	617

Wave period (τ) as a function of the speed (c)

c, m/sec	0.0	0.1	0.2	0.3	0.4	0.5	0.6	0.7	0.8	0.9
					τ, sec					
0	—	0.1	0.1	0.2	0.3	0.3	0.4	0.4	0.5	0.6
1	0.6	0.7	0.8	0.8	0.9	1.0	1.0	1.1	1.2	1.2
2	1.3	1.4	1.4	1.5	1.5	1.6	1.7	1.7	1.8	1.8
3	1.9	2.0	2.0	2.1	2.2	2.2	2.3	2.4	2.4	2.5
4	2.6	2.6	2.7	2.8	2.8	2.9	2.9	3.0	3.1	3.1
5	3.2	3.3	3.3	3.4	3.5	3.5	3.6	3.6	3.7	3.8
6	3.8	3.9	4.0	4.0	4.1	4.2	4.2	4.3	4.4	4.4
7	4.5	4.5	4.6	4.7	4.7	4.8	4.9	4.9	5.0	5.0
8	5.1	5.2	5.2	5.3	5.4	5.4	5.5	5.6	5.6	5.7
9	5.8	5.8	5.9	6.0	6.0	6.1	6.1	6.2	6.3	6.3
10	6.4	6.5	6.5	6.6	6.7	6.8	6.8	6.9	7.0	7.0
11	7.1	7.2	7.2	7.3	7.4	7.4	7.5	7.6	7.6	7.7
12	7.7	7.8	7.8	7.9	7.9	8.0	8.1	8.1	8.2	8.3
13	8.4	8.4	8.4	8.5	8.6	8.6	8.7	8.8	8.8	8.9
14	9.0	9.0	9.1	9.2	9.2	9.3	9.4	9.4	9.5	9.6
15	9.6	9.7	9.7	9.8	9.9	9.9	10.0	10.0	10.1	10.2
16	10.2	10.3	10.4	10.4	10.5	10.6	10.6	10.7	10.7	10.8
17	10.9	10.9	11.0	11.0	11.1	11.2	11.2	11.3	11.4	11.4
18	11.5	11.5	11.6	11.7	11.8	11.8	11.9	12.0	12.0	12.1
19	12.2	12.2	12.3	12.3	12.4	12.5	12.5	12.6	12.7	12.7

The speed at which the particles traverse their orbits is given by

$$v = \frac{2\pi r}{\tau} = \frac{2\pi r}{\lambda}\, c = \frac{r}{R}\, c,$$

(1.6.28)

or

$$v = \frac{\pi h}{\tau} = \frac{\pi h c}{\lambda} = \omega r,$$

(1.6.29)

and also

$$\frac{v}{c} = \pi\,\frac{h}{\lambda}\,.$$

(1.6.30)

The angular velocity of a particle in the orbit is given by

$$\omega^2 = \frac{2\pi g}{\lambda}\,,$$

(1.6.31)

$$\omega = \frac{2\pi}{\tau}\,.$$

(1.6.32)

The steepness of a trochoidal wave varies along its length, the wave being shallower at the trough and steeper at the crest. The angle of maximum wave steepness is found from

$$\sin\alpha = \frac{r_0}{R}\,.$$

Multiplying the numerator and denominator of this fraction by 2π and making use of the fact that $2\pi R$ is equal to the length of the circumference of a rolling circle, or to the wave length λ, we obtain

$$\sin\alpha = \frac{2\pi r_0}{\lambda} = \frac{\pi h}{\lambda}\,.$$

(1.6.33)

The average wave steepness can be characterized approximately by the ratio of the height h to half the wave length. Hence the average steepness is given by

$$\operatorname{tg}\alpha = \frac{2h}{\lambda}\,.$$

(1.6.34)

The values of trochoidal-wave steepness for various ratios $\frac{h}{\lambda}$ are listed in Table 1.6.3.

TABLE 1.6.3. Average and maximum steepness (in degrees) of a trochoidal wave

Wave steepness	Ratio of wave height to length						
	1:7	1:10	1:15	1:20	1:25	1:30	1:40
Average	16	11	8	6	5	4	3
Maximum	26	18	12	9	7	6	4

It is seen from profiles of an undulated sea surface that the form of wind waves in certain cases is quite different from a trochoid. Sometimes the wave form is close to a cycloid, and in other cases it is quite shallow (Figure 1.6.6). At the same time the average profile of a wind wave, obtained from a large number of wave profiles, was found to agree satisfactorily with a trochoid, differing from it somewhat at the crest slopes. Swell profiles conform to the trochoid better and more frequently. Since there exists a large variety of wind-generated wave forms, it is quite useless to seek rigorous conformance to a trochoid.

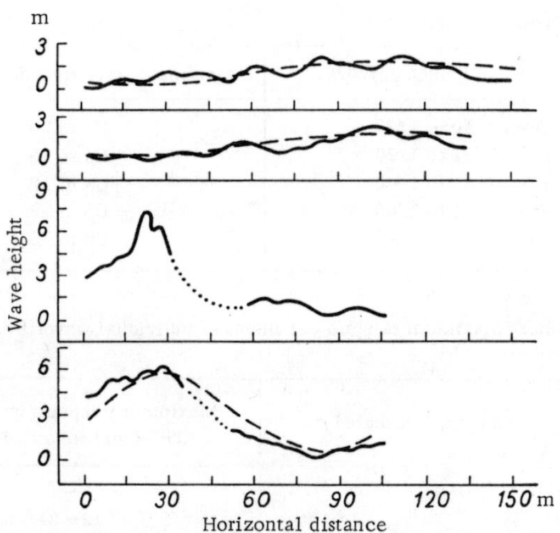

FIGURE 1.6.6.   Wind-wave profiles obtained by stereo-photography.

The dashed line indicates the trochoidal-wave profiles.

The steepness of wind-wave crests is highly variable from point to point; this follows from the aforementioned arbitrariness of the wave forms. For a large number of wind waves the steepness of the windward slope is greater than that of the leeward.

Wind wave profiles are very frequently asymmetric; the projection of the windward is usually in a ratio of 2:3 to that of the leeward slope. The maximum steepness of small, secondary waves usually does not exceed 30°, while that of major waves lies within the limits of 15 — 16° for a force 2 — 3 wind and increases to 20° for stronger winds. In individual cases the wave steepness of small sections is as high as 45°. Maximum steepness over the entire crest slope can be 20° or somewhat higher; this excludes individual, highly interfered-with waves, the crests of which may be even steeper.

Table 1.6.4 lists maximum average wave steepness values obtained by stereophotography. Doubtlessly, the steepness of trochoidal waves (Table 1.6.3) is much smaller than that of real waves (Table 1.6.4).

Even greater differences are found when comparing individual wave slopes. Then the true slope of the windward crest may exceed 4—5 times, and in individual cases even 7 times, the steepness of the trochoidal profile, while the steepness of the leeward slope may exceed it by a factor of 3—4. This difference is attributable to the fact that the wind-wave profile is asymmetric. It should also be remembered that the fact that the crest steepness is measured on the basis of the profile in itself underestimates it. An even higher steepness could be obtained by measuring the slopes of individual waves (Table 1.6.5).

TABLE 1.6.4.

Ratio of wave height to length	Maximum wave steepness, degrees
From 1/10 to 1/20	Up to 20
From 1/21 to 1/30	Up to 15
From 1/31 to 1/40	Up to 12
From 1/41 to 1/50	Up to 9
From 1/51 to 1/60	Up to 6

TABLE 1.6.5. Maximum steepness of slopes of individual waves from stereophotography data

Wave strength [sea scale]	Maximum steepness of slopes of individual waves, degrees
I—III (swell)	1—4
IV—V	12—24
V—VI	13—14
VI—VII	9—20
VII	~ 20

The vertex angles of the crest increase the more rapidly, the faster the wind waves are transformed to swell. Observations show that these angles lie between 149° and 154°. The vertex angle of crests of breakers are found to equal 130° and not to exceed 140°. These observation results agree with theoretical conclusions, according to which the angle at the crest vertex cannot be higher than 120°, after which the crest breaks and should spill over.

The limiting ratio (1/7) of wave height to its length followed from theoretical conclusions and was not established for any clearly delineated wind-wave profile. However, according to stereophotography in open sea, the limiting ratio of wind height to its length is 1/9—1/10, which is quite close to the theoretical value. Steeper waves were not observed.

FIGURE 1.6.7. Profile of wave-crest tops (after Shuleikin).

The actual variety of wave forms and dimensions is responsible for the fact that the actual ratios of wave elements vary within $\pm 20\%$ from the ratios given by trochoidal wave theory. The difficulty in checking the theory is amplified by the fact that presently used experimental techniques are such that no single instrument is available for recording simultaneously, and with the same accuracy, the length, period and phase velocity of waves. It may be assumed in the first approximation that the difference between the values of wave elements calculated from formulas of the trochoidal theory and observed values is $\pm 20\%$.

The profile of real wind waves was assessed in laboratory studies, carried out in a special annular tank (Shuleikin, 1956), in which waves travel along a circle as long as necessary, sustained by winds up to $19\,\mathrm{m/sec}$. The waves may be as high as $1.5\,\mathrm{m}$ and as long as $18\,\mathrm{m}$. It became possible to express the profile of such a wave analytically in a form (Shuleikin, 1956) similar to the equation of a trochoidal wave:

$$x = R\theta + a \sin\theta,$$
$$z = b \cos\theta. \qquad (1.6.35)$$

The first equation of this system contains the major semiaxis (a) of some ellipse, and the second contains the minor semiaxis (b) of this ellipse. The profile of such a wave is shown in Figure 1.6.7. The wave crest cuts a segment AB on the horizontal axis which is shorter than segment BC, cut by the wave trough. Denoting the ratio AB/BC by α, we have

$$a = \frac{1-\alpha}{1+\alpha}\frac{\lambda}{4},$$
$$b = \frac{h}{2}.$$

The actual form of wind-wave profile is sharper than the trochoidal wave profile. The cause of this in swell is the wave current (Chapter 1, Section 5), the velocity of which, however, is not constant, but varies continuously within a single period (Shuleikin, 1956). The velocity of this translational motion is

$$u = \omega\frac{r_0^2}{R}(1 + \cos\theta). \qquad (1.6.36)$$

This expression indicates that this velocity varies along the wave.

On the basis of (1.6.36) the average velocity of translational motion over the segment $\theta = 0$ to $\theta = 2\pi$ is given by

$$\bar{u} = \omega\frac{r_0^2}{R} = r_0^2\left(\frac{2\pi}{\lambda}\right)^2 c = \pi^2\delta_0^2 c,$$

which agrees with (1.5.3) obtained on the basis of purely formal postulates of the theory of finite-amplitude waves for the surface, if $z = 0$ in (1.5.3). Naturally, equation (1.6.36) is applicable to any water depth, if equation (1.6.6) is used. The higher the ratio of the wave height to its length, the stronger the fluctuations of this flow and the sharper the wave peaks. This pattern is complicated by the wave, but the substance of the phenomenon remains unchanged. However, alongside the wave current, the wind now starts generating drift currents. This additional, but much stronger current, upon being superimposed on the wave current, makes the peaks of waves (which are now wind waves) even sharper. This process is expressed by the following simple formula (Shuleikin, 1956):

$$\frac{a}{b} = 1 + \frac{r_0}{R}\left(1 + \frac{\overline{u}_{dr}}{v}\right), \tag{1.6.37}$$

where $R = \dfrac{\lambda}{2\pi}$.

The sharpness of wave peaks, typified by the ratio $\dfrac{a}{b}$, can also be expressed by the ratio $\dfrac{r_0}{R}$, i.e., wave steepness, and by the ratio of $\overline{u}_{dr}$, the drift-current velocity, to v, the orbital velocity of water particles. When there is no wind, the sharpness of wave peaks depends only on the wave steepness. A continuous increase in sharpness of wind-wave peaks should in the final count result in breaking of the wave, which formally can occur when its profile has a vertex angle of 120° and its steepness is $\delta_{max} = 0.143$ (Chapter 1, Section 5). The critical velocity $u_{dr.cr}$ of the drift current at which the crest breaks is given by

$$\left(\frac{\overline{u}_{dr}}{v}\right)_{cr} = 0.65\left(\frac{R}{r_0}\right)^2 - \frac{R}{r_0} - 1. \tag{1.6.38}$$

When the current velocity exceeds this critical value, foamy whitecaps appear on the waves.

A wind wave breaks before its steepness reaches $\frac{1}{7}$, as is required by the theory (Chapter 1, Section 5), i.e., at $\delta = \frac{1}{10}$ and vertex angle 140°. At this steepness $\frac{u_{dr}}{v} \simeq 2.4$. Whitecaps appear in all these cases.

7. Energy of a trochoidal wave

Water particles on the surface of a trochoidal wave (Shuleikin, 1959) not only travel in circular orbits, but also rise above the calm-water level (1.6.24) by an amount

$$d_0 = \frac{\pi h^2}{4\lambda} = \frac{\pi r_0^2}{\lambda}.$$

Hence the energy of a particle (the mass m of which is taken as unity) is composed of the kinetic energy produced by the motion of particles in a circular orbit with linear velocity v and equal to

$$\Delta E_\kappa = \frac{1}{2}v^2, \tag{1.7.1}$$

and of the potential energy, produced by raising the water particles to some height d and equal, for surface particles, with consideration of (1.6.24), to

$$\Delta E_p = g d_0 = \frac{g \pi r_0^2}{\lambda} .$$ (1.7.2)

It follows from (1.6.28) that

$$v^2 = \frac{2\pi r_0^2 g}{\lambda} ,$$

and hence expression (1.7.1) is replaced by

$$\Delta E_\kappa = \frac{g \pi r_0^2}{\lambda} .$$ (1.7.3)

The latter expression is the same as (1.7.2). Consequently, the kinetic energy of a surface water particle with unit mass is equal to its potential energy.

The total energy of the particle is equal to the sum of the kinetic and potential energies:

$$\Delta E = \Delta E_p + \Delta E_\kappa = \frac{2\pi g r_0^2}{\lambda} ,$$ (1.7.4)

or, making use of (1.6.6), for a water layer at any distance from the surface

$$\Delta E = \frac{2\pi g}{\lambda} r_0^2 e^{-2kb} .$$ (1.7.5)

To determine the total energy in the entire layer participating in the wave motion, per unit wave surface, (1.7.5) should be integrated from $b = 0$ to $b = \infty$:

$$\int_0^\infty \Delta E \, db = \frac{\rho 2\pi g}{\lambda} r_0^2 \int_0^\infty e^{-2kb} \, db .$$

Hence the total wave energy per unit of its surface is

$$E = \frac{\rho g r_0^2}{2} = \frac{\rho g h_0^2}{8} .$$ (1.7.6)

To determine the wave energy in a layer, for example, from b_1 to b_2, corresponding integration yields

$$E_{b_1} - E_{b_2} = \frac{\rho g}{8} h_0^2 (e^{-2kb_1} - e^{-2kb_2}) .$$ (1.7.7)

If the energy is calculated for a wave length λ, then

$$E_\lambda = \frac{\rho g}{8} h_0^2 \lambda .$$ (1.7.8)

It should be noted that neither the expression for the kinetic, nor for the potential energy per unit sea surface contains the wave length. Consequently, it is natural that it is the increase in wave height which is related to external energy supply.

The kinetic energy is constant along the wave length, since it depends on the orbital velocity of particles, and the latter does not change. As to the potential energy, it varies as a function of the positions of particles in the orbit.

When considering the energy-distribution pattern with depth (Kondrat'ev, 1953) h in expression (1.7.6) is replaced by $h = 2r_0 e^{-\frac{2\pi g}{\lambda}}$. Then (1.7.6) can be rewritten in the form

$$\frac{E}{\lambda} = \frac{1}{2} g\rho r_0 e^{-2kb}.\tag{1.7.9}$$

From this expression one may find the vertical energy distribution, i.e., the amount of energy referred to unit volume. Denoting this energy by ϑ, we derive the expression

$$\theta = \frac{1}{\lambda}\frac{dE}{db} = \frac{2\pi\rho g r_0^2 e^{-2kb}}{\lambda}.\tag{1.7.10}$$

The variation of θ as a function of r_0 is given by

$$d\theta_{r_0} = \frac{\partial\theta}{\partial r_0} dr_0 = \frac{4\pi g\rho r_0}{\lambda} e^{-2kb} dr_0,\tag{1.7.11}$$

and making use of (1.7.10) we have

$$d\theta_{r_0} = \frac{2\theta}{r_0} dr_0.\tag{1.7.12}$$

The variation of θ with λ is given by the expression

$$d\theta_\lambda = \frac{\partial\theta}{\partial\lambda} d\lambda = \frac{2\pi g\rho r_0^2 e^{-\frac{4\pi b}{\lambda}}}{\lambda} d\lambda,\tag{1.7.13}$$

from which it follows that

$$d\theta_\lambda = -\frac{\theta}{\lambda}\left(1 + \frac{4\pi b}{\lambda}\right) d\lambda.\tag{1.7.14}$$

Comparison of (1.7.12) and (1.7.14) indicates (Figure 1.7.1) that the wave energy referred to unit volume increases over the entire vertical with increasing wave amplitude (curves 1 and 2 in Figure 1.7.1). However, this quantity also varies with increasing wave length, but in a different direction, depending on the value of b. For b in the interval from $b = 0$ to $b = -\frac{\lambda}{4\pi}$ the term in parentheses in (1.7.14) has a positive sign, while when $b < -\frac{\lambda}{4\pi}$ it is negative. This means that the energy in the upper part of the vertical above the elevation

$$b = -\frac{\lambda}{4\pi} \simeq -0.08\lambda,\tag{1.7.15}$$

decreases with increasing wave length, while the energy below this elevation increases (curve 3 in Figure 1.7.1). Consequently, with increasing r_0 (wave height) the energy increases, primarily in the upper part of the water layer, while with increasing λ the energy is redistributed so as to equalize its distribution in the water layer participating in the wave oscillations.

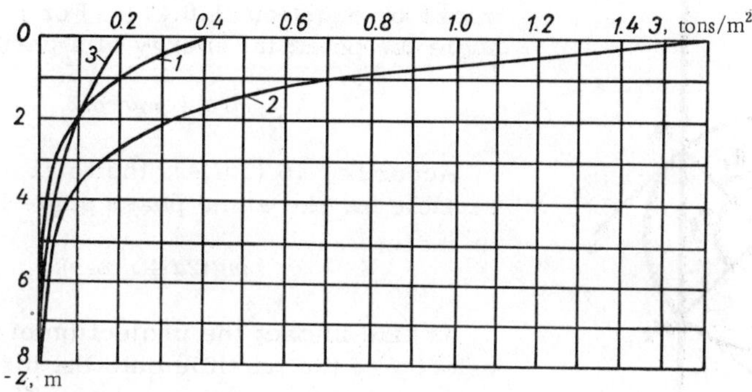

FIGURE 1.7.1. Vertical energy distribution for waves of different size (after Kondrat'ev):

$1 - h = 1.0$ m, $\lambda = 16$ m; $2 - h = 2.0$ m, $\lambda = 16$ m; $3 - h = 1.0$ m, $\lambda = 32$ m.

Expressions for the energy of a trochoidal wave are also valid for a progressive wave with infinitesimal amplitude, including tidal waves. The energy of a standing wave is given by

$$F = \frac{\rho g h_{st.}^2 \lambda}{16},$$ (1.7.16)

where $h_{st} = 2h$.

8. Energy transport by wave motion. Group velocity of waves

To gain insight into the question of energy transmission by waves (Shuleikin, 1956) we must again consider water-particle motion in a wave motion consisting of trochoidal waves. The wave-form displacement is denoted in Figure 1.8.1 by arrows. Vectors v denote the linear velocity of the motion of particle M along an orbit corresponding to phase angles θ_2 and θ_1. Here the moving particle intersects the same vertical plane AB, first when moving from left to right (in the direction of wave-form displacement), and later when moving from right to left (opposite to the wave direction). The kinetic energy of the particle is the same in both cases, while the potential energy is different because the particle at point M_1 is higher than in position M_2. Hence a water particle traveling in a circular orbit from that part of space to the left of plane AB transmits a part of its potential energy to the right. This results in continuous

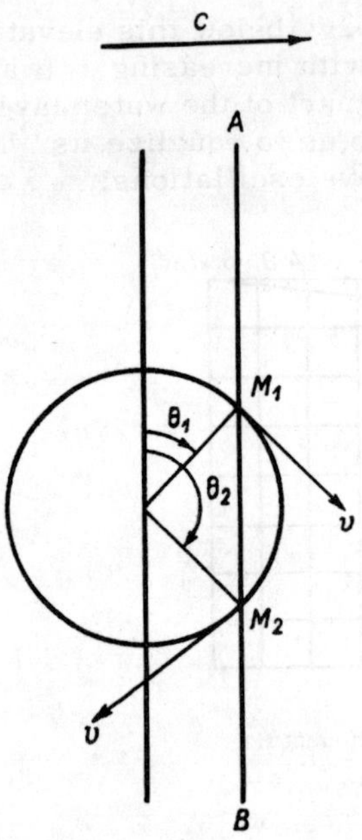

FIGURE 1.8.1.

transmission of the energy of oscillatory motion in the direction of wave propagation.

The kinetic energy of fluid particle M, according to results of the preceding section, is equal to its potential energy. The oscillations of potential energy for a trochoidal wave are determined by equation (1.6.11). For any given phase angle the potential energy of a particle is

$$mgd + mgr \cos \theta.$$

According to (1.7.4), the total energy of the particle for the same phase angle is

$$mg(2d + r \cos \theta).$$

At this instant the projection of the linear velocity of the particle onto the direction of wave propagation (the horizontal axis) is expressed by the product of v and $\cos \theta$. Hence the energy flux transmitted by the particle in the direction of wave motion equals the product of the particle energy and the horizontal component of linear velocity:

$$\Delta \Phi = mg(2d + r \cos \theta) v \cos \theta. \qquad (1.8.1)$$

According to the above, the energy flux in the direction of motion is greater than in the opposite direction. Usually when considering energy transmission by waves reference is had to the energy flux averaged over a wave period. This averaging is carried out by integrating (1.8.1) from $\theta = 0$ to $\theta = 2\pi$, and then dividing the result obtained by 2π:

$$\Delta \overline{\Phi} = \frac{mgv_0}{2\pi} \int_0^{2\pi} (2d + r_0 \cos \theta) \cos \theta \, d\theta = \frac{1}{2} mgv_0 r_0.$$

However, using (1.6.28),

$$v_0 = \frac{r_0}{R} c,$$

$$\Delta \overline{\Phi} = \frac{mg\pi r_0^2}{\lambda} c,$$

or, assuming $m = 1$,

$$\Delta \overline{\Phi} = \frac{g\pi r_0^2}{\lambda} c. \qquad (1.8.2)$$

The coefficient of the phase velocity in (1.8.2) represents, according to (1.7.2), the averaged potential energy of the same particle. Denoting the

total energy of the water particle by ΔE, expression (1.8.2) may be
replaced by

$$\Delta \overline{\Phi} = \Delta E \frac{c}{2}.$$

Proceeding in the same manner the entire wave-energy flux transmitted
over the depth of the undulated sea is found to be given by the similar
expression

$$\overline{\Phi} = E \frac{c}{2}. \tag{1.8.3}$$

It appears at first sight that the energy should be transmitted at the wave
velocity. It may be imagined that there propagates in the medium (water in
our case) an ensemble of waves having a dispersion, with the wave frequencies,
and consequently periods, differing from one another. The term dispersion
denotes variation in the wave velocity as a function of the wave length. The
dispersion of waves results in the appearance of wave groups, i.e., the
existence of successive wave trains.

As a result of interference the propagation pattern of a wave group with
close velocities assumes the form shown in Figure 1.8.2.

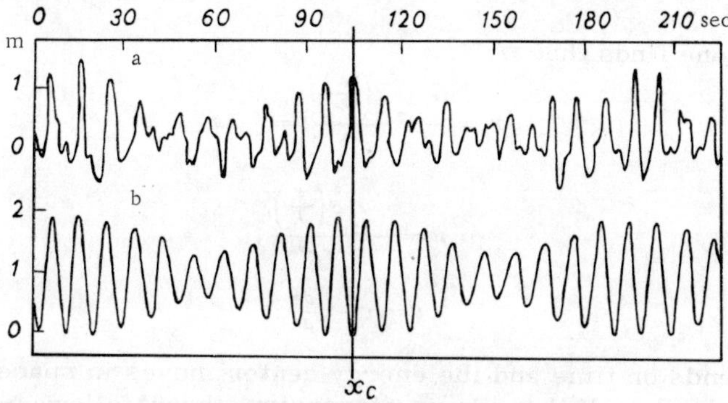

FIGURE 1. 8. 2.

a—record of wave oscillations obtained by a wave recorder; b—result of
adding two wave systems with periods close to those expressed by the wave
record; wave groups are clearly seen.

Due to differences in velocity, the waves will in general arrive at a given
point of the wave field out of phase and, at any instant, the resulting advance
at a given point in space will depend on the relationship between phases of
individual oscillations. If the waves propagate in a large number of phases,
then the resulting oscillations are moderate; if the phases are approximately
equal, then the oscillations add and attain a maximum. Here the energy
density at this point will be maximum, which is also true of the amplitude.
However, at a subsequent instant the phase relationship will definitely
change and this involves a reduction in the energy density. Points with
maximum energy density are termed the energy center of the wave group.
It is obvious that due to the continuous variation in the relationship between
phases of individual oscillations the energy center will move in the space

embraced by the wave field.   The rate at which this center moves can be determined.   For this it is assumed that at some time (t) the energy center lies at a point with coordinate x_c (Figure 1.8.2).   The motion corresponding to each individual oscillation can be obtained from the expression

$$z = r \sin \frac{2\pi}{\tau}\left(t - \frac{x_c}{c}\right) = r \sin \omega \left(t - \frac{x_c}{c}\right),$$

where

$$\omega t - \omega \frac{x_c}{c} = \frac{2\pi}{\tau} t - \frac{2\pi}{\tau} \frac{x_c}{c} = \theta.$$

Since the phases of oscillations of waves of different, but close length coincide at the energy center, here the oscillation phases do not depend on the wave length and hence the derivative of the phase along the wave is zero. Consequently

$$\theta = 2\pi \left(\frac{c}{\lambda} t - \frac{x_c}{\lambda}\right),$$

$$\frac{d\theta}{d\lambda} = 2\pi \left[t \frac{d\left(\frac{c}{\lambda}\right)}{d\lambda} - \frac{d\left(\frac{x_c}{\lambda}\right)}{d\lambda} \right] = 2\pi \left[t \frac{d\left(\frac{c}{\lambda}\right)}{d\lambda} + \frac{x_c}{\lambda^2} \right] = 0,$$

from which one finds that

$$x_c = -\lambda^2 \frac{d\left(\frac{c}{\lambda}\right)}{d\lambda} t = -ut.$$

Hence

$$u = -\lambda^2 \frac{d\left(\frac{c}{\lambda}\right)}{d\lambda},$$

or

$$u = c - \lambda \frac{dc}{d\lambda}, \tag{1.8.4}$$

i.e., x_c depends on time and the energy center moves in space with speed u, the group velocity.   If there is no dispersion, then it follows from equation (1.2.14) that

$$\frac{dc}{d\lambda} = 0,$$

and consequently (1.8.4) implies that

$$u = c. \tag{1.8.5}$$

If, on the other hand, dispersion is present and the velocity, for example, is given by (1.2.13), then it is easily found that

$$u = \frac{1}{2} c. \tag{1.8.6}$$

It is possible to obtain an expression for u in general form. Let the relationship for the phase velocity have the form

$$c = m\lambda^n,$$

where m is a dimensional constant, while n is an exponent which has a characteristic value for different types of waves.

Dividing the last expression by λ, we obtain

$$\frac{c}{\lambda} = m\lambda^{n-1},$$

in which case

$$\frac{d\left(\frac{c}{\lambda}\right)}{d\lambda} = (n-1)\,m\lambda^{n-2}.$$

Multiplying both parts of the above equation by $-\lambda^2$, replacing $m\lambda^n$ by c, and making use of the fact that

$$u = -\lambda^2 \frac{d\left(\frac{c}{\lambda}\right)}{d\lambda},$$

it is found that

$$u = (1-n)\,c.$$

Consequently, as n decreases u approaches c.

According to (1.8.6), the group velocity of waves (u) is related to the phase velocity c for deep-water conditions by the expression

$$c = 2u.$$

As a result (1.8.3) may be replaced by

$$\overline{\Phi} = Eu, \qquad\qquad\qquad (1.8.7)$$

where E is the total wave energy per unit wave length. Denoting the amount of energy transmitted through a vertical section of unit width per unit time by E_Φ, and using (1.8.3) and (1.8.7), we may write

$$\frac{E_\Phi}{E} = \frac{u}{c},$$

i. e., the quantity of transmitted wave energy is in the same ratio to its total energy, as the group velocity is to the phase velocity. In terms of the notation

$$\frac{E_\Phi}{E} = n,$$

we have

$$n = \frac{u}{c}. \tag{1.8.8}$$

For infinite depth

$$\frac{u}{c} = \frac{1}{2}$$

and consequently

$$n = \frac{1}{2}. \tag{1.8.9}$$

It follows from the above that the energy lags behind the geometric wave form. This is the reason why any local excitation of the surface of deep water always leaves "residual" motions in the wake of its main part. These motions are the more pronounced the deeper the water.

For a small-depth fluid (long waves in shallow water)

$$\frac{u}{c} = 1; \tag{1.8.10}$$

hence

$$n = 1.$$

It follows that in this case the entire energy is transmitted by the waves and consequently the dynamic characteristics do not lag behind the geometric characteristics of motion. It is for this reason that long waves propagate without changing the wave form.

The energy transport for

$$0.1 < \frac{H}{\lambda_\infty} < 0.3$$

is given by

$$u = \frac{1}{2}\,c\left(1 + \frac{2kH}{\operatorname{sh} 2kH}\right). \tag{1.8.11}$$

The previously examined transmission of wave-motion energy can be explained by the following example. Consider a wave group. As soon as the first wave of this group advances by one wave length, it will induce corresponding velocities in previously calm water. The kinetic energy corresponding to these velocities should be subtracted from the energy transmitted at the phase velocity, i.e., together with the wave form. Suppose the energy in the wave is distributed such that half the potential energy moving with the wave is converted into kinetic energy. Consequently, the energy of the entire wave decreases and this means a reduction in wave height. When the wave at hand, i.e., the first wave of the group advances by still another length, it will again lose half the potential energy it presently has. This again results in a reduction in wave height. This process will continue until the first wave under study becomes so small as to be imperceptible. The waves following the first wave will not advance over water disturbed by the first wave and hence the reduction in their height will be less pronounced. Consequently, the potential energy at the forward edge of the wave group "advances," while the rear of the group contains half the entire energy in the kinetic form.

TABLE 1.8.1.

Wave serial number n	1	2	3	4	5	6	7	8	9	10
1	$\frac{1}{2}E$									
2	$\frac{3}{4}E$	$\frac{1}{4}E$								
3	$\frac{7}{8}E$	$\frac{4}{8}E$	$\frac{1}{8}E$							
4	$\frac{15}{16}E$	$\frac{11}{16}E$	$\frac{5}{16}E$	$\frac{1}{16}E$						
5	$\frac{31}{32}E$	$\frac{26}{32}E$	$\frac{16}{32}E$	$\frac{6}{32}E$	$\frac{1}{32}E$					
6	$\frac{63}{64}E$	$\frac{57}{64}E$	$\frac{42}{64}E$	$\frac{22}{64}E$	$\frac{7}{64}E$	$\frac{1}{64}E$				
7	$\frac{127}{128}E$	$\frac{120}{128}E$	$\frac{99}{128}E$	$\frac{64}{128}E$	$\frac{29}{128}E$	$\frac{8}{128}E$	$\frac{1}{128}E$			
8	$\frac{255}{256}E$	$\frac{247}{256}E$	$\frac{219}{256}E$	$\frac{163}{256}E$	$\frac{93}{256}E$	$\frac{37}{256}E$	$\frac{9}{256}E$	$\frac{1}{256}E$		
9	$\frac{511}{512}E$	$\frac{502}{512}E$	$\frac{466}{512}E$	$\frac{328}{512}E$	$\frac{256}{512}E$	$\frac{130}{512}E$	$\frac{46}{512}E$	$\frac{10}{512}E$	$\frac{1}{512}E$	
10	$\frac{1023}{1024}E$	$\frac{1013}{1024}E$	$\frac{968}{1024}E$	$\frac{848}{1024}E$	$\frac{638}{1024}E$	$\frac{386}{1024}E$	$\frac{170}{1024}E$	$\frac{56}{1024}E$	$\frac{11}{1024}E$	$\frac{1}{1024}E$

It is assumed that the water in a sufficiently long wave channel was initially at rest. A wave generator is provided at one of the channel ends and the former initiates harmonic wave oscillations. The first push of the wave generator produces a wave with energy $\frac{1}{2}E$. When the generator hits the water for the second time, the first wave moves through a wave length, but in the process it loses (leaves behind) half of its energy at the point of origin, i.e., at the paddle. This energy is expended upon oscillatory motion of the water particles, i.e., it is converted into kinetic energy. The second push of the paddle again produces a wave with energy $\frac{1}{2}E$. However, this energy is imposed upon an energy $\frac{1}{4}E$ left behind by the previous wave. Consequently, the energy of the second wave is now $\frac{1}{2}E + \frac{1}{4}E = \frac{3}{4}E$. But half of this energy is again left behind, and only $\frac{3}{8}E$ advances. This will continue with each push of the paddle.

After a time interval equal to n periods we obtain n waves. The position of the first wave of each group is denoted by m, i.e., for this wave

$$m = 1.$$

The position of the wave at the group center is

$$m = \frac{n+1}{2},$$

and the position of the wave at the end of the group, i.e., of the wave which advanced further than the others, is

$$m = n.$$

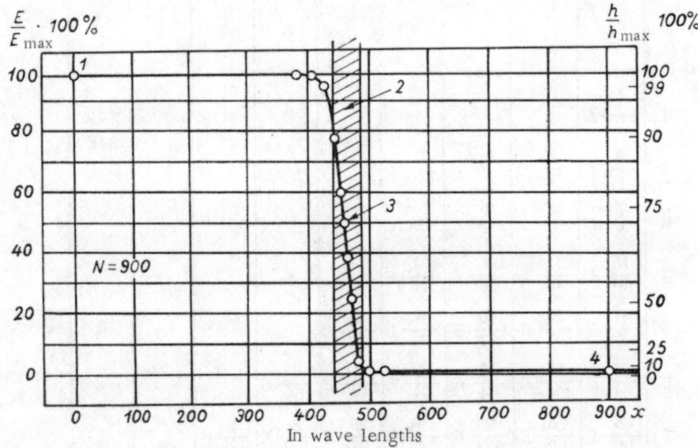

FIGURE 1.8.3. Variation in wave height with increasing distance from the point of origin:

1—waves just leaving the point of origin; 2—region of sharp reduction in wave height; 3—central wave; 4—waves which advanced through the greatest distance from the origin.

Table 1.8.1 gives the distribution of wave energy in a wave channel after ten pushes of the paddle ($n = 10$). It follows that the last, tenth wave of the group has negligible energy, only $1/1024 E$. The middle number in the rows of Table 1.8.1 corresponding to odd wave groups is always $1/2 E$. This characterizes the advance of half the energy at the phase velocity (or of all the energy at half the phase velocity). It is also clear from this table that the quantity of energy in waves advancing ahead of the central wave decreases sharply in each group. This is most clear from Figure 1.8.3 in which the right vertical axis shows the percent of the maximum wave height $\left(\frac{h_{max}}{h} 100\% \right)$, while the left axis shows the percent of maximum energy $\left(\frac{E_{max}}{E} 100\% \right)$; the horizontal axis represents the distance in wave lengths.

There is a total of 900 waves. The central wave is located at a distance of 450.5 wave lengths from the wave generator, and its height is 70.7% of the maximum height. The wave height (energy) decreases sharply only in 50 waves. For them the wave height decreases from 90 to 10% of the maximum. These waves are located at $1/2$ of the distance traversed by the first of the 900 waves. Consequently, the energy center (energy front), which is the region of maximum disturbance, moves at a velocity equal to half or somewhat more of the velocity of each individual wave.

Chapter 2

FUNDAMENTALS OF THE THEORY OF WIND WAVES

1. Generation of wind waves

Observations of the generation of wind waves are carried out best in laboratory experiments: a wind flux, generating oscillatory motion in water, is produced in small tanks (wave channels). It is found (Kononkova, 1953) that even relatively weak winds, with velocities of $85 - 100$ m/sec, generate small waves (ripples) on the initially smooth water surface. These waves have amplitudes of the order of only $10^{-3} - 10^{-2}$ cm, and are produced by turbulent fluctuations in wind pressure. Figure 2.1.1 shows the recorded fluctuations of wind pressure (p) at different wind velocities. The maximum of these fluctuations increases with increasing wind speed. Figure 2.1.2 illustrates the spectra of amplitudes (a) of wind-generated waves at approximately the same wind speeds as those given in Figure 2.1.1. As the wind speed increases, the maximum of the spectra also increases. Both types of spectra pertain to the same frequency range and vary in the same direction with increasing wind speed. The frequency range expands at the same time. Figure 2.1.3 shows the variation in wave length as a function of wind speed at points located at different distances from the end of the wave channel, i. e., from the point of wave generation.

Before discussing experimental results it should be recalled (Kochin, Kibel' and Roze, 1948) that the velocity of quite small waves cannot be calculated from (1.2.12), since for these waves one must allow for the effect of capillary forces. These forces arise from interaction between the fluid molecules. This interaction is appreciable only in a very thin surface fluid layer. The velocity of such waves is determined from the expression

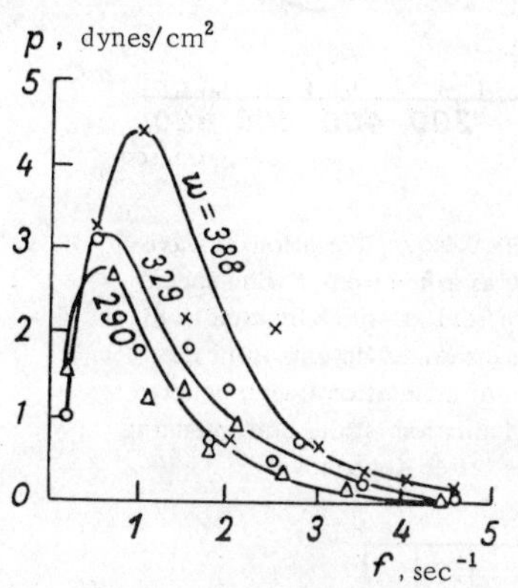

FIGURE 2.1.1. Spectra of turbulent wind pressure fluctuations for different wind speeds (in cm/sec), indicated on the corresponding curve (after Kononkova).

$$c^2 = \left(\frac{g}{k} + \frac{ak}{\rho}\right) \operatorname{th} kH,$$

$$(2.1.1)$$

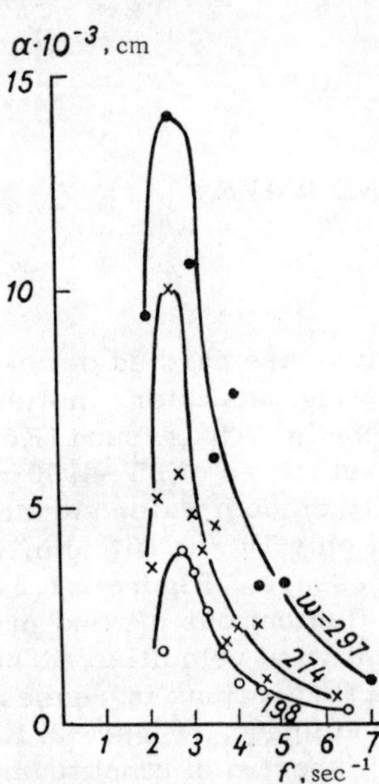

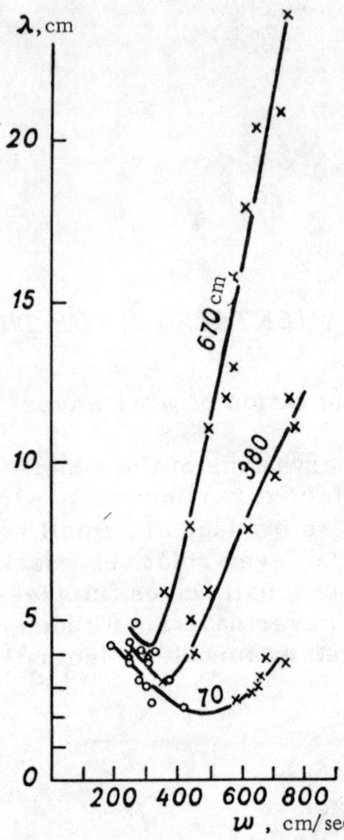

FIGURE 2.1.2. Spectra of amplitudes of wind-generated waves with wind speeds (in cm/sec) indicated on the corresponding curves (after Kononkova).

FIGURE 2.1.3. Variation in wavelength as a function of wind speed (in cm/sec) at points located at different distances (in cm) from the region of generation (wave generator), indicated on the corresponding curves (after Kononkova).

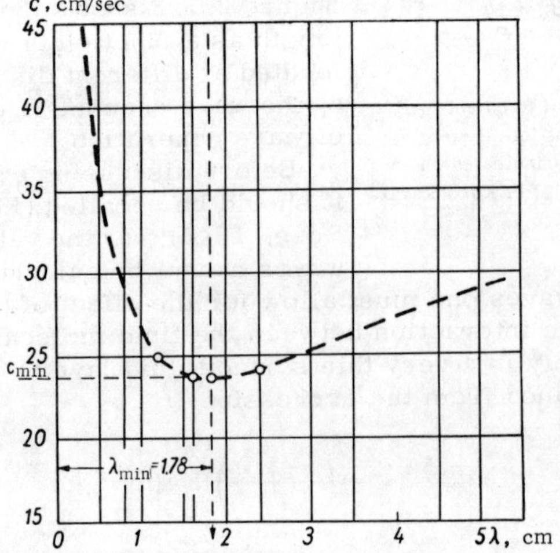

FIGURE 2.1.4.

66

where α is the surface tension of the fluid.

For deep-water conditions $(H \to \infty)$ one may assume that in (2.1.1) th $kH \approx 1$; hence

$$c^2 = \frac{g}{k} + \frac{\alpha k}{\rho},$$

or

$$c^2 = \frac{g\lambda}{2\pi} + \frac{2\pi\alpha}{\rho\lambda}. \tag{2.1.2}$$

Figure 2.1.4 shows the curve of $c = f(\lambda)$ given by (2.1.2). At very small and very large λ the velocity c can be quite appreciable. The derivative

$$\frac{dc^2}{d\lambda} = \frac{g}{2\pi} - \frac{2\pi\alpha}{\rho\lambda^2}$$

vanishes when

$$\lambda_{min} = 2\pi \left(\frac{\alpha}{\rho g}\right)^{\frac{1}{2}}.$$

When $\lambda \ll \lambda_{min}$, $\dfrac{dc^2}{d\lambda}$ is negative; and when $\lambda > \lambda_{min}$ it is positive. Consequently the rate of propagation (c) first decreases, as λ increases from 0 to ∞, from infinity to some minimum (c_{min}), and then increases from c_{min} to ∞. This minimum c_{min} is attained at $\lambda = \lambda_{min}$ and is given by

$$c_{min}^2 = 2 \left(\frac{\alpha g}{\rho}\right)^{\frac{1}{2}}.$$

It is also possible to use the expression

$$c^2 = \frac{g\lambda_{min}}{\pi} = \frac{4\pi\alpha}{\rho\lambda_{min}}.$$

Using the latter two expressions, we may express (2.1.2) in the form

$$\frac{c^2}{c_{min}^2} = \frac{1}{2}\left(\frac{\lambda}{\lambda_{min}} + \frac{\lambda_{min}}{\lambda}\right). \tag{2.1.3}$$

Comparison of expressions (2.1.2) and (2.1.3) show that for $\lambda \gg \lambda_{min}$ the only term of substance in (2.1.2) is the first. In this case capillary forces may be neglected. When $\lambda \ll \lambda_{min}$, on the other hand, one should neglect the effect of gravity forces and the first term in (2.1.2) can be discarded. Waves of length $\lambda < \lambda_{min}$ are termed capillary. It follows from the above that gravitational waves cannot be shorter than some minimum wave length (λ_{min}). It should also be mentioned that two waves of different wave lengths may propagate at the same speed: one of these will be capillary (λ_α), and the other gravitational (λ_g). In all cases $\lambda_\alpha < \lambda_{min} < \lambda_g$.

The numerical values of λ_{min} and c_{min} for water and air, assuming $\alpha = 74$ dynes/cm, $\rho = 1$ gram/cm, and $g = 981$ cm/sec^2, are as follows:

$$\lambda_{min} = 1.78 \text{ cm}; \quad c_{min} = 23.5 \text{ cm/sec}.$$

Experiment shows that during wave development the wave length first decreases, which agrees with (2.1.2). The continuing increase in wave amplitude (Figure 2.1.2) results in increasing wave steepness. After attaining some minimum, the wave length starts increasing (Figure 2.1.3). This occurs at some known critical wind speed, which is different for points lying at different distances from the start of the wave channel, i.e., from the origin of wave generation. It can consequently be assumed that at these points capillary waves are converted into gravity waves and their steepness starts decreasing. When the wind speed overcomes the critical effect of turbulent fluctuations predominance is had by increasing average air-flow velocity. Each point of the water surface will have waves which arrive from further removed sections (i..e., from sections closer to the start of the wave channel), and waves which were generated in the immediate vicinity of this point and which did not have sufficient time for significant development or damping. If the wind speed is insufficient to maintain the waves against the viscosity forces, i.e., if the energy supplied to the wave by the wind per unit time is smaller than the dissipated energy, then each point will have only waves generated in its immediate vicinity, since the "longer" waves are damped on the way. Thus, when the wind speed is quite insufficient for maintaining the waves, the wave field at the given point depends only on the structure of turbulent pressure fluctuations in the air flow. However, at high wind speeds the initial perturbations of the water surface will not be damped, but rather will develop due to the wind-supplied energy.

The initial amplitudes and lengths of waves forming near a given point can be denoted by a_0 and λ_0; x is the longitudinal coordinate in the direction of wave propagation, with $x=0$ at the beginning of the wave channel; w is the wind speed.

At any point one may have waves with amplitudes a_0 and [lengths] λ_0, as well as waves which arrived from points closer to the start of the wave channel, with amplitude a and length λ:

$$a = a_0 + \int_0^x \frac{\partial a\,(w,\,x)}{\partial x}\,dx, \tag{2.1.4}$$

$$\lambda = \lambda_0 + \int_0^x \frac{\partial \lambda\,(w,\,x)}{\partial x}\,dx. \tag{2.1.5}$$

The change in wave length at the given point from the start of the wave channel is composed, for increasing wind speed, of two factors: of the reduction in λ_0, the initial wave length, which is due to increasing the wave frequency upon increasing frequency of turbulent wind oscillations; and from the increase in increment of the wave length over the path from x to x_0 up to the given point x_1, i.e.,

$$\Delta \int_0^{x_1} \frac{\partial \lambda\,(w,\,x)}{\partial x}\,dx,$$

which results from increasing the derivative $\dfrac{\partial \lambda (w,\, x)}{\partial x}$, namely, the rate of wave length increase as a function of the traversed path (x) with increasing wind speed:

$$\frac{\partial \lambda\,(w,\,x)}{\partial w} = \frac{\partial \lambda_0}{\partial w} + \int_0^{x_1} \frac{\partial^2 \lambda\,(w,\,x)}{\partial w\,\partial x}\,dx. \qquad (2.1.6)$$

Experiments show that $\dfrac{\partial}{\partial w}\left[\dfrac{\partial \lambda\,(\omega,\,x)}{\partial x}\right]$ is positive, in other words, at high wind speeds the wave length of a traveling wave increases more rapidly than at low speeds. The derivative $\dfrac{d\lambda_0}{dw}$ is negative (Figure 2.1.3). Hence the wave length at the given point will either increase or decrease with increasing wind speed, depending on which of the two terms in the right-hand side of (2.1.6) has higher absolute magnitude.

If

$$\left|\frac{d\lambda_0}{dw}\right| > \int_0^{x_1} \frac{\partial^2 \lambda\,(w,\,x)}{\partial w\,\partial x}\,dx, \qquad (2.1.7)$$

then

$$\frac{d\lambda\,(w,\,x)}{dw} < 0,$$

which means that the wave length at the given point will decrease with increasing wind speed. Experimental results (Figure 2.1.3) show that the wave length decreases until the wind speed attains a value which is fixed for a given point, but differs for different points. In these cases the wave length remains constant for small changes in wind speed. However, this implies that $\dfrac{d\lambda\,(w,\,x)}{dw}$ becomes zero, i.e.,

$$\frac{d\lambda_0}{dw} = - \int_0^{x_1} \frac{\partial^2 \lambda\,(w,\,x)}{\partial w\,\partial x}\,dx. \qquad (2.1.8)$$

The wind speed for which (2.1.8) is satisfied at each point x_1 is termed the critical wind speed (w_{cr}). The reason for reduction in w_{cr} with increasing x can be interpreted (Kononkova, 1953) as follows. Let the wind speed w_2 be smaller than wind speed w_1, which is critical for the given point x_1. This means that

$$\left|\frac{d\lambda_0}{dw}\right|_{w\,=\,w_1} > \left[\int_0^{x_1} \frac{\partial^2 \lambda\,(w,\,x)}{\partial w\,\partial x}\,dx\right]_{w\,=\,w}. \qquad (2.1.9)$$

Since $\dfrac{\partial^2 \lambda\,(w,\,x)}{\partial w\,\partial x} > 0$, quantity $\displaystyle\int_0^{x_1} \frac{\partial^2 \lambda\,(w,\,x)}{\partial w\,\partial x}$ will increase with increasing x, and there exists a value $x = x_2$ such that

$$\left[\int_0^{x_2} \frac{\partial^2 \lambda\,(w,\,x)}{\partial w\,\partial x}\,dx\right]_{w\,=\,w_1} = \left[\frac{d\lambda_0}{dw}\right]_{w\,=\,w_1}. \qquad (2.1.10)$$

Consequently, the critical wind speed w_{cr} at point x_1 farther removed from
the start of the wave channel is smaller than at the closer point x_1. For
wind speeds above critical, the wave length at the given point will increase
with increasing wind speed. This implies that

$$\left|\frac{d\lambda_0}{dw}\right|_{w>w_1} < \left[\int_0^{x_1} \frac{\partial^2\lambda\,(w,\,x)}{\partial w\,\partial x}\right]_{w>w_1}, \qquad (2.1.11)$$

i. e., if the wind speed attains its critical value, then as it increases even
more the rate of increase in quantity $\int_0^{x_1} \frac{\partial\lambda(w,\,x)}{\partial x}\,dx$ is greater than the rate
at which the initial wave length (λ_0) decreases.

In the above discussion of the wave development at point x_1 consideration
is given to waves arriving from the origin of wind action. Waves develop
along the entire water surface of the wave channel. Hence the waves
arriving at a given point x_1 are not only those originating at the very start
of the channel, but also at closer points. According to (2. 1. 4) and
(2. 1. 5), the amplitudes and wave lengths of these waves will naturally
be smaller than those which traversed greater distances. At the same
time, it is possible to assume at subcritical wind speeds (Kononkova,
1953) that the amplitudes of waves forming near the point of measurement
can be commensurable with the amplitude of waves arriving from the
start of the wave channel, i. e.,

$$\frac{d\lambda}{dw} = \frac{d\lambda_0}{dw_0}.$$

In view of (2. 1. 6) it follows from the above expression that the lengths
of waves generated near the point of measurement decrease more rapidly
with increasing wind speed than those arriving at this point from the start
of the wave channel. This results in reducing the average wave length
measured at the given point with increasing wind speed (Figure 2. 1. 3).

It is thus seen from experimental results that the appearance of initial
wave spectra is related primarily to turbulent wind-pressure fluctuations.
Subsequent development of wind waves is a result of the average wind-flux
velocity. The first developing wind waves appear at different critical
wind speeds; here these may have different initial wave lengths, related
to the wind speed. This initial length is smaller, the higher the wind
speed. For example, the length of waves in the initial development
stage at wind speeds of 300 cm/sec at a distance of 670 cm from the start
of the wave channel was found to be close to 4.5 cm (Figure 2. 1. 3), and
at a wind speed of 500 cm/sec at a distance of 70 cm from the channel start
the length was found to be close to 2.5 cm.

The initial disturbances at the water surface can be represented
(Uspenskii, 1937) as a train of waves with length λ, propagating at
velocities c. The profile of the water-air interface can be represented
in the form of potential waves (Chapter 1, Section 3)

$$z = a\cos(kx - \sigma t).$$

The energy corresponding to a bandwidth of unit width λ is E_λ (Chapter 1, Section 7). The variation of this energy per unit time can be expressed (Uspenskii, 1937) in the form

$$\frac{dE_\lambda}{dt} = -\frac{1}{2}\, p\, \frac{\partial}{\partial t_0} \int_0^\lambda z^2\, dx. \qquad (2.1.12)$$

It is assumed that water at the crests of these initially-formed waves is subjected to an above-atmospheric pressure, while the pressure at the wave troughs is below atmospheric, due to the surface-layer curvature produced by surface tension. The symbol p_2 denotes the pressure per unit water surface. It is found (Uspenskii, 1937) that the pressure fluctuates together with the oscillation of the z coordinate of the surface point under study according to the relationship

$$p_2 = p_z. \qquad (2.1.13)$$

Here

$$p = \left\{ -g\,\frac{2\rho_1\rho_2}{\rho_1+\rho_2} + \rho_2\,\frac{k^2\alpha}{\rho_1+\rho_2} - \frac{k\rho_1\rho_2\,(\rho_2-\rho_1)}{(\rho_1+\rho_2)^2}\,w^2 + \right.$$
$$+ \frac{2k\rho_1\rho_2}{\rho_1+\rho_2}\,w\left[\frac{g\rho_2-\rho_1}{k\,(\rho_1+\rho_2)} + \frac{k\alpha}{\rho_1+\rho_2} - \right.$$
$$\left.\left. - \frac{\rho_1+\rho_2}{(\rho_1-\rho_2)^2}\,w^2\right]^{\frac{1}{2}}\right\} r\cos(kx-\sigma t), \qquad (2.1.14)$$

where ρ_1 and ρ_2 are the air and water densities; α is the surface tension at the interface; w is the wind speed.

The wave development depends on the sign of the multiplier of p in (2.1.12). The latter quantity, as follows from (2.1.14), depends on three factors: the gravity force (g), surface tension (α) and wind speed (w). These three factors determine the pressure distribution at the water surface. If p is positive, then $\dfrac{dE_\lambda}{dt}$ is negative (expression (2.1.12)), which means that the wave energy cannot decrease. A positive value of p is obtained when the predominant term in (2.1.14) is the positive term $\rho_2\dfrac{k^2\alpha}{\rho_1+\rho_2}$, which depends on the surface tension. The limit at which surface tension is no longer significant depends on the wind speed, because the effect of the wind is opposite to that of the surface tension. The term depending on wind speed, i. e., $-\dfrac{k\rho_1\rho_2(\rho_2-\rho_1)}{(\rho_1+\rho_2)^2}w^2$, has a negative sign, opposite to that of the term containing α. At positive p no waves can develop, since then $\dfrac{dE_\lambda}{dt}$ is negative. If $p = 0$, the variable part of the pressure vanishes and the pressure at the interface is constant everywhere; here the wave energy is also constant.

If p is negative, $\dfrac{dE_\lambda}{dt}$ is positive. The energy of wave oscillations increases due to the air-flux energy with a simultaneous increase in time of the wave amplitude. Consequently, when p has this sign one observes the origination and development of gravity wind waves. The variation of p as a function of wave length (λ) is given in Figure 2.1.5 for wind speeds of 200, 250 and

300 cm/sec. The region of positive p, when there can be no wave development, is separated for each wind speed in the region of negative values of p by the critical point at which the (p, λ, w) curve intersects the λ axis. This critical point defines those minimum critical values starting with which development of waves is possible. Consequently, it is impossible to have under actual conditions some single minimum wave length. Wind waves may originate and develop at any wind velocity; generation requires only that there exist at the water surface some initial disturbances, which owe their origin to the air flux turbulence.

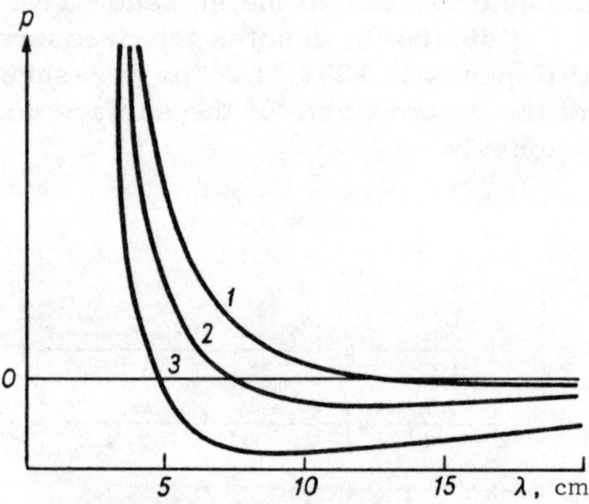

FIGURE 2.1.5. Variation in p as a function of wave length at different wind speeds (after Uspenskii):

1 — wind speed of 200 cm/sec; 2 — 250 cm/sec; 3 — 300 cm/sec.

These results are in sufficiently satisfactory agreement with results of experiments described at the start of the present section. In fact, the length of capillary waves which pass into wind waves decreases with increasing wind speed (Figure 2.1.3). At wind speeds of 380 cm/sec the length of the initial gravity waves is equal to approximately 4 cm (at $x = 670$ cm); according to Figure 2.1.5 it should be ~ 5 cm.

2. Transfer of wind energy to waves and dissipation of this energy

The appearance of initial disturbances at the surface water layer is attributable to pressure fluctuations in the turbulent air flow (Chapter 2, Section 1). However, when the wind speed exceeds the critical effect of these fluctuations, the decisive role in wave development is being acquired by the velocity of the air flow above the initial gravity wind waves (Chapter 2, Section 1). Now the increase in the latter's size is due to direct transfer to them of the wind energy. In this process increasing importance is acquired by those changes in the wind flow which occur as it passes the wave profile.

If we consider for simplicity a sinusoidal wave, then the air flow above
such a wave does not have the same velocity above different points of its
profile (Figure 2.2.1). The wind speed above the crest is ~80% of its value
at the 12 cm level above water. The wind speed at a point near the bottom
of the trough is only 20% of this velocity. It is appreciably smaller at the
leeward slope of the crest than at the windward slope (Figure 2.2.1). The
variation in wind speed above the profile of a sinusoidal wave has associated
with it variations in the aerodynamic pressure above it (Figure 2.2.2). The
pressure drops above those points of the profile where the wind speed is
higher. Hence the minimum aerodynamic pressure lies above the wave crest
(Bonchkovskaya, 1954) also in accordance with the wind-speed distribution,
i.e., at points lying at the same level to both sides of the crest (Figure 2.2.2)
the pressure above the windward slope is higher than the leeward slope. This
phenomenon is amplified, during wind-energy transfer to waves, by separation of the wind flow and formation of an eddy above the leeward slope of the
wave. The latter in turn is related to the wave profile form. The wind flux
above a relatively shallow wave (Figure 2.2.3a) is quite laminar, but departure from laminarity occurs as the crest becomes sharper (Figure 2.2.3b).
At some crest steepness (Figure 2.2.3c) an eddy with a horizontal axis
perpendicular to the wind flow direction is formed above the leeward part of
the slope as a result of the separation in the air flow above the wave crest.
These eddies, in addition to reducing the atmospheric pressure above the
leeward slope, are also responsible for fluctuations in the aerodynamic
pressure. Among the entire ensemble of these fluctuations some are also
produced which are in resonance with oscillations of the water surface.
These will be eddies whose transport velocity c and distance λ between
them are related by

$$c^2 = \frac{g\lambda}{2\pi},$$

i.e., the relationship between wave elements. Consequently resonance sets
in between disturbances in the water-surface layer of the atmosphere and
oscillations of the latter's surface.

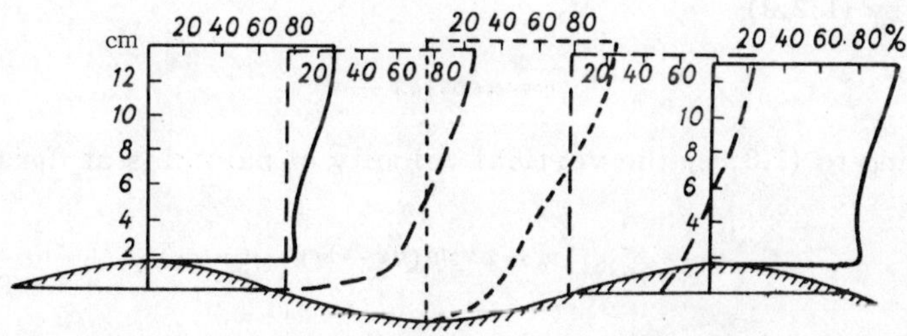

FIGURE 2.2.1. Relative (%) wind speed above the profile of a sinosoidal wave
(rigid profile).

The horizontal lines emanating from the corresponding verticals to the curves
show the relative wind speed as a percent of its total velocity, which it has at
the level of 14 cm from the surface of the profile (according to laboratory
experiments).

The difference in wind pressures at the leeward and windward slopes is responsible for the fact that the total work done by the wind is not zero, but some positive quantity, equal to exactly the energy imparted by the wind to the waves. However, subsequent transfer of wind energy to waves does not occur by means of turbulent energy fluctuations, but by energy influx from some average air-flow velocity.

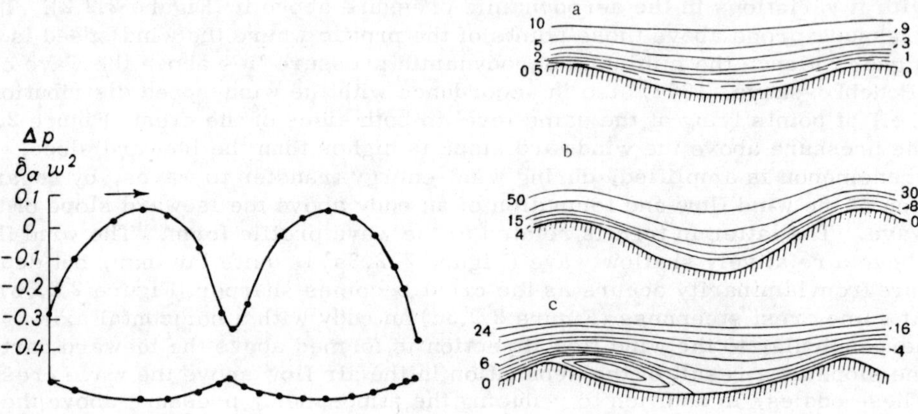

FIGURE 2.2.2. Pressure distribution above a sinusoidal wave (after Bonchkovskaya).

FIGURE 2.2.3. Streamlines above sinusoidal waves with different profiles.

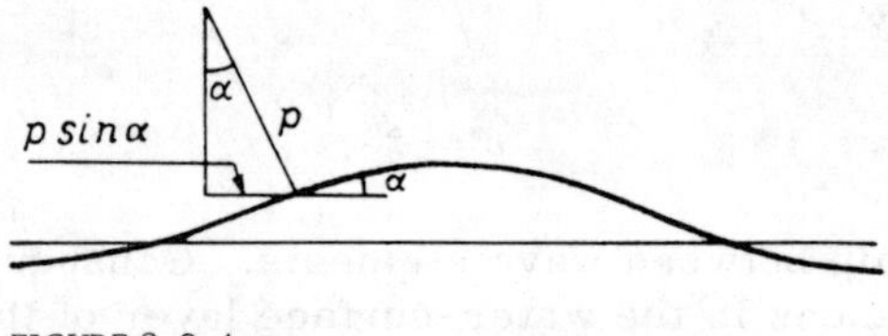

FIGURE 2.2.4.

This energy is assessed quantitatively (Krylov, 1954) for a wave profile described by (1.2.9):

$$\eta = a \cos(kx - \sigma t).$$

According to (1.3.8), the vertical velocity of particles at the free surface is

$$v_0 = a\sigma \sin(kx - \sigma t).$$

Functions v_0 and η are interrelated by the expression

$$v_0 = -c \frac{\partial \eta}{\partial x}.$$

If α is the angle made by the tangent to the wave profile with the horizontal at any point (Figure 2.2.4),

$$\frac{\partial \eta}{\partial x} \approx \operatorname{tg} \alpha \approx \sin \alpha,$$

74

and consequently, approximately,

$$v_0 = -c \sin \alpha.$$

The work done by the wind pressure per unit wave length is approximately

$$N_w = \frac{1}{\lambda} \int_0^\lambda \Delta p\, v_0\, dx.$$

Substitution for v_0 yields

$$N_w = \frac{c}{\lambda} \int_0^\lambda \Delta p \sin \alpha\, dx = c p_x.$$

The energy expended for wave development can be expressed in the form

$$N_w = p_x \frac{dx}{dt} = c p_x, \qquad (2.2.1)$$

where c is the phase velocity of the wave, while p_x is the horizontal force per unit wave length, produced by the pressure difference at the windward and leeward slopes of the waves. In its turn, it follows from numerous experiments with models that

$$p_x = c_p(\delta) \frac{\rho'}{2} (w - c)^2, \qquad (2.2.2)$$

where $c_p(\delta)$ is the drag coefficient of the wave form, depending in some manner on wave steepness; ρ' is the density of air; $(w-c)$ is the relative wind speed.

The wave-form drag coefficient can be expressed as

$$c_p = n\delta^2, \qquad (2.2.3)$$

by assuming that the wind pressure is proportional to the inclination of the free surface. Using (2.2.3) and (2.2.2), expression (2.2.1) is replaced by

$$N_w = \frac{\rho'}{2} n\delta^2 (w - c)^2 c. \qquad (2.2.4)$$

Formula (2.2.4) contains some constant factor n, the magnitude of which is defined differently by different investigators. Kapitsa (1949) takes

$$n \approx 2\pi\gamma, \qquad (2.2.5)$$

where γ is a constant factor, close to unity. Jeffreys (1924) assumes

$$n = s\pi^2, \qquad (2.2.6)$$

where s is a proportionality constant, expressing the ratio of the wave area subjected to direct, normal wind action, to the entire wave area. It is estimated variously at from 0.27 to 0.03.

Substitution of (2.2.5) into (2.2.4) yields the following expression for the energy transmitted by the wind to the wave (per unit area):

$$N_w = \frac{\gamma \rho' g^2}{4\pi} h^2 (w - c)^2 c^{-3},$$ (2.2.7)

and using (2.2.6),

$$N_w = \frac{s \rho' g^2}{8} h^2 (w - c)^2 c^{-3}.$$ (2.2.8)

The latter two formulas differ by the values of the constant coefficients. If $s = 0.27$ and $\gamma = 1$, then N_w calculated from (2.2.7) is approximately three-fold greater than the value calculated from (2.2.8).

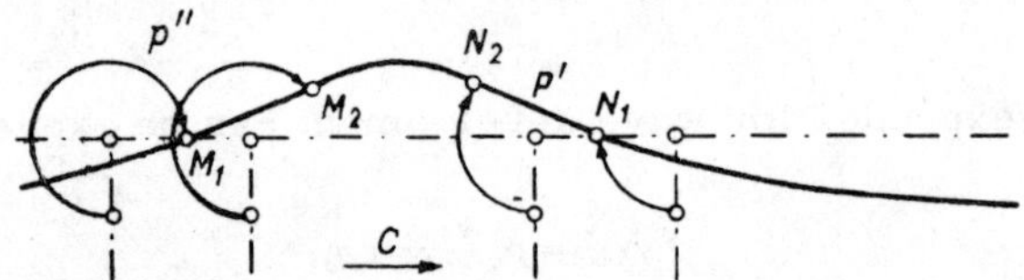

FIGURE 2.2.5. Diagram of wind-energy transfer to waves (after Shuleikin).

The above expressions for estimating the wind energy fed to the waves make allowance for the effect of the horizontal wind-pressure force on the profile of a wave moving at phase velocity. The mechanism of energy transfer by the water particles in orbital motion is disregarded. It is possible to consider the action of the wind not on the wave form, but directly on the water particles participating in the wave motion (Shuleikin, 1956). Here it should be remembered, as before, that the atmospheric pressure distribution over the wave profile is not symmetrical relative to the crest (Figure 2.2.2).

Figure 2.2.5 depicts schematically the orbital motion of water particles in the surface layer of an undulated surface. Particles M_1 and M_2 at the left, windward side are in descending motion phase, while particles N_1 and N_2 on the leeward side are in ascending motion phase along their orbits. But the aerodynamic pressure per unit area at the windward slope is higher than at the leeward slope. Hence each of the two particle pairs intersecting the same horizontal plane will be subjected to a higher pressure from the top when they move downward, through this plane, and a smaller pressure from the top when they move upward through the same plane. If in both cases the particles travel a distance dz in the vertical direction, then, as a result of these opposing elementary motions, an energy increment of

$$(p_z'' - p_z') \cos \alpha \, dz$$

should be produced per unit sea surface. In the above expression α is the angle between a sea-surface element and the horizontal. The total increase in particulate energy due to motion of the water surface from the wave trough to its crest and back again to the trough is expressed by the integral from zero to h:

$$E_0 = \int_0^h (p_z'' - p_z') \cos \alpha \, dz. \qquad (2.2.9)$$

Since $\cos \alpha$ differs little from unity, particularly at the uppermost and lower-most parts of the crest and trough, it is assumed that $\cos \alpha \approx 1$. An energy increment occurs during a single wave period. Hence the average energy $(\overline{N_w})$ transmitted by the wind to the wave per unit sea surface can, using (2.2.9), be expressed

$$\overline{N}_w = \frac{dE}{dt} = \frac{1}{\tau_0} \int_0^h (p_z'' - p_z') \, dz. \qquad (2.2.10)$$

According to (2.2.10) the amount of energy put into the waves by the wind is determined by the pressure at the different points of the wave profile, due to which the wind transmits energy $\overline{N}_w$ through unit sea surface area to water particles moving in their orbits. It is assumed that the part of the kinetic energy that is proportional to the square of the orbital velocity of particles which is expended for developing the "wave" current, can be neglected. As to the drift current which is produced at the sea surface by the wind simultaneously with its generation of waves it is induced by tangential wind forces, and not by the energy transmitted to waves by means of the above mechanism.

A quantitative estimate of the energy defined by (2.2.10) can be obtained by estimating the value of $(p_z'' - p_z')$ corresponding to each value of z. For this purpose Shuleikin carried out special wind-tunnel experiments of wave models with profiles constructed according to (1.6.35). The models were constructed from thin transformer iron in two versions: $\delta = 0.6$ and $\delta = 0.12$. The results obtained with these models in the wind tunnel are shown in Figure 2.2.6. The abscissa corresponds to the dimensionless ratio

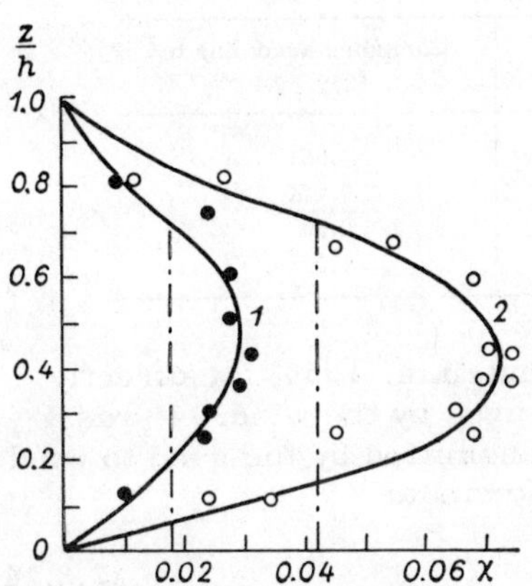

FIGURE 2.2.6. Wind-tunnel results for wave models (after Shuleikin).

$$\chi = \frac{p_z'' - p_z'}{\rho' w^2}, \qquad (2.2.11)$$

which corresponds to some elevation of the pair of points above the wave trough. The ordinate represents ratio $\frac{z}{h}$, where z is the vertical distance from the wave trough. Each point on Figure 2.2.6 corresponds to experiments with a model of a given steepness. For the model with $\delta = 0.6$ the value of χ (average value of χ from the trough to the crest) is 0.018 (curve 1), while for the model with $\delta = 0.12$ it is $\chi = 0.042$ (curve 2). Consequently, χ is proportional to the wave steepness

$$\chi = \gamma \pi \delta,\tag{2.2.12}$$

where ν is some aerodynamic coefficient which, under laboratory conditions, equals ~ 0.11 for $c = 89\,\mathrm{cm/sec}$, $\delta = 0.72$ and $\lambda = 1.24\,\mathrm{m}$. Doubtlessly, for the complex waves observed under actual conditions this coefficient should have a value higher than that given by laboratory experiments for the wave models. Apparently this coefficient is quite a variable quantity.

Substitution of (2.2.11) into (2.2.10) yields

$$N_w = \chi \rho' \frac{h}{\tau} (w - c)^2 \ \mathrm{W/m^2}.\tag{2.2.13}$$

Here w in (2.2.11) is replaced, with respect to progressive waves, by the relative wind speed.

TABLE 2.2.1. Energy (N_w, tons/m^2) transferred to waves by the wind (after Shuleikin)

Wind speed, m/sec	Laboratory results	Computed according to (2.2.13)
8	0.059	0.051
10	0.172	0.120
13	0.93	0.60
17	2.9	1.2

Laboratory experiments were employed (Shuleikin, 1956) for directly estimating the amount of energy put into the waves by the wind. It was established as a result that the energy (N_w) transmitted by the wind to wind waves is expressible by the following simple formula:

$$N_w = Ah (w - c)^2,\tag{2.2.14}$$

where A is some constant. The numerical values of N_w are listed in Table 2.2.1.

Subsequently the energy transmitted by the wind to the waves was measured directly in the sea (Efimov, 1966). The results obtained for winds of from 5 to 10 m/sec were found to be sufficiently close to the data tabulated in Table 2.2.1.

Expression (2.2.14) contains the relative wind speed $(w - c)^2$, and can be written as

$$N_w = Ah (1 - \bar{\beta})^2 w^2,\tag{2.2.15}$$

where $\bar{\beta} = \dfrac{\bar{c}}{w}$.

Comparison of (2.2.14) and (2.2.13) shows that the values of N_w for wind speeds of $8-10\,\mathrm{m/sec}$ are close to one another. But as the wind speed increases the difference becomes more pronounced, and at a wind speed of

17 m/sec the value of N_w given by (2.2.13) is approximately 2.5 times smaller
than that obtained experimentally (Table 2.2.1). This difference is probably
due to the appearance of very steep secondary waves at the crests of the
primary waves. This facilitates the separation of air jets, which should
increase the asymmetry of the pressure field above the waves. This hypothesis was checked in the laboratory and found to be correct. It sufficed to
put on the crests of the previously described wave models small prisms
whose height was 7.5 times smaller than the basic wave height in the model,
in order for resulting values of N_w in the wind tunnel to increase approximately 6-fold as compared with those obtained previously. It may hence be
assumed that (2.2.14) correctly reflects the substance of wind energy transfer
to waves. Substitution of (2.2.12) into (2.2.13) yields

$$N_w = \gamma \pi \rho' \delta^2 (w - c)^2 c. \tag{2.2.16}$$

This may also be expressed in the form

$$N_w = \frac{\gamma \rho' g^2}{4\pi} h^2 (w - c)^2 c^{-3}, \tag{2.2.17}$$

which is the same as (2.2.7). Expression (2.2.17) can also be written in the
form

$$N_w = \gamma p \delta^2 \beta (1 - \beta)^2 w^3, \tag{2.2.18}$$

where

$$p = \pi \rho'.$$

The aerodynamic coefficient γ in (2.2.16) should be determined from
reliable observations under natural conditions.

The above expressions for estimating the energy put into the waves by the
wind were obtained on the assumption that we are concerned with the work
done by the wind due to its pressure on the windward wave slopes. The
mechanism of wind energy transfer can be associated with its tangential
pressure (Makkaveev, 1951).

Shuleikin (1956) concludes from a number of theoretical considerations
and experiments that the energy transmitted by tangential wind pressures
is approximately an order of magnitude smaller than the energy due to the
normal wind pressure.

In classical hydrodynamics one determines the amount of energy dissipated by the wave motion in the entire column of fluid participating in wave
motion per unit area, from the expression

$$N_\mu = \frac{1}{2} \rho \nu \left(\frac{2\pi}{\lambda} \right)^3 h^2 c^2 = 2\pi^2 \rho \nu g \delta^2, \tag{2.2.19}$$

where ν is the kinematic viscosity coefficient defined by

$$\nu = \frac{\mu}{\rho},$$

μ being the molecular viscosity coefficient, equal to 0.018 g/cm · sec for
water, and ρ being the density of water, equal to 1.013 g/cm^3.

However, the use of ν in (2.2.19), even if permissible, is limited to the earliest stages of wave generation, i.e., until capillary waves are converted into wind waves (Chapter 2, Section 1). The latter may attain large values upon a known combination of wave-forming factors. The velocity of water particles in these waves becomes appreciable and the development of wind waves proper is accompanied by phenomena which impart a certain randomness to the particle motion. The laminar motion in the water layer undergoing such a process becomes turbulent.

It should be emphasized that the subdivision of fluid motions, on the one hand, into potential, without rotation of the fluid macroparticles, and vortex flow, i.e., flow with rotation of the fluid macroparticles (Chapter 1, Section 1), and on the other hand, into laminar (layered) and turbulent flow, pertains to different aspects of the same phenomenon. In turbulent flow the primary translational motion may be potential, but there may exist vortices which are superimposed on the principal potential motions and result in mixing of the fluid, i.e., in transmission to it of some momentum.

The fluid-particle trajectories in turbulent motion consist of complicated space curves. By virtue of the same fact the velocity components normal to the flow axis are not zero, and the particle trajectories proper are unstable in time. The conditions of fluid motion at some space element will vary continuously, irrespective of the state of the process as a whole. If the external effect which induced the fluid flow is permanent, then the variation in the water-particle velocity will be in the nature of oscillations about some average value with fluctuational velocity v_l. In this case reference is had to steady-state turbulent motion of the fluid. The study of turbulent motion which is accompanied by continuous transfer and exchange of momentum between the fluid masses is carried out by introducing the (Prandtl) concept of mixing length (l) in a direction transverse to the entire fluid flow. The liquid volumes mix along this length until they collide or mix with other volumes.

Hence the turbulence coefficient (ν_τ) can be expressed as

$$\nu_\tau \sim v_l l. \tag{2.2.20}$$

Consequently, ν_τ depends on the scales (l) of the averaged motion in wave oscillations and on the velocity of turbulent fluctuations (v_l) of these oscillations, related to changes in the average velocity over distance l. These latter velocities (v_l) in wave motion should at the same time be limited, since the turbulence scales (l) are also limited. If the velocities of turbulent fluctuations were commensurable with the velocities of the average wave motion, then wind wave motion would be accompanied by random fluid ejection from the surface, similar to that observed at the surface of boiling water. This is not observed because the velocities of turbulent fluctuations in wave motion are limited. The scales of turbulent fluctuations in wave motion should be limited similarly. Near the surface they should be smaller than the wave dimensions. The condition (Levich, 1955)

$$\rho g a \gg \rho v_l, \tag{2.2.21}$$

where a is the wave amplitude and v_t is the maximum velocity of turbulent fluctuations, must be satisfied. This inequality expresses the stability condition for the surface of a fluid participating in wave motion in a gravitational field. Hence the velocity fluctuation scales are much smaller than the entire region participating in wind wave motion, and the velocity of the fluctuations is also smaller than the velocity of particle motion in the waves. These considerations enable one to use the semiempirical theory of turbulent transport. It is here assumed that the particle motions in wind waves are described with sufficient accuracy by the linear wave theory, and the wave dimensions are estimated on the basis of their statistical characteristics (Chapter 2, Section 3), referring all estimates to the average wave.

Coefficient v contained in (2.2.19) should be replaced by the turbulence coefficient v_T when applying this relation to wind wave processes. Expression (2.2.19) thus becomes

$$N_v = \frac{1}{2}\, \rho v_T g \left(\frac{2\pi}{\lambda}\right)^3 h^2 c^2 = 2\pi^2 v_T \rho g \delta^2. \qquad (2.2.22)$$

In effect, coefficient v_T (Zhukovets, 1963) is the generalized coefficient of turbulent and kinematic viscosity, since the turbulence of wave motion is associated with a velocity gradient and some randomness of wave particle motion. This randomness apparently exists primarily in the topmost water layer. This combination is justified by the fact that the energy losses for turbulent and kinematic viscosity can be determined only together. Some investigators maintain the view that (2.2.22) may contain only v_T pertaining to energy losses only in the surface layer (v_{T_0}) which is contradicted by the physical meaning of (2.2.22) (Zhukovets, 1963).

The amount of dissipated energy thus depends, in addition to the wave elements, also on the value of v_T. The determination of this coefficient is the subject matter of many studies.

Starting with the premise that turbulence in wave motion is determined by the wave dimensions, and having reference to the average value of these elements for the entire ensemble of waves (Chapter 2, Section 3), we may express the turbulence coefficient as some function of the elements of water waves:

$$v_T = f\left(\bar{h},\ \bar{\lambda},\ \bar{\tau}\right). \qquad (2.2.23)$$

Bearing in mind that the dimensions of v_T are cm^2/sec, we should use parameters h, λ and τ to construct the following expressions:

$$v_T = f_1\left(\frac{\bar{h}, \bar{\lambda}}{\bar{\tau}}\right) \qquad (2.2.24)$$

or

$$v_T = f_2\left(\frac{\bar{h}^2}{\bar{\tau}}\right). \qquad (2.2.25)$$

Using relationships pertaining to the theory of trochoidal waves (Chapter 2, Section 6) the latter expressions assume the form

$$v_{\tau} = f_3(\overline{h},\ \overline{c})$$

or

$$v_{\tau} = f_4(\overline{h},\ \overline{v}). \tag{2.2.26}$$

Here the latter equation appears to be more convincing, since the turbulent nature of wave motion should more likely depend on the fluctuation in the orbital velocity of particles (v) than on the fluctuation in the phase velocity (c).

Equations (2.2.25) and (2.2.26) can be expressed in the following general form (Kitaigorodskii, 1961):

$$v_{\tau} = k\,(\delta)^a\, \frac{w^3\beta^3}{g}, \tag{2.2.27}$$

where k is a constant, while exponent a will depend on the form of function (2.2.23), i.e., on the selection of wave elements typifying the turbulent fluctuations of the wave motion.

In equation (2.2.27) δ is replaced by some function

$$\delta = f(\beta) \tag{2.2.28}$$

and then equation (2.2.27) becomes

$$v_{\tau} = k\,[\delta(\beta)]^a\, \frac{w^3\beta^3}{g} \tag{2.2.29}$$

or

$$v_{\tau} = k\,[\varphi(\beta)]\, \frac{w^3}{g}, \tag{2.2.30}$$

where $\varphi(\beta)$ depends on the form of function (2.2.28) and on the value of a in equation (2.2.27). Equation (2.2.30) shows that v_{τ} depends on the wind velocity w and on the wave development stage $\beta = \dfrac{c}{w}$ (Chapter 2, Section 5); the wave amplitude and its other elements increase with increasing β. Hence also v_{τ} should increase with increasing β, namely,

$$v_{\tau}(\beta_2) > v_{\tau}(\beta_1), \tag{2.2.31}$$

where

$$\beta_2 > \beta_1.$$

This can serve as a criterion for selecting the value of a when function (2.2.23) is known.

The conclusions from the semiempirical theory of turbulent exchange employing the von Karman similitude hypothesis are based (Kitaigorodskii, 1961) on the use of equations defining the motion of liquid particles in progressive waves of infinitesimal magnitude (Chapter 1, Section 3).

An item of interest in the study of turbulent exchange in the sea are the characteristics of vertical exchange associated with the averaged velocity along the x axis. Hence the average motion is regarded as one with velocity components

$$v_x = u = \frac{2}{\tau}\frac{2}{\lambda} \int_0^{\tau/2} \int_0^{\lambda/2} v_x \, dt \, dx,$$

$$v_z = w = \frac{2}{\tau}\frac{2}{\lambda} \int_0^{\tau/2} \int_0^{\lambda/2} v_z \, dt \, dz. \tag{2.2.32}$$

The averaging scales used in (2.2.32) are based on the fact that velocity v_x is a periodic function of t and x; here for a time period τ and wave length λ, obviously $v_x = 0$. For the selected averaging scales the fluctuations in velocities u and w are taken for a volume element which is small compared with $\dfrac{\lambda}{2}$, and for a time which is small compared with $\dfrac{\tau}{2}$.

Substitution of (1.3.7) and (1.3.8) into (2.2.32) yields

$$\bar{u} = u(z) = \frac{4h}{\pi\tau}\, e^{-\frac{2\pi z}{\lambda}}, \tag{2.2.33}$$

$$w = 0.$$

The principal expressions for the case

$$\bar{u} = u(z), \quad \bar{v} = \bar{w} = 0 \tag{2.2.34}$$

take the form

$$l = \varkappa \, \frac{\dfrac{d\bar{u}}{dz}}{\dfrac{d^2 u}{dz^2}}, \tag{2.2.35}$$

$$v_\tau = l^2 \frac{d\bar{u}}{dz} = \varkappa^2 \frac{\left(\dfrac{d\bar{u}}{dz}\right)^3}{\left(\dfrac{d^2 u}{dz^2}\right)^2}, \tag{2.2.36}$$

$$N_v = \rho \int_{-\infty}^0 v_\tau \left(\frac{d\bar{u}}{dz}\right)^2 dz = \rho\varkappa^2 \int_{-\infty}^0 \frac{\left(\dfrac{d\bar{u}}{dz}\right)^3}{\left(\dfrac{d^2 u}{dz^2}\right)^2}\, dz, \tag{2.2.37}$$

where l is the Prandtl "mixing length" which, as was noted, denotes the distance traveled by individual water volumes without change in the average velocity until their collision; $\varkappa$ is the universal turbulent von Karman constant; v_τ is the coefficient of vertical turbulent momentum exchange.

Using equations (2.2.33), (2.2.35) and (2.2.37), we have

$$l = \frac{\varkappa\lambda}{2\pi}, \tag{2.2.38}$$

$$v_{\tau_0} = \frac{2\varkappa^2}{\pi^2}\frac{h\lambda}{\tau}\, e^{-\frac{2\pi z}{\lambda}}. \tag{2.2.39}$$

At the surface this expression, with the aid of (2.2.27), reduces to

$$\nu_{\tau_0} = k\delta\,\frac{w^3\beta^3}{g},$$ (2.2.40)

where

$$k = \frac{4\varkappa^2}{\pi}.$$

Consequently,

$$N_\nu = \frac{64\rho\varkappa^2}{3\pi^3}\,\frac{h^3}{\tau^3}.$$ (2.2.41)

An alternative form for the latter expression is

$$N_\nu = \frac{64\rho\varkappa^2}{3\pi^3}\,\delta^3\beta^3 w^3.$$ (2.2.42)

Equating the last expression with (2.2.22) we derive an expression for the coefficient of integral eddy viscosity $(\overline{\nu_\tau})$, since (2.2.27) defines the dissipation of energy over the entire fluid column participating in wave motion:

$$\overline{\nu_\tau} = k\delta\beta^3\,\frac{w^3}{g},$$ (2.2.43)

where

$$k = \frac{32\varkappa^2}{3\pi^5}.$$

The above solution yields an expression for the mixing length as a function of the wave length, shows that exponent a in (2.2.28) is unity, and that identifying the value of ν_{τ_0} at $z = 0$ in (2.2.39) with the value of $\overline{\nu_\tau}$ in (2.2.43) is inappropriate.

The description of eddy viscosity in trochoidal waves by using the von Karman similitude hypothesis (Dobroklonskii, 1947) is based on the following expression for the wave-length-averaged average-velocity gradient:

$$\frac{d\bar{u}}{dz} = \frac{2\pi^3\,\dfrac{\delta^2}{\tau}\,e^{\frac{4\pi z}{\lambda}}}{1 + \pi^2\delta^2 e^{\frac{4\pi z}{\lambda}}}.$$ (2.2.44)

On the basis of (2.2.44) and using (2.2.35) we obtain the following expressions for the turbulence scale l and coefficient ν_τ:

$$l = \frac{\varkappa\lambda}{6\pi}\left(1 - \pi^2\delta^2 e^{\frac{4\pi z}{\lambda}}\right),$$

$$\nu_\tau = l^2\,\frac{d\bar{u}}{dz} = \frac{\varkappa^2\pi}{18}\,\frac{h^2}{\tau}\,e^{\frac{4\pi z}{\lambda}}\left(1 - \pi^2\delta^2 e^{\frac{4\pi z}{\lambda}}\right).$$

The second term in the latter formula can be neglected due to its small-ness. Hence

$$\nu_\tau \cong \frac{\varkappa^2 \pi}{18}\,\frac{h^2}{\tau}\,e^{\frac{4\pi z}{\lambda}}. \qquad (2.2.45)$$

Reducing the last expression to the form of (2.2.28), approximation (2.2.45) at the surface can be written as

$$\nu_{\tau_0} \approx k\delta^2\,\frac{w^3\beta^3}{g}, \qquad (2.2.46)$$

where

$$k = \frac{\varkappa^2\pi^4}{g}.$$

Using expressions (2.2.44) and (2.2.37), the following may be obtained (Kitaigorodskii, 1961) for N_ν:

$$N_\nu = \frac{\rho\varkappa^2\pi^6}{54}\,\delta^6\,\frac{\lambda^3}{\tau^3}, \qquad (2.2.47)$$

which is reduced to the form

$$N_\nu = \frac{\rho\varkappa^2\pi^6}{54}\,\delta^6\beta^3 w^3. \qquad (2.2.48)$$

It is seen by equating (2.2.48) and (2.2.22) that the integral eddy viscosity coefficient $(\bar\nu)$ is given by

$$\bar\nu_\tau = k\delta^4\beta^3\,\frac{w^3}{g}, \qquad (2.2.49)$$

where

$$k = \frac{\varkappa^2\pi^4}{108}.$$

The last expression differs markedly from (2.2.46), which gives the turbulent motion coefficient for the surface. By virtue of the same fact it is incorrect to use it in (2.2.22).

It is found by using the von Karman similitude hypothesis that the turbulence scale (or mixing length) l is proportional to the wave length λ (2.2.38) and (2.2.45). If it is assumed (Kitaigorodskii, 1961) that

$$l \sim h \quad \text{or} \quad l \sim he^{-2\pi z/\lambda},$$

one derives an expression for $\bar\nu_\tau$ corresponding to $a = 3$ in (2.2.29). Consequently, the use of different mixing lengths l results in different numerical values of a in (2.2.29). If (2.2.28) is taken in the form

$$\delta = \frac{0.023}{\beta^{0.5}},$$

then, according to (2.2.30), expressions (2.2.40), (2.2.46), (2.2.43) and (2.2.49) for v_{T_0} and $\bar{v}_T$ become, respectively,

$$v_{T_0} = k\,(0.023)\,\frac{\beta^{2.5}w^3}{g}\,, \qquad (2.2.50)$$

$$v_{T_0} = k\,(0.023)^2\,\frac{\beta^2 w^3}{g}\,, \qquad (2.2.51)$$

$$\bar{v}_T = k\,(0.023)\,\frac{\beta^{2.5}w^3}{g}\,, \qquad (2.2.52)$$

$$\bar{v}_T = k\,(0.023)^4\,\frac{\beta w^3}{g}\,. \qquad (2.2.53)$$

Coefficients k in these expressions correspond to the values presented in (2.2.40), (2.2.46), (2.2.43) and (2.2.49).

To calculate the values of v_T or v_{T_0} from these expressions one must estimate the value of constants k in the above expressions, which in turn requires estimating $\varkappa$, the von Karman constant.

It was found from experiments in pipes and channels that $\varkappa$ lies between 0.31 and 0.44. According to a number of experiments, the value of $\varkappa$ for conditions of motion in the sea should be different; for example, $\varkappa = 0.12$ and 0.065 (Rossby), $\varkappa = 0.071$ (Bowden). More recent studies (Shuleikin, 1962) yield a value of $\varkappa = 0.1$. This is also the value of $\varkappa$ for sea currents. If we assume this latter value of $\varkappa$, then the constants k in (2.2.40), (2.2.46), (2.2.43) and (2.2.49) take the values listed in Table 2.2.2.

TABLE 2.2.2. Numerical values of k for $\varkappa = 0.1$

Formula	k
(2.2.40)	$13 \cdot 10^{-3}$
(2.2.46)	$11 \cdot 10^{-3}$
(2.2.43)	$35 \cdot 10^{-6}$
(2.2.49)	$89 \cdot 10^{-4}$

TABLE 2.2.3. Values of v_{T_0} and v_T given by different formulas

Formula	$v_T\,\mathrm{cm}^2/\mathrm{sec}=f\,(w,\beta)$	$\beta = 0.8$			$\beta = 1.0$		
		w m/sec					
		10	20	30	10	20	30
(2.2.40)	$3 \cdot 10^{-7}\ \beta^{2.5}w^3$	$2 \cdot 10^2$	$16 \cdot 10^2$	$54 \cdot 10^2$	$3 \cdot 10^2$	$24 \cdot 10^2$	$81 \cdot 10^2$
(2.2.46)	$6 \cdot 10^{-9}\ \beta^2 w^3$	4	32	108	6	48	142
(2.2.43)	$8 \cdot 10^{-9}\ \beta^{2.5}w^3$	$5 \cdot 10^{-1}$	$40 \cdot 10^{-1}$	$135 \cdot 10^{-1}$	$8 \cdot 10^{-1}$	$64 \cdot 10^{-1}$	$216 \cdot 10^{-1}$
(2.2.49)	$3 \cdot 10^{-12}\ \beta w^3$	$2 \cdot 10^{-3}$	$16 \cdot 10^{-3}$	$54 \cdot 10^{-3}$	$3 \cdot 10^{-3}$	$24 \cdot 10^{-3}$	$81 \cdot 10^{-3}$

The use of (2.5.24) enables one to calculate values of v_{T_0} and v_T (Table 2.2.3) from the previously mentioned formulas for different w and for conditions of limiting wave development, assuming $\beta = 0.8$ or $\beta = 1.0$ (Chapter 2, Section 5). It follows from these results that the values of v_T differ by 2 or 3 orders of magnitude. It is also noted that $\bar{v}_T$ from expression (2.2.49) for a wind speed of 10 m/sec is smaller than μ, which obviously does not match the facts. The above results show that the question of actual values of v_τ for sea waves requires further study, in which a proper place should be assigned to the determination of v under natural conditions.

3. Statistical regularities of wind waves

With the advent of stereophotography of the sea surface (Titov, 1951) it was found that wind waves in each case of wind wave motion have very different dimensions. Subsequent measurements of wind wave elements by wave recorders (Morozov, 1953; Vilenskii and Glukhovskii, 1955; and others) made it possible to establish statistical regularities in the observed variety of wind waves.

A number of variable values of wave elements can be described with accuracy sufficient for practical purposes by several numerical character- istics of the statistical distribution. The first of these is the arithmetic mean (M), denoted by the corresponding symbol of the wave element with a bar and calculated from the expression

$$M = \frac{\sum v}{n}, \tag{2.3.1}$$

where $\sum v$ is the sum of all the values (v) of series of the wave element under study, and n is the number of values in the series $M = \overline{h}, \overline{\lambda}, \overline{c}, \overline{\tau}, \overline{\delta}$, etc.

The variability of the given wave element is described by the average values of the squares of deviations of the terms of the series from their arithmetic mean, namely, the standard deviation (σ), and is calculated from the expression

$$\sigma = \left(\frac{\sum x^2}{n}\right)^{1/2}, \tag{2.3.2}$$

where $\sum x^2$ is the sum of squares of deviations of all the variables from the arithmetic mean; for example, for the wave height $\sum x^2 = \sum (h_i - \overline{h})^2$; n is the number of values in the series.

The variability of various series is estimated by the ratio of the standard deviation to the arithmetic mean. This ratio is termed the coefficient of variation of the series or the coefficient of variance (C_v), which is usually expressed in percents:

$$C_v \% = \frac{100\sigma}{M}. \tag{2.3.3}$$

Analysis of observations of wind wave elements, carried out by wave recorders, gives contradictory results as to the value of C_v. Some results show that C_v depends to some extent on whether the waves develop or are being damped (Morozov, 1953). In the former case C_v is higher than in the latter. Thus, the coefficient of variation for the wave height (C_{vh}) is ~ 0.49 for developing waves and ~ 0.37 for damped waves. For swell it decreases to ~ 0.20. On the other hand, C_v depends on the average value of the wave

elements. The value of C_v increases with increasing wave size. The
coefficient of variation for the wave period, $C_{v\tau}$, is more stable. Other
results (Vilenskii and Glukhovskii, 1961) do not point to variations in
C_{vh} and $C_{v\tau}$ at different wave-development stages.   According to others
(Korneva and Liverdi, 1964), $C_{v\tau}$ ranges from 0.25 to 0.47. On the average,
for wind waves it equals 0.33, for mixed waves 0.35, and for wind swell
0.33. The values of C_{vh} according to different observations range from 0.30
to 0.68. On the average, for wind waves it is 0.47, for wind swell 0.46, and
for mixed waves 0.53. It is assumed in the majority of cases that the coefficient of variation is virtually constant for the height of waves. Under sea
conditions it is assumed to be $0.50-0.52$, and ~ 0.43 for natural lakes and
water storage lakes.

The average error of the arithmetic mean (m) is

$$m = \pm \frac{\sigma}{\sqrt{n}}. \qquad (2.3.4)$$

It is expressed in the same units as the arithmetic mean and the standard
deviation. The average error can be expressed as a percent of its corresponding arithmetic mean. This accuracy index $(p\%)$ is calculated from the
formula

$$p = \frac{100m}{M}. \qquad (2.3.5)$$

In resolving the majority of questions requiring consideration of the size
of wind waves, the accuracy index should not exceed 5%. If the coefficient
of variation and the accuracy index are known, the number of required
observations (n) is

$$n = \frac{C_v^2}{p^2}. \qquad (2.3.6)$$

When approaching absolute reliability in the numerical characteristics of
the statistical distribution, the number of observations should be calculated
from the expression

$$n = 10 \frac{C_v^2}{p^2} + 5. \qquad (2.3.7)$$

The coefficient of variation of wind waves in the sea may be taken as
$\sim 50\%$. For a permissible error of 5% the number of observations according to (2.3.6) and (2.3.7) should be between 100 and 1005. Experience shows
that for $n = 150-200$ only the results of estimates of statistical characteristics of wind wave elements can be regarded as more or less reliable.

When a sufficiently long continuous record (150 or more observations) of
some wind wave element is available, the variation in the quantity under
study can be represented graphically. Here it is assumed that the value of
the given element is a random continuous quantity, which can take on any
value within some interval. On this assumption the observational record is
subdivided into groups (classes), which differ from one another by the same
value. The average value of the group is indicated in increasing order on
the abscissa, and the number of variables in the group is indicated, also in
increasing order, on the ordinate; the resulting series of points is joined
by a smooth curve. This distribution curve is termed in statistics the

probability density curve, while in oceanography it is termed the frequency curve. The frequency curve can be represented in integral form, by sequentially summing the number of variables along the ordinate and the average values of classes along the abscissa. Such a curve allows one to estimate the probability of appearance of a wave element with a value which is higher than any given value. These integral frequency curves are termed in oceanography cumulative probability curves.

Distribution curves may have a variety of shapes. Asymmetrical curves are frequently observed. Asymmetry of a distribution points to the fact that the majority of variables in the series has accumulated in a region where their values are high (negative asymmetry or skewness), or where their values are low (positive asymmetry or skewness). The asymmetry of a series is characterized by a coefficient of skewness (C_s), given by

$$C_s = \frac{\sum x^3}{n\sigma^3},$$

(2.3.8)

where $\sum x^3$ is the sum of the cubed deviations of individual variables from the arithmetic mean; n is the number of variables; σ is the standard deviation. The value of C_{sh} for sea waves is assumed to be approximately equal to $0.6-0.8$, i.e., the wave height distribution has positive asymmetry: $C_{s\tau} = 0$; in other words, the wave-period distribution is symmetrical.

When processing results of observations of wave elements under actual conditions one usually does not use absolute magnitudes of the elements but rather their dimensionless value, expressed as the ratio (k) of the given magnitude to the arithmetic mean of the series. Consequently, one computes

$$k_h = \frac{h_i}{\bar{h}},$$

(2.3.9)

$$k_\tau = \frac{\tau_i}{\bar{\tau}}$$

(2.3.10)

and similarly for any other wave elements.

Correlation of a large number of wave recordings taken in different seas and oceans under a large variety of meteorological conditions yielded a number of regularities (Vilenskii and Glukhovskii, 1955) in the statistical distribution of wind-wave elements in terms of their relative values (Table 2.3.1 and Figure 2.3.1).

It is found that only 5% of all the waves has a height exceeding approximately two-fold the average wave height, and only 1% of the total number of measured waves exceeds the average height of waves by a factor of ~ 2.5 (Table 2.3.1, and curve 1 in Figure 2.3.1). The greatest wave height, close to the maximum possible, which has a cumulative probability of 0.1%, apparently does not exceed three times the average height of the recorded waves. The upper part of the cumulative probability of wave elements must be extrapolated to small cumulative probabilities (less than 1%), since the available observational records usually do not contain more than $100-150$ measurements of wave elements. This makes it impossible to estimate reliably the corresponding $k_{\bar{h}}$ for cumulative probabilities less than 1%. Hence it is assumed that the minimum possible cumulative probability is

$\sim 0.1\%$, regarding it as corresponding to the limiting value of the given wave element. The cumulative probability of average wave height is about 46% (Table 2.3.1, and curve 1 in Figure 2.3.1). Curve 2 in Figure 2.3.1 shows the wave height frequency. The maximum of the curve corresponds to $\sim 0.8\,\bar{h}$. This means that among all the waves most have relative heights of the order of $0.8\,\bar{h}$. Curve 3 in Figure 2.3.1 shows the cumulative probability of wave periods constructed from the corresponding data of Table 2.3.1. It is found that the maximum period does not exceed twice the average period. The cumulative probability of the average period is 50%. Curve 4 in Figure 2.3.1 represents the frequency of wave periods. The maximum of curve 4 corresponds approximately to $\bar{\tau} \approx 1$. Consequently, most waves possess the average period. It should be also noted that the variety of wave periods is not as wide as that of wave heights.

TABLE 2.3.1. Cumulative probability of wind wave elements according to observational results (after Vilenskii and Glukhovskii)

	Cumulative probability, %							
	0.1	1	2	3	5	10	20	30
$k_{\bar{h}}$, $k_{\bar{\lambda}}$ . .	2.96*	2.52	2.28	2.10	1.91	1.69	1.38	1.21
$k_{\bar{\tau}}$, $k_{\bar{c}}$ . .	1.80*	1.65	1.57	1.52	1.47	1.37	1.23	1.15
$k_{\bar{\delta}}$	—	6.50	—	—	2.75	1.90	1.30	0.97

	Cumulative probability, %						
	40	50	60	70	80	90	99
$k_{\bar{h}}$, $k_{\bar{\lambda}}$ . .	1.05	0.93	0.81	0.69	0.54	0.37	0.1
$k_{\bar{\tau}}$, $k_{\bar{c}}$ . .	1.07	1.00	0.93	0.85	0.76	0.66	0.44
$k_{\bar{\delta}}$	0.77	0.62	0.51	0.40	0.30	0.20	0.15

* Extrapolated data.

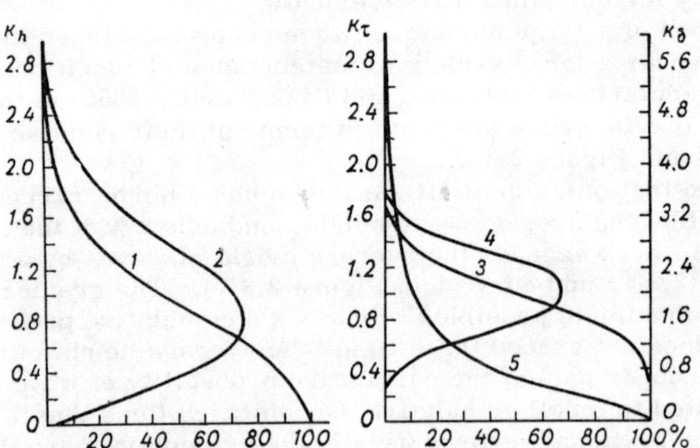

FIGURE 2.3.1. Distribution of wind-wave elements from results of wave recordings (after Vilenskii and Glukhovskii):

1 — cumulative probability (in %) of relative wave heights (k_h); 2 — frequency (in %) of relative wave heights (k_h); 3 — cumulative probability (in %) of relative wave periods (k_τ); 4 — frequency (in %) of relative wave periods (k_τ); 5 — cumulative probability (in %) of relative wave steepness (k_δ).

A fixed relationship exists between the corresponding ordinates of cumulative probability curves of wave heights and periods (curves 1 and 3 in Figure 2.3.1). Denoting the ordinate of curve 1 by k_{h_i}, and the ordinate with the same cumulative probability on curve 3 by k_{τ_i}, we find that

$$k_{\tau_i} = \left(k_{h_i} \right)^{1/2}.$$

On the other hand, according to (1.6.26)

$$\tau \approx (\lambda)^{1/2}.$$

Hence the distribution curves of wave heights reflect at the same time also the wave length distribution. It follows that the most probable wave length is $0.8\bar{\lambda}$. The wave speed is related functionally to the wave period. Consequently, the distribution curves of wave periods and velocities should be identical. .

The data presented in Table 2.3.1 on the cumulative probability of wave elements express some of their average values. Individual measurements exhibit large deviations. Thus, in processing 80 wave recordings Vilenskii and Glukhovskii (1961) found that $k_{h_{1\%}}$ ranges from 2.01 ($n = 108$) to 2.69 ($n = 176$); $k_{h_{50\%}}$ ranges from 0.87 ($n = 112$) to 1.02 ($n = 91$); $k_{\tau_{1\%}}$ ranges from 1.61 ($n = 108$) up to 2.13 ($n = 112$); $k_{\tau_{50\%}}$ ranges from 1.06 ($n = 82$) to 0.92 ($n = 206$). It can be established from these data that values of $k_{h_{1\%}}$ from 2.21 to 2.70 pertain to 80% of all the measurements (80 wave recordings), and values of $k_{\tau_{1\%}}$ from 1.65 to 1.90 also pertain to approximately 80% of the same measurements.

Using the previously presented distribution curves of wave elements, from which one can gain an insight into the variety with respect to any given feature, for example, height or period, it is possible to construct multidimensional frequency or cumulative probability curves. This can be done on the basis of the multiplication rule for probabilities, according to which the probability of the simultaneous occurrence of two independent random events can be expressed as the product of the probabilities of these separate events:

$$F(h, \tau) = F(h)\,F(\tau). \tag{2.3.11}$$

Here the probability of the appearance of waves with height $\geqslant h$ is denoted by $F(h)$, the probability of appearance of waves with period $\geqslant \tau$ is denoted by $F(\tau)$, while the probability of appearance of waves possessing simultaneously the above features, i.e., corresponding values of h and τ, is denoted by $F(h, \tau)$. This approach can be followed by assuming in the first approximation that the wave elements referred to their average values are random independent variables. This also involves the assumption that, for example, the relative height $k_{\bar{h}}$ is in no way related to the relative period of these waves or their length. This assumption is obviously correct only for those wave elements which are not functionally related.

Figure 2.3.2 shows the distribution of wave heights and lengths, constructed from corresponding calculations using (2.3.11). It follows from this figure that the probability of the appearance of waves with simultaneous maximum height and length is negligible; it is virtually impossible to detect such waves. For example, the probability of occurrence of waves $\geqslant 1.8\bar{h}$ and $\geqslant 1.8\bar{\lambda}$ can be approximately 0.5% of the total number of waves, i.e., 5 waves out of a thousand. Naturally, the probability of higher and at the same time longer waves will be even smaller. Figure 2.3.3 (Morozov, 1953) shows the distribution of a number of wave features. It contains a straight line which bisects the coordinate axes. This line represents the relationship $\frac{\delta_i}{\bar{\delta}} = \frac{h_i/\lambda_i}{\bar{h}/\bar{\lambda}}$. Waves for which $\frac{\delta_i}{\bar{\delta}} > 1$ are counted among waves with above-average steepness, while those with $\frac{\delta_i}{\bar{\delta}} < 1$ have below-average steepness. The former can be conditionally classified as "steep" and the others as "shallow" waves. This figure also shows three more lines. One parallel to the abscissa, corresponds to $\frac{h_i}{\bar{h}} = 1$, while the other, parallel to the ordinate, corresponds to $\frac{\lambda_i}{\bar{\lambda}} = 1$. Waves for which $\frac{h_i}{\bar{h}} > 1$ will be those with above-average heights, conditionally termed "high" waves. Waves with $\frac{h_i}{\bar{h}} < 1$ are "low" waves. Similarly, waves with $\frac{\lambda_i}{\bar{\lambda}} > 1$ are "long" and those with $\frac{\lambda_i}{\bar{\lambda}} < 1$ are "short." In Figure 2.3.3 the entire distribution surface is subdivided into eight sectors. In each of them is listed the number of waves as a function of their total number on the sea surface, belonging by virtue of their length, height and steepness to the given sector. Thus, for example, there are 35% of "low," "short," "steep" and "small" waves, and 10% of "low," "short," "shallow" and "small" waves. There are only 10% of "high," "long" and "large" waves and so on. As can be seen, there is a general predominance of "low," "short," "long" and "small" waves, comprising more than half of all the waves (58%). Conversely, "long" and at the same time "high" waves are encountered much less frequently.

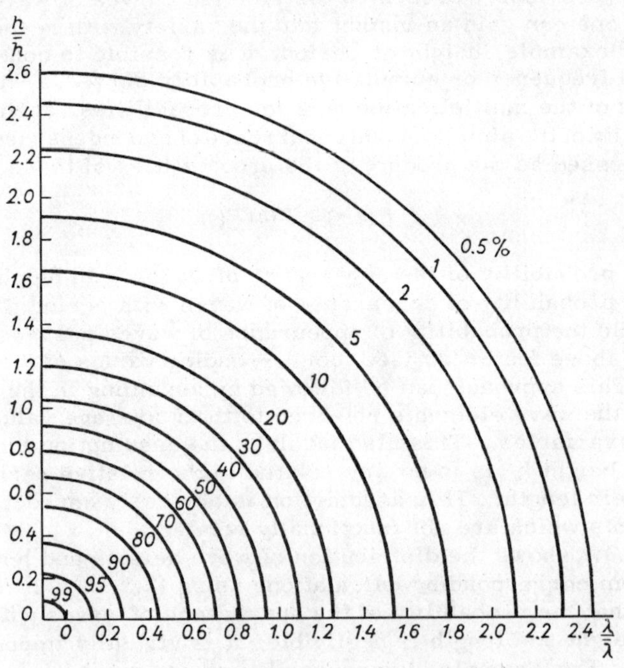

FIGURE 2.3.2. Distribution (in %) of relative wave height (k_h) and relative wave length (k_λ) (after Vilenskii and Glukhovskii).

Figure 2.3.4 can be employed for following the manner in which the
number of waves (in percents) with given relative height varies as a function
of the relative period, and vice versa. For example, one virtually does not
encounter waves in which simultaneously the relative height exceeds twice
the average height ($k_{\bar{h}} > 2$), and the relative period is above 1.6 ($k_\tau > 1.6$).

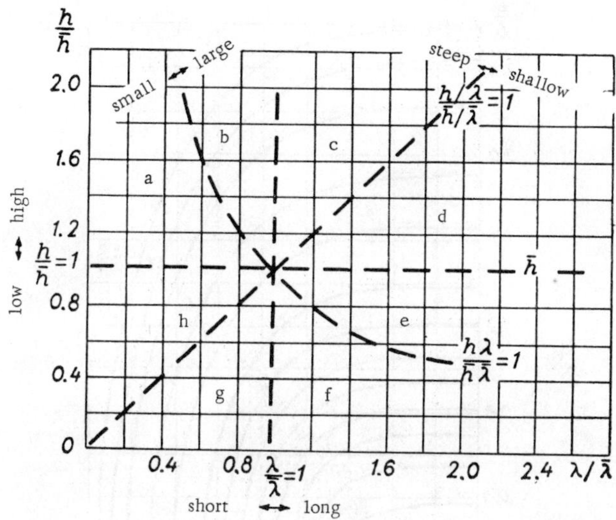

FIGURE 2.3.3. Distribution of various wave features (after Morozov):

a — high, short, steep, small (9%); b — high, short, steep, large (5%);
c — high, long, steep, large (10%); d — high, long, shallow, large
(19%); e — low, long, shallow, large (8%); f — low, long, shallow,
small (4%), g — low, short, shallow, small (10%); h — low, short,
steep, small (35%).

The distribution curves of wave elements plotted from observed data can
be approximated by equations developed in mathematical statistics. The
feasibility of replacing an actual series by a mathematical curve very much
facilitates the study of any varying phenomenon and, in particular, of the
variety of wind-wave elements. At the same time approximation of the
observed wave train by some theoretical distribution curves does not provide
insight into the causes and mechanisms of the wave-variety phenomenon, i.e.,
its physical substance. An item of importance in using theoretical curves
is the feasibility of extrapolating empirical distribution curves into a region
of extremely rare cases, i.e., to wave elements in the range of waves with
low cumulative probability, which means extremely large waves. This extrapolation is to some extent arbitrary, but it is still permissible for relatively
small cumulative probabilities, of the order of 1 — 0.1 %, since it yields quite
real values of wave elements. The basic criterion in selecting curves is
their agreement with experimental data. This agreement should in practice
be acceptable in the first place in the upper part of the distribution curves,
i.e., for cumulative probability less than 50 %, since the most important
information is that on the appearance of waves with large height, length and
period, and in particular for waves with small cumulative probability, i.e.,
1 % and less, down to 0.1 %. From these considerations a less rigorous

approach is also possible to the agreement between theoretical and experimental curves for the lower part of the curves, i.e., for cumulative probabilities in excess of 50%. Of greatest interest are those theoretical distribution curves which can be constructed and used with the smallest number of parameters.

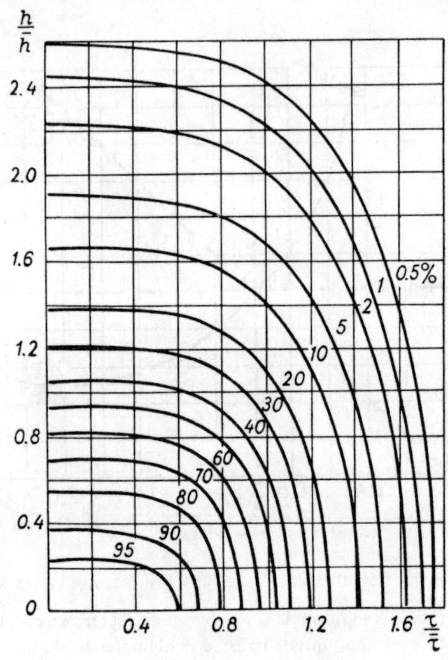

FIGURE 2.3.4. Distribution in (%) of the relative height (k_h) and relative period (k_τ) of waves (after Vilenskii and Glukhovskii).

From the physical essentials of wind-wave motion and from considerations of the possible distribution of wave elements under natural conditions, the theoretical distribution curves should correspond at their lower boundary to the zero value of variable x with a cumulative probability of 100%. The upper branch of curves may be without a limit and allow for the probability of existence of an as-large-as-desired value of the variable under study, i.e., a given wave element. The zero cumulative probability for these curves corresponds to an infinite value of a wave element. The dimensions of wind waves are in the majority of cases limited by conditions of their development, i.e., wind speed and other factors. Hence from this point of view they cannot be infinitely large. However, the appearance of extremely large waves can be physically interpreted as the result of interference between simultaneously existing different wave systems. From this point of view one may encounter a wave with as-large-as-desired dimensions, but its probability is zero, i.e., actually such waves do not exist. However, there is no consensus on this point and in some cases, for example, for the distribution of wind wave elements over natural and water-storage lakes, theoretical distribution curves are selected with a restriction on the upper limit.

Using specified values of the average wave element and the corresponding C_s and C_v, it is possible to select theoretical distribution curves (with consideration of the above conditions) by employing normal distribution curves described by binomial-type equations (Selyuk, 1961). In practice these distribution curves can be constructed from tables used in mathematical physics.

Approximation of empirical distribution curves of wind-wave elements by theoretical distribution curves can be replaced by treating the wind-wave field as a complex oscillatory process, produced by variability in the air flow over the sea. The instantaneous ordinates of elevation of the undulated surface above the still-water level are treated as random functions of coordinates and time. They can be treated as steady, i.e., as stationary, only during individual time intervals, for which reason wind-wave processes are termed quasistationary. The average values of coordinates of the undulated sea surface are taken as constant only for such time intervals during which the waves are treated as invariable, i.e., stationary.

If a float is placed on the undulated water surface (Krylov, 1956), then its trajectory in the case of sinusoidal waves can be written in the form

$$x = r(t)\cos\theta(t),$$
$$z = r(t)\sin\theta(t), \qquad (2.3.12)$$

where $r(t)$ is the distance of the float from the state of rest, i.e., from the coordinate origin at time t, while $\theta(t)$ is the polar angle, formed by the radius vector r and the horizontal axis x. Obviously, $2r = h$. For simplification we consider the two-dimensional case. The coordinate origin is placed at the rest position of the float. The complex wind-wave process will set the float into oscillatory motion, deflecting it over different distances r. Some method, for example, photography, can be employed for recording the location of the float at equal time intervals. These latter should, naturally, substantially exceed the average period of the traveling waves. If this recording is carried out over a sufficient time period, we will get a statistical ensemble of radial deviations (r) of the float from its position of rest. It may be assumed, due to the random nature of wave-height variation, that the probability density function for radial deflections of the float $f(r)$ will not depend on angle θ. Hence $f(r)$ is a function of r only and does not depend on θ.

One may single out on the xoz plane an element of area bounded by concentric arcs of radii r and $r + dr$ and rays θ and $\theta + d\theta$ (Figure 2.3.5). The probability that the float position will lie between concentric circles r and $r + dr$ is $f(r)\,dr$, and the probability that it will lie between rays θ and $\theta + d\theta$ is $\dfrac{d\theta}{2\pi}$.

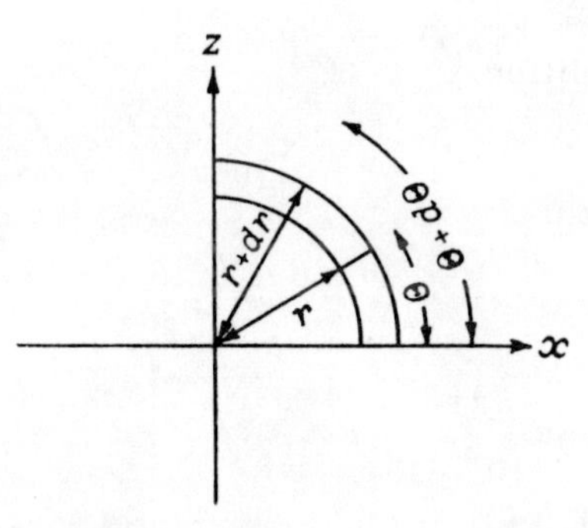

FIGURE 2.3.5.

Using (2.3.11), the probability that the float will be located in the chosen element of area can be written in the form

$$f(r)\,dr\,\frac{d\theta}{2\pi}.$$

On the other hand, this probability is also equal to the product of the probability that the float will lie within the unit-area element, i.e.,

$$\varphi(r^2)\, r\, dr\, d\theta.$$

Hence functions f and φ are interrelated as follows:

$$f(r) = 2\pi r \varphi(r^2).$$

Function φ is sought by assuming that it depends on $r^2 = x^2 + z^2$, i.e., it is the product of two functions, one depending only on x and the other only on z:

$$\varphi(r^2) = \varphi_1(x)\, \varphi_2(z). \tag{2.3.13}$$

This means that components with radius r on axes ox and oz are random, independent quantities. Consequently, when recording the float position at equal time intervals it cannot be determined beforehand in which part of the plane xoz the float will be located. This is due to the fact that the succession of waves and their size are random and independent of one another.

Taking logarithms in (2.3.13) and differentiating, one derives the following differential equation for φ:

$$\varphi' + k^2 \varphi = 0,$$

whence

$$\varphi(r^2) = C \exp\left[-k^2 r^2\right]. \tag{2.3.14}$$

Substitution of function (2.3.14) into the previously obtained expression for $f(r)$ yields

$$f(r) = 2\pi C r \exp\left[-k^2 r^2\right],$$

where constant C is determined from the condition

$$\int_0^\infty f(r)\, dr = 1,$$

which yields

$$\pi C = k^2.$$

Hence

$$f(r) = 2k^2 r \exp\left[-k^2 r^2\right]. \tag{2.3.15}$$

Constant k^2 is expressed in terms of the average value of radius vector $\bar{r}$ of the float by making use of the fact that

$$\bar{r} = \int_0^\infty f(\bar{r})\, \bar{r}\, d\bar{r}. \tag{2.3.16}$$

Substitution of function (2.3.15) into the above expression yields

$$k^2 = \frac{\pi}{4\bar{r}^2}.$$

Now the probability density function of radius vector of the float, i.e., function (2.3.15), assumes the final form

$$f(r) = \frac{\pi}{2\bar{r}} \frac{r}{\bar{r}} \exp\left[-\frac{\pi}{4}\left(\frac{r}{\bar{r}}\right)^2\right]. \qquad (2.3.17)$$

As was previously mentioned, probability theory (Venttsel', 1962) utilizes not only distributions in the form of probability density, but also their integrals, expressing the probability of the fact that the quantity under study, i.e., the wave element (for example, height), takes on a value not below some specified value. In the case at hand

$$F(r) = \int_r^\infty f(r)\,dr;$$

consequently,

$$F(r) = \int_r^\infty \frac{\pi}{2\bar{r}} \frac{r}{\bar{r}} \exp\left[-\frac{\pi}{4}\left(\frac{r}{\bar{r}}\right)^2\right] dr = \exp\left[-\frac{\pi}{4}\left(\frac{r}{\bar{r}}\right)^2\right].$$

Recalling that $2r = h$, the above expression can be written in the form

$$F(h) = \exp\left[-\frac{\pi}{4}\left(\frac{h}{\bar{h}}\right)^2\right]. \qquad (2.3.18)$$

After simple transformations, the above integral probability density function can be also written in the form

$$\frac{h}{\bar{h}} = 1.712(-\lg F)^{1/2}. \qquad (2.3.19)$$

For the distribution described by function (2.3.18) one may determine C_{rh} and C_{sh}, which are found to be respectively ~ 0.52 and ~ 0.635. These values are in general agreement with results obtained by analysis of observations. Function (2.3.17) has a maximum at $\dfrac{h}{\bar{h}} = 0.8$. The computed integral distribution function (2.3.18) is tabulated in Table 2.3.2. It is seen that the agreement with observational results listed in Table 2.3.1 is excellent.

Each time the location of the float in the plane is recorded it is also possible to simultaneously measure the wave length or the length of the crest. These measured elements should be plotted in the direction of the corresponding radius vector of the float. This will yield a statistical ensemble of linear quantities with the same properties as the ensemble of radial displacements of the float. Hence the distribution of wave or crest lengths is a distribution defined by (2.3.17). Also the integral distribution function for wave and crest lengths has the same form as (2.3.18). Instead of h one should now use the random values of the wave length (λ) or of the crest length (L). It was pointed out above that function (2.3.17) attains its maximum at $\dfrac{h}{\bar{h}} = 0.8$. Consequently, the most probable wave and crest lengths are those equal to 4/5 of the average values of these quantities.

TABLE 2.3.2. Distribution of wind-wave elements from theoretical relationships (after Krylov)

	Cumulative probability, %							
	0.1	1	2	3	5	10	20	30
$k_{\bar{h}}$; $k_{\bar{\lambda}}$; $k_{\bar{l}}$ (by 2.3.18)	2.96	2.42	2.23	2.11	1.95	1.71	1.43	1.24
$k_{\bar{\tau}}$; $k_{\bar{c}}$ (by 2.3.21)	1.78	1.61	1.55	1.52	1.45	1.35	1.24	1.15

	Cumulative probability, %						
	40	50	60	70	80	90	100
$k_{\bar{h}}$; $k_{\bar{\lambda}}$; $k_{\bar{l}}$ (by 2.3.18)	1.08	0.94	0.81	0.67	0.53	0.37	0.00
$k_{\bar{\tau}}$; $k_{\bar{c}}$ (by 2.3.21)	1.07	1.00	0.93	0.85	0.76	0.62	0.00

The distribution functions of the wave height, its length and the crest length, assumed in the above derivation, describe these functions for defining the distribution of these elements at a point, since the oscillations of the float on the wave were recorded at each given point of the undulated sea. In fact, the location of the float was recorded in terms of polar coordinates r and θ at plane xoz; consequently, functions (2.3.17) and (2.3.18) describe the distribution of the above elements of two-dimensional waves. Under natural conditions one frequently encounters three-dimensional waves. It is found from theoretical analysis of the distribution of heights of three-dimensional waves (Krylov, 1956) that these have a greater variety than wave heights at a fixed point. It was found that the ratio of the average height of a three-dimensional wave $(\bar{h}_T)$ to the average wave height at a point $(\bar{h})$ is constant and equal to 1.27. It is also possible to obtain a statistical relationship between the heights of three-dimensional waves and of waves at a point (two-dimensional). This relationship is shown in Table 2.3.3, from which it follows that ratio $\dfrac{h_T}{h}$ decreases with a reduction in cumulative probability of wave height and, as the latter probability tends to zero, this ratio tends to unity. For example, for a cumulative probability of 1 % of a wave measured from a wave recording, i.e., close to the maximum possible height, it will on the average differ by less than 0.1 from the height of a three-dimensional wave of the same cumulative probability. This conclusion, which as yet awaits experimental verification, has important practical implications.

Probability theory (Venttsel', 1962) provides a method for determining
the distribution of a random variable, functionally related to another random
variable with known distribution. Hence the distribution of wave periods
can be found by using the relationhsip between the wave period and length,
for which distribution (2.3.18) is known.

TABLE 2.3.3. Statistical relation between height (h_τ) of three-dimensional wave and height (h) at a point (after Krylov)

$p\%$	$\dfrac{h_\tau}{h}$	$p\%$	$\dfrac{h_\tau}{h}$	$p\%$	$\dfrac{h_\tau}{h}$
0.1	1.07	20	1.20	60	1.34
1.0	1.10	30	1.23	70	1.42
5.0	1.14	40	1.27	80	1.51
10	1.18	50	1.30	90	1.73

If $f(x)$ is the probability density function of a continuous random variable
x, and another random variable y is related to it functionally,

$$y = \varphi(x),$$

then the probability density of the latter is expressed as

$$F(y) = \int_0^{\varphi(y)} f(x)\,dx. \tag{2.3.20}$$

In the case under study $x = \lambda$, $y = \tau$, and $f(x)$ is determined from (2.3.18) in
which $\dfrac{h}{h}$ is replaced by $\dfrac{\lambda}{\lambda}$, using the above postulate. The relationship
between λ and τ is given by (1.6.25):

$$\lambda = \frac{g\tau^2}{2\pi}.$$

Now the cumulative probability function for the period is

$$F(\tau) = \int_0^{\frac{g}{2\pi}\tau^2} \frac{\pi}{2}\frac{\lambda}{\bar\lambda}\exp\left[-\frac{\pi}{4}\left(\frac{\lambda}{\bar\lambda}\right)^2\right]d\lambda.$$

As a result one obtains the following cumulative probability function of
the wave period:

$$F(\tau) = e^{-\Gamma^4\left(\frac{5}{4}\right)\left(\frac{\tau}{\bar\tau}\right)^4}, \tag{2.3.21}$$

where Γ is the gamma function. Taking logarithms to base 10 and making
use of the fact that $\Gamma\left(\dfrac{5}{4}\right) = 0.9064$, one obtains

$$\frac{\tau}{\bar\tau} = 1.36(-\lg F)^{1/4}.$$

Here, $C_{vt} = 0.3$ and $C_{st} = 1.0$.

Utilizing the fact that the rate of wave propagation is directly proportional to the period, the distribution function for the wave speeds can be obtained from (2.3.21) by simply replacing τ by c. The numerical values of function (2.3.21) are tabulated in Table 2.3.2. Comparing these with data of Table 2.3.1, one sees that agreement with the empirical distribution is excellent.

The fact that observations and theoretical analysis yield the distribution functions of wind-wave elements and that these also expose their random nature, is of substantial significance in the analysis of processes of wind-wave development and propagation, in improving methods of observing them, and for practical utilization of the results of these observations and studies.

All the above data on the large variety of waves pertain to wind waves propagating on the surface of a sufficiently deep sea, i.e., when the wave size is unaffected by the depth. For wave motion in shallow waters, observations show (Vilenskii and Glukhovskii, 1957) that the statistical distribution of their elements begins to diverge from that in deep waters. It starts to depend on the ratio of the average wave height $(\bar{h})$ to the sea depth (H), i.e., on $\dfrac{\bar{h}}{H}$.

Figure 2.3.6 shows the variation in the dimensionless ratio $\dfrac{h}{\bar{h}}$ as a function of $\dfrac{\bar{h}}{H} = H^*$. These data were obtained from wave recordings taken in two coastal regions of the Caspian Sea and one coastal region of the Baltic Sea. It is found that the higher H^*, the closer, in general, the curves of ratios $\dfrac{h}{\bar{h}}$ to one another. Consequently, the variety of waves decreases.

The highest waves, which have a cumulative probability of less than 30%, decrease in size when moving in shallow waters, and the reduction is the greater, the higher the waves, i.e., the smaller their cumulative probability. Wave heights with cumulative probabilities of 0.1 and 1% when passing from deep into highly shallow water ($H^* \geqslant 0.4$) decrease in size by respectively ~ 60 and $\sim 40\%$. Wave heights with cumulative probability close to $30-50\%$ in deep water do not change when moving into shallow waters. Waves with cumulative probability of $\sim 50\%$ increase in height with increasing H^*. For example, the heights of waves with cumulative probability of 70 and 90% increase, when passing to a depth of H^*, respectively by 20 and 60%. The most significant changes in heights of waves with low as well as high cumulative probability is observed for values of H^* from 0.08 to 0.3.

The observed height distribution in coastal, shallow waters was approximated with sufficient accuracy by a formula (Glukhovskii, 1966), describing the distribution of relative wave height $\left(k_h = \dfrac{h}{\bar{h}} \right)$ in integral form:

$$F(k_h) = \exp\left[-\frac{\pi}{4\left(1 + \dfrac{H^*}{\sqrt{2\pi}}\right)} \left(\frac{h}{\bar{h}}\right)^{\frac{2}{1-H^*}} \right], \qquad (2.3.22)$$

where $\bar{h}$ is the mean wave height, and $H^* = \dfrac{\bar{h}}{H}$, H being the sea depth.

Assuming in expression (2.3.22) that $H^* = 0$, we can reduce it to the form

$$F(k_h) = \exp\left[-\frac{\pi}{4}\left(\frac{h}{\bar{h}}\right)^2\right],\qquad(2.3.23)$$

which is identical with (2.3.18).

At the boundary of the crest breaking zone, when it may be assumed that $H = 2h$ and consequently $H^* = 0.5$, function (2.3.22) reduces to

$$F\left(\frac{h}{\bar{h}}\right) = \exp\left[-\frac{\pi}{4.8}\left(\frac{h}{\bar{h}}\right)^4\right],\qquad(2.3.24)$$

or

$$\frac{h}{\bar{h}} = 1.37(-\lg F_h)^{1/4},\qquad(2.3.25)$$

which is identical with (2.3.21). Consequently, in this case the distribution of wave heights coincides with the distribution of wave periods in deep water. Function (2.3.22) can be rewritten in a form convenient for computation:

$$\frac{h}{\bar{h}} = [-2.932(1 + 0.4H^*)\lg F_h]^{\frac{1-H^*}{2}}.\qquad(2.3.26)$$

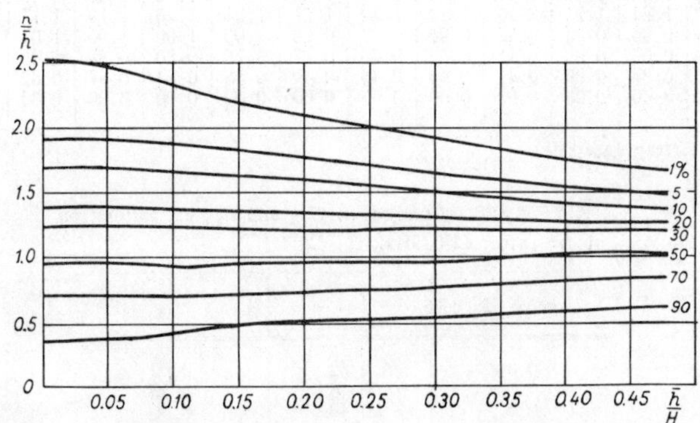

FIGURE 2.3.6. Variation (in %) of relative wave heights (k_h) as a function of relative sea depth h/H (after Vilenskii and Glukhovskii).

This form was used for computing the values in Table 2.3.4.

The coefficients of variation (C_{vh}) and skewness (C_{sh}) of wave heights undergo certain changes as the waves enter shallow waters (Table 2.3.5).

It follows from Table 2.3.5 that before breaking, the variability in wave height decreases, the skewness of its distribution becomes smaller, and the wave height distribution is symmetrical.

The distribution of wave periods does not, in the first approximation, change in shallow waters, and remains the same as in deep water, i.e., it is described by (2.3.21) and, consequently, also for deep water $C_{vt} = 0.28$, while $C_{st} = 0$.

The wave length distribution coincides with that of the wave heights, i.e., is given, similarly to (2.3.26), by the expression

$$\frac{\lambda}{\bar{\lambda}} = [-2.932(1+0.4\lambda^*)\lg F_\lambda]^{\frac{1-\lambda^*}{2}}. \qquad\qquad (2.3.27)$$

Here

$$\lambda^* = f\left(\frac{2\pi H}{\bar{\lambda}}\right)$$

is tabulated in Table 2.3.6 (after Glukhovskii), while Table 2.3.7 lists the results obtained with (2.3.27) using the data of Table 2.3.6.

TABLE 2.3.4. Cumulative probability of relative wave heights $\left(\dfrac{h}{\bar{h}}\right)$

$p\%$	Deep sea	Intermediate zone									Breaking zone
	\multicolumn — $H^* = \bar{h}/H$										
	0	0.05	0.1	0.15	0.20	0.25	0.30	0.35	0.40	0.48	0.5
0.1	2.96*	2.84*	2.71*	2.59*	2.47*	2.34*	2.23*	2.12*	2.01*	1.85*	1.81*
1	2.42	2.34	2.26	2.17	2.09	2.01	1.93	1.85	1.78	1.66	1.63
5	1.95	1.89	1.86	1.81	1.76	1.71	1.66	1.61	1.56	1.48	1.46
10	1.71	1.68	1.65	1.62	1.59	1.55	1.52	1.48	1.44	1.38	1.37
20	1.43	1.42	1.41	1.39	1.37	1.36	1.34	1.32	1.30	1.26	1.25
30	1.24	1.23	1.23	1.23	1.22	1.21	1.21	1.20	1.19	1.17	1.16
50	0.94	0.94	0.95	0.96	0.97	0.98	0.99	1.00	1.00	1.01	1.01
70	0.67	0.69	0.71	0.73	0.75	0.77	0.79	0.80	0.82	0.85	0.86
90	0.37	0.39	0.41	0.44	0.46	0.49	0.52	0.54	0.57	0.62	0.63
100	0.00	0.00	0.00	0.00	0.00	0.00	0.00	0.00	0.00	0.00	0.00

* Extrapolated data.

TABLE 2.3.5. Changes in C_{vh} and C_{sh} as a function of H^* (after Glukhovskii)

$H^* = \dfrac{\bar{h}}{H}$	C_{vh}	C_{sh}
0.0	0.52	0.62
0.1	0.48	0.50
0.2	0.43	0.36
0.3	0.38	0.22
0.4	0.33	0.07
0.5	0.28	0.00

TABLE 2.3.6.

$\dfrac{2\pi H}{\bar{\lambda}}$	0	0.5	1.0	1.5	2.0	2.5	3	3.5	4	4.5	5	5.5	$\geqslant 6$
λ^*	0.5	0.44	0.37	0.30	0.225	0.16	0.12	0.09	0.065	0.045	0.02	0.01	0.00

The above observed data and expressions (2.3.22)—(2.3.27) obtained on their basis, as well as the resulting conclusions, are valid for bottom slopes from 0.01 to 0.1 for a soil consisting of sand and gravel. The application of these results to other shore conditions with different bottom slopes naturally requires verification.

All that was said about the variety of wind waves apparently applies to wind waves proper. The variety of waves in the case of swell most likely differs in its quantitative assessment from the above relationships. This question has not until now been provided with an explicit quantitative formulation.

All that was said in this section pertains, as noted previously, to statistica characteristics of wind waves, reflecting their distribution over a very short time. Consequently, we have been discussing the so-called quasistationary distribution functions of wind-wave elements. However, in the study of wind waves one encounters other distribution functions, which are constructed on the basis of long-term systematic observations of wave dimensions in seas or oceans. Such functions will be discussed in Section 6 of this chapter.

TABLE 2.3.7. Cumulative probability of relative wave lengths $\left(\dfrac{\lambda}{\bar{\lambda}}\right)$ (after Glukhovskii)

$F\%$	Deep Sea	Intermediate zone							Breaking zone
		$2\pi H/\bar{\lambda}$							
	∞	5.0	4.0	3.0	2.0	1.5	1.0	0.5	0
1	2.42	2.37	2.30	2.22	2.02	1.92	1.79	1.68	1.63
2	2.23	2.19	2.15	2.08	1.91	1.82	1.70	1.60	1.56
3	2.11	2.09	2.06	2.01	1.85	1.75	1.65	1.56	1.51
5	1.95	1.94	1.91	1.88	1.75	1.67	1.58	1.49	1.46
10	1.71	1.71	1.70	1.69	1.59	1.55	1.46	1.39	1.37
20	1.43	1.43	1.44	1.44	1.40	1.36	1.32	1.27	1.25
30	1.24	1.25	1.26	1.27	1.25	1.24	1.21	1.18	1.16
50	0.94	0.95	0.96	0.98	1.00	1.01	1.02	1.02	1.02
70	0.67	0.68	0.69	0.71	0.74	0.78	0.81	0.84	0.86
90	0.37	0.37	0.38	0.39	0.42	0.45	0.50	0.60	0.63
100	0.00	0.00	0.00	0.00	0.00	0.00	0.00	0.00	0.00

4. Wave-generating factors

The development of large wind waves is aided most by strong winds of stable direction and speed, blowing for an extended period of time over large water spaces. These conditions will most frequently be encountered in the path of cyclones or, as these are called, depressions. Figure 2.4.1 shows schematically the passage of such a cyclone in the Northern Hemisphere from point M to points M_1, M_2 and finally M_3. The approximate wind directions in the cyclone are shown by arrows. The winds blowing in the southern part of the cyclone are southwesterly, those in the eastern part are southeasterly, those in the northern part are northeasterly, and in the western part, northwesterly. The development and propagation of wind waves for this cyclone motion pattern will be approximately the following: the northeasterly winds in the northern part of the cyclone cannot produce very heavy wind waves, since the waves, propagating in the southwesterly direction as the depression moves to the southeast, will soon be outside the range of the wind which generated them. Hence, as the cyclone moves to the southeast, they will gradually become a weak swell, propagating from the northeast.

The growth and propagation of wind waves is promoted most by northwesterly winds, blowing in the western part of the depression. The direction of these winds is the same as the direction in which the minimum proper is moving. Hence, as the depression moves to the southeast, the

northwesterly winds will continuously affect the wave growth. Thus, the
part of the depression most favorable for wave development is that more
to the west, and the least favorable part is the northern part. Here, relatively weak wind waves and weak swell will appear.

When the depression, moving from point M_1 to M_2, turns to the west, the
northwesterly winds will have a smaller effect on the wave growth. Wind
waves from the northwest, which developed appreciably when the depression
moved to the southeast, will, after the latter turns to the west, propagate in
a straight line and transform into heavy swell. After the depression turns
to the east, the swell will continue to move from the northwest. However,
this swell will no longer be as heavy, since the northwesterly winds will not
induce such a strong wind motion as was generated when the depression
moved to the southeast.

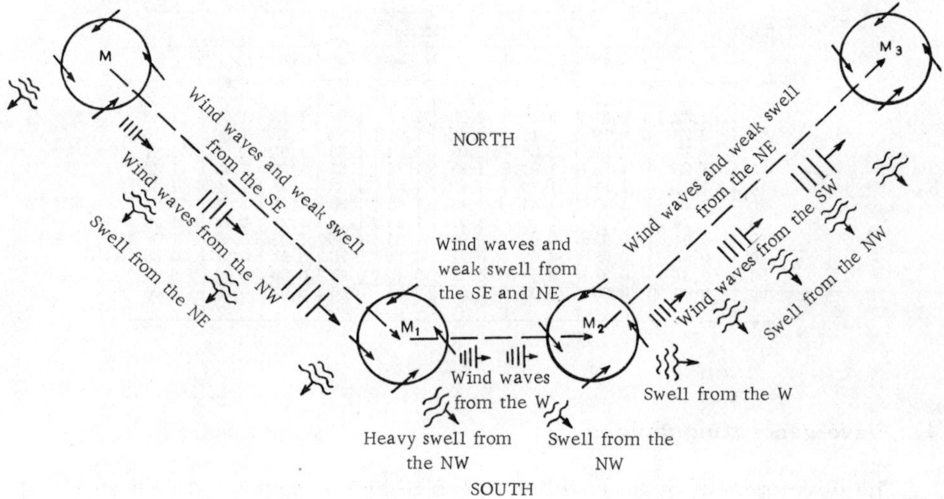

FIGURE 2.4.1. Schematic diagram of the propagation of wind waves in a region of the ocean over which
a barometric depression is moving.

With the depression moving to the northeast, the wind waves will grow
most rapidly due to southwesterly winds. The wind waves produced by
northwesterly winds will rapidly move out from under the wind and transform
into swell. However, this swell will not be as heavy as that generated by
winds from the same direction over the first segment of the depression's
path, when these winds were capable of prolonged inducement of wave growth
over large distances. The strong effect of northwesterly winds on wind-wave
development is also attributable to the fact that the generation of these winds
in depressions of the Northern Hemisphere is induced by the passage of cold
fronts. Usually this amplifies the winds from the northwest, and results in
the appearance of squalls, which taken together promotes the growth of
waves.

If the depression is sufficiently deep and, consequently, the winds are
quite strong, then the motion of such a depression over a sufficiently long
period of time over large water spaces may generate waves in the forward
part of the depression which travel at speeds higher than the minimum

proper. In this case swell will precede the passage of the depression.

It frequently happens that the strong wind which generated wind waves abates. Then these waves are transformed into swell. However, after the wind strength increases again, the presence of the swell aids in a more rapid development of new wind waves, and much less time is needed by the wind to again induce heavy wind waves.

The above occurs particularly frequently in those regions of seas and oceans where storms follow one another rapidly. Then not enough time is available for the sea to abate and each successive storm quite rapidly induces strong waves. These conditions are encountered, for example, in the North Atlantic and North Pacific, in the Barents Sea and Sea of Okhotsk, and especially frequently in the southern parts of the Atlantic, Indian and Pacific oceans, where storms follow one another frequently and reach tremendous strengths.

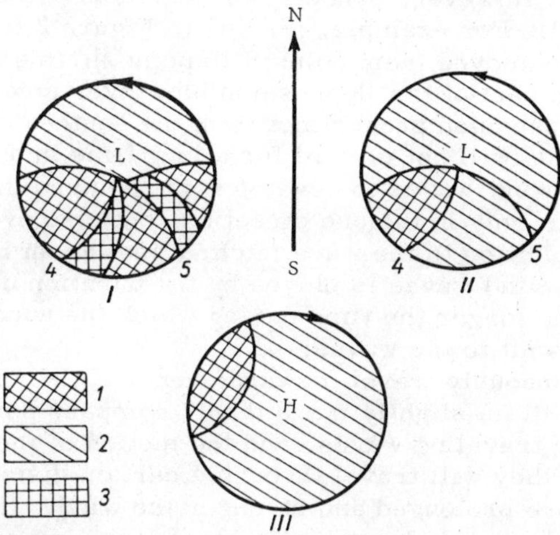

FIGURE 2.4.2. Schematic distribution of wind waves in cyclones (L) and anticyclones (H) of the Northern Hemisphere (after Efimov):

I — deep immobile cyclone; II — immobile cyclone with poorly delineated fronts; III — anticyclone; 1 — region of heavy wind waves (wave height ~ 5—7 m); 2.— region of moderate wind waves (wave height ≤ 3-5 m); 3 — region where swell appears; 4 — cold front; 5 — warm front.

Figure 2.4.2 depicts schematically the distributions of heavy (wave heights of 6 — 8 m and more with a 3 % cumulative probability) and weak (wave heights up to 3 — 4 m with a 3 % cumulative probability) waves in cyclonic and anticyclonic regions of the Northern Hemisphere. These schemes have been constructed from calculations of wind-wave elements employing daily synoptic maps and observations of weather ships in the North Atlantic. These figures supplement the explanations provided above for Figure 2.4.1.

It is seen from the above discussion of the manner in which wind waves develop over spaces covered by a moving depression, that an item of importance in the development of wind waves, in addition to the wind speed, is the distance over which there blow winds of relatively stable direction and such strength at which generation of wind waves is possible.

Wind fluxes may possess different curvature, but observations show that changes of $2-3$ compass points $(20-30°)$ in the direction of wind do not appreciably change the development and propagation of waves (Titov, 1951). The path traveled by this wind flow over the sea determines the wave fetch (x).

For example, stable northwesterly winds blow in the western part of a depression (Figure 2.4.1) as it moves from point M to point M_1. Consequently, the wave fetch in this case is the distance MM_1. As the depression continues moving the wave fetch for wave development in its southeastern part is the distance M_1M_2. Hence the wave fetch may be highly variable. It depends on synoptic conditions, i.e., on the shape, dimensions and nature of the motion of baric formations. The wind in deep depressions in moderate latitudes of both hemispheres may blow with a substantial change in direction over many hundreds of miles. However, synoptic conditions are also possible when the wave fetch is small. For example, turning to Figure 2.4.1 it can be seen that when the depression moved from point M to point M_1, the northeasterly winds in the northwestern part of the depression blow over small distances, due to the fact that the depression proper moves in the southeasterly direction. In addition, these winds will not prevail for a long time in a given region. In fact, in order for wind waves to develop it is required that they be acted upon by the wind not only in the one direction, but also over an extended time. Consequently, in addition to the wave fetch, a just as an important role in the development of wind waves is played by the duration during which the wind blows (t). The longer the time during which the wind blows, the more energy will it transmit to the waves.

The waves continuously travel forward; hence if the wind which blows in a stable direction within slightly more than 2 compass points $(\sim 22°)$ acts continuously on the traveling waves, and the motion of the latter is not in any way retarded, they will travel through a certain distance, in general the larger, the more prolonged and stronger the wind. However, the wave fetch is determined not only by synoptic conditions, but also by the size of the free water surface, i.e., the size and shape of the wave channel. If the shape and size of the latter are such that the wind fluxes do not prevail over the entire body of water, irrespective of prevailing synoptic conditions, then the waves do not only fully develop under the action of the wind, but also travel large distances in the form of swell, until they reach the shores of this body of water. Such conditions are encountered in oceans. Conversely, wind fluxes over seas usually prevail over the entire free space above the sea. Hence swell occurs in seas much less frequently than in oceans. In smaller bodies of the water, for example, natural or large water-storage lakes, the wave fetches usually do not exceed the size of the water surface, for which reason swell is usually not observed in such bodies of water.

Thus the dimensions of wind waves undergoing development depend primarily on the wind speed (w), length of air flow, i.e., the wave fetch (x) and the time during which the wind blows (t):

$$h = f(w, x, t). \tag{2.4.1}$$

Obviously, wave development is affected by other factors, for example, gravitational acceleration (g), density of the water (ρ) and of the air (ρ'), etc. Of appreciable importance in shallow waters is the sea depth (H). However, assuming that g, ρ and ρ' are constant, and considering the development of waves in deep seas, it is possible to assume that (2.4.1) satisfies, in the first approximation, conditions of wave development.

The wave fetch, in general, should begin at the shore line (leeward shore), from which the wave development starts, which is very suitable for small bodies of water (natural and water-storage lakes). In open seas, particularly in oceans, the start of the wave fetch is usually located at the point where marked changes in wind direction and speed occur. This is observed, for example, at the boundary between cold and warm meteorological fronts (Figure 2.4.2).

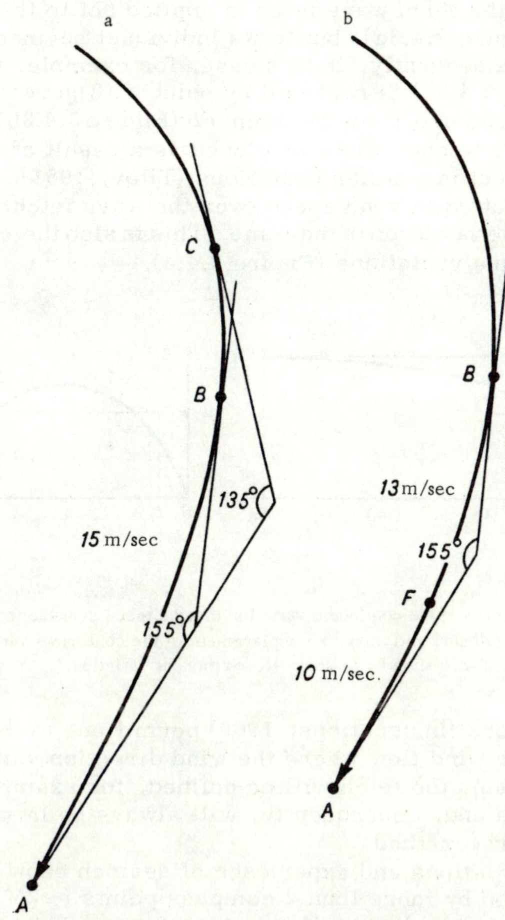

FIGURE 2.4.3. Schematic diagram clarifying the definition of wave fetch (see the text):

a — identical wind velocity over wave fetch; b — nonidentical wind velocity over wave fetch.

According to the above, the wave fetch is estimated by the extent of the open water across which the wind blows relative to the point for which it has to be determined (the terminal point of the wave fetch, point A in Figure 2.4.3a), opposite to the wave direction.

It is here assumed that the wind direction over the entire wave fetch should not vary by more than 25°. This condition defines the location of the terminal point of the required length of the wave fetch (point B in Figure 2.4.3a). This section of the wind flow (AB in Figure 2.4.3a) is taken as the wave fetch (x). However, consideration is given to yet another condition: the wind speed over the entire wave fetch thus measured should be constant within ±2 m/sec. In the opposite case the measured wave fetch is separated into such segments over which the wind speed is constant within the above range (±2 m/sec). Then the term wave fetch is applied not to the entire wind-flow segment (AB in Figure 2.4.3a), but to its individual segments (AF, FB in Figure 2.4.3b). Consequently, in this case, for example, the initial point of the fetch B (Figure 2.4.3a) is replaced by point F (Figure 2.4.3b), while point B is the initial point of the wave fetch FB (Figure 2.4.3b). These methods of determining wave fetches were developed as a result of many such measurements under specific weather conditions (Titov, 1951). It follows that the continuous variation in wind speed over the wave fetch is replaced by continuous, stepwise variation of the same. This is also the case when the wind speed undergoes time variations (Figure 2.4.4).

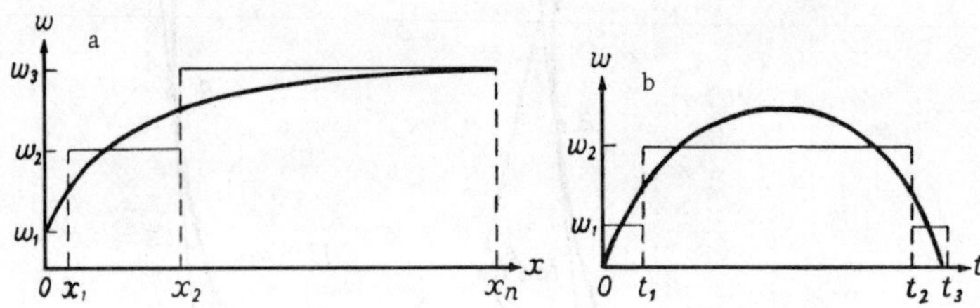

FIGURE 2.4.4:

a — replacement of the continuous variation in wind speed over the fetch by its stepwise, continuous variation; b — replacement of the continuous variation in the wind speed with time by its stepwise, continuous variation.

Some investigators (Instructions, 1960) permit one to take as the fetch such sections of the wind flow where the wind direction varies within 45°. Under these conditions the fetch will be defined, for example, by segment AC in Figure 2.4.3a, and, consequently, will always be larger than the fetch measured by the first method.

Numerous observations and experience of seamen show that if the wind changes its direction by more than 2 compass points (~25°), this produces a marked reduction in the height of existing waves. When the wind direction changes by approximately 4 compass points (~45°), then, provided the wind strength does not decrease, a new wave system starts developing and the previously existing waves rapidly lose their outlines. Some numerical estimates of the effect of changes in wind direction on wave size show that the height of waves moving at an angle of 45° to the wind direction is approximately 55% of the height they had when propagating in the direction of the

wind (Arthur, 1949). By virtue of the same fact it is more rational to take
the limiting angle in estimating the wave fetch as 25°. The wave fetch is
determined by this method from synoptic maps, from which the distance
over which the wind blows and its velocity can be obtained.

By virtue of the variety in distributions of wind directions and strength
over the ocean or sea surface, which is due to given synoptic processes,
the latter and (by virtue of the same fact) wave fetches may be markedly
different. The variability in air flows can be estimated for the duration of
the storm, i.e., for several days, as well as, naturally, for longer time
periods (month, season, year). A study of this variety (Titov and Zubova,
1967) showed that it can be expressed by the following distribution function

of the integral frequency of relative wave fetches $\dfrac{x}{x^*}$:

$$F\left(\frac{x}{x^*}\right) = \exp\left[-\frac{x}{x^*}\right] = \exp\left[-k_x\right],\tag{2.4.2}$$

where x is the wave fetch, measured in each case by the methods described
above, while x^* is the average wave fetch, computed as the arithmetic mean
of the total of measured fetches (x), corresponding to the same wind speed;

$$k_x = \frac{x}{x^*}.$$

The standard error in determining $F\left(\dfrac{x}{x^*}\right)$ from function (2.4.2) is $\pm 12\%$,
but it increases for $F\left(\dfrac{x}{x^*}\right) \approx 0.01$ and smaller.

These results were obtained (Titov and Zubova, 1967) by measuring the
extent of ocean over which wind blows by methods presented above (Figure
2.4.3a), from daily synoptic maps for individual storms, continuously for
different seasons and for individual years, over a region in the North
Atlantic bounded by 30 to 65°N and 10°E to 60°W; approximately 10,000 such
measurements for wind speeds from 4 to 30 m/sec were taken.

The distribution of $k_x = \dfrac{x}{x^*}$ given by (2.4.2) does not depend on wind speed
(Titov and Zubova, 1967) and is applicable over periods of from several days,
i.e., for the duration of the storm and longer, provided the number of
variables in the distribution under study is at least 100—150. Values of
$F\left(\dfrac{x}{x^*}\right)$ given by (2.4.2) are listed in Table 2.4.1. Here, if $F\left(\dfrac{x}{x^*}\right) < 1\%$,
the values of $\dfrac{x}{x^*}$ should be regarded as approximate, since in general
$\dfrac{x}{x^*} \to \infty$ does not have physical meaning as applied to the extent over which
the wind blows under natural conditions. By virtue of the same fact, function
(2.4.2) is inexact for values of $\dfrac{x}{x^*}$ with low cumulative probability.

If the average fetch is determined for the duration of a storm and longer
(month, season, year), x^* can be calculated from the expression (Figure 2.4.5a)

$$x^* = \frac{50 \cdot 10^5}{w},\tag{2.4.3}$$

where w is the wind speed in m/sec, x^* is in meters, and $50 \cdot 10^5$ is a constant
with dimensions of m^2/sec. The above formula is applicable for $w \geqslant 10$ m/sec.

If $w < 10 \, \text{m/sec}$, then $x^* = \text{const} \cong 500 \, \text{km}$ (Figure 2.4.5a). The standard error of expression (2.4.3) is approximately $\pm 8\%$.

TABLE 2.4.1. $F\left(\dfrac{x}{x^*}\right)$, in percent, according to function (2.4.2)

$F\%$	$\dfrac{x}{x^*}$	$F\%$	$\dfrac{x}{x^*}$	$F\%$	$\dfrac{x}{x^*}$
0.01	~ 9.2	10	2.30	45	0.80
0.1	~ 6.9	11	2.20	50	0.70
1	4.6	12	2.12	55	0.60
2	3.91	13	2.04	60	0.51
3	3.50	14	1.96	65	0.43
4	3.21	15	1.90	70	0.36
5	3.00	20	1.60	75	0.29
6	2.81	25	1.38	80	0.22
7	2.65	30	1.20	85	0.16
8	2.52	35	1.05	90	0.10
9	2.40	40	0.91	95	0.05
				99	0.01

Expression (2.4.3) was obtained on the condition that the number of fetches (x), from which the arithmetic mean value (x^*) was determined, includes only air flows with uniform wind velocity, i.e., with a velocity variation not exceeding 2 m/sec (Figure 2.4.3a). The value of x^* can be calculated also from all the measured values of air flows, meaning also those with non-uniform wind velocity (Figure 2.4.3b), which also means that the wind direction varied by less than 25°. Then

$$x^* = \frac{35 \cdot 10^5}{w},\qquad (2.4.4)$$

and if $w \leqslant 8 \, \text{m/sec}$, x^* is constant and equal to $\sim 450 \, \text{km}$. Here again x^* is in meters, and w is in m/sec (Figure 2.4.5a). The distribution function of $\dfrac{x}{x^*}$ given by (2.4.2) is valid also when x^* is determined from (2.4.4).

Using (2.4.2) and (2.4.3), the distribution of the integral frequency of air flows, i.e., fetches (x), with uniform wind speed is calculated from

$$F(x) = \exp\left[-\frac{xw}{50 \cdot 10^5}\right].\qquad (2.4.5)$$

This distribution is shown in Figure 2.4.5b, from which it is seen that the probability of the existence of fetches is substantially different for each wind speed. The higher the latter, the more probable are small fetches, and the probability that they will have large values decreases.

It should be noted that (2.4.2) can be conveniently represented in graphical form after taking logarithms twice. Then

$$\lg\left[-\lg\left(F\left(\frac{x}{x^*}\right)\right)\right] = \lg\frac{x}{x^*} + \lg(\lg e),$$

and consequently, if $\lg \lg F\left(\dfrac{x}{x^*}\right)$ is marked on the vertical axis and
$\lg \dfrac{x}{x^*}$ on the horizontal axis, function (2.4.2) will be represented by a straight
line. This can also be done for function (2.4.5).

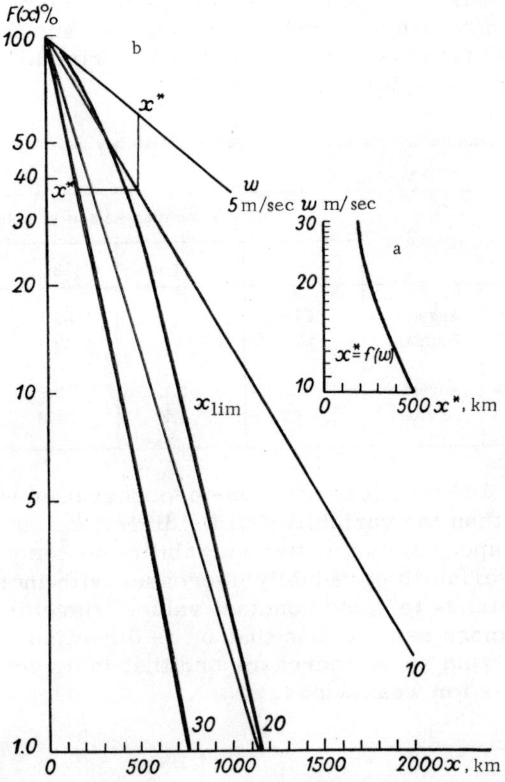

FIGURE 2.4.5:

$a - x^*=f(w)$ from (2.4.3); $b - F(x)\% = f(w)$ from (2.4.5);
$x_{\lim}$ is the limiting fetch given by (2.5.41).

The wind speed is determined by measurement or is estimated from
synoptic maps. The measured wind speed must be corrected (Chapter 3,
Section 2) before it can be used to estimate its effect on the wave development. The wind speed, as its direction, which is related to the definition
of the fetch, varies in time as well as in space. Hence the measured wind
direction and speed represent some averages over the measurement period
(usually 100 sec) or over more prolonged time intervals, determined by the
dates of measurements or dates of compilation of synoptic maps. The first
variations, i.e., variations during the period of observation, are usually
insignificant, while the others, i.e., during the time from one observation
to another, may be appreciable. They are greater in storm regions and

smaller in regions with relatively weak winds. Table 2.4.2 gives the average observation-to-observation variability in wind speed and direction from observational data in one of the storm regions of the North Atlantic. The limiting variations in wind speed of 2 m/sec and of ∼ 25° with respect to direction, which do not appreciably affect the wave development, are averages over 6 hour intervals between measurements (Table 2.4.2). In estimating the duration of wind action, defined as the period during which its velocity does not change by more than 2 m/sec and the direction remains within 25°, consideration should be given to the variability in the wind between the times of observation (Chapter 3, Section 2).

TABLE 2.4.2. Average absolute observation-to-observation variability in wind direction and speed. Weather ship J (Petrov, 1967)

Characteristic	Month	Time elapsed between observations, hours				
		6	12	18	24	48
Variations in wind direc-tion, degrees	August	27	41	55	59	66
	February	27	41	49	53	60
Variations in wind speed, m/sec	August	2.2	3.1	3.6	4.0	4.4
	February	3.2	4.2	4.9	5.0	5.4

Here (Petrov, 1967) the observation-to-observation wind speed variability is much smaller than the variability in its direction. In addition, a rapid increase in wind speed and direction variability does not continue indefinitely. The increase in variability gradually decreases with increasing time between observations and tends to some constant value. Here the variability in wind speed stabilizes more rapidly than that in its direction. For strong winds the variability in wind speed increases and that in direction decreases, the opposite being true for weak winds.

5. Dependence of wind-wave elements on the wave-generating factors

In this section we consider the results of determining empirical relationships between wave-generating factors (Chapter 2, Section 4) and wind-wave elements. The search for relationships of this kind has long been the subject of studies of many investigators in the USSR and elsewhere (Introduction, Section 1). The first publications on this topic appeared in the middle of the last century. The use recently of automatic recorders of wave elements, which replaced previous naked-eye estimates, instrumental measurements of wind speed, as well as a more objective estimate of wave fetches from synoptic maps, have markedly improved the reliability of the data.

Below we present the results of studies carried out during the past several years on the basis of instrumental observations of wave elements and of a sufficiently objective determination of wave-generating factors.

The relationship for a given wind-wave element, for example, height, can, in the first approximation for deep-sea conditions, be expressed as a function of three independent variables (2.4.1):

$$h = f(w, \, x, \, t).$$

Wave elements and wave-generating factors, contained in this or similar relationships, are expressed in dimensionless form (Sverdrup and Munk, 1947). These dimensionless expressions are:
  a) dimensionless wave height

$$\tilde{h} = \frac{g\overline{h}}{w^2}; \tag{2.5.1}$$

  b) dimensionless wave period

$$\tilde{\tau} = \frac{g\overline{\tau}}{w}; \tag{2.5.2}$$

  c) dimensionless wave speed

$$\tilde{\beta} = \frac{\overline{c}}{w}; \tag{2.5.3}$$

  d) dimensionless wave steepness

$$\tilde{\delta} = \frac{\overline{h}}{\overline{\lambda}}; \tag{2.5.4}$$

  e) dimensionless wave fetch

$$\tilde{x} = \frac{gx}{w^2} \tag{2.5.5}$$

(it should be noted that, by virtue of the variety of wave fetches discussed in Section 4 of the present chapter, it would have been more reasonable to calculate the dimensionless fetch from expression (2.5.5) with consideration of the probability of the given fetch);
  f) dimensionless duration of wave growth

$$\tilde{t} = \frac{gt}{w}. \tag{2.5.6}$$

The bars over the symbols of wave elements in the right-hand sides of the above expressions imply that the dimensionless quantities contain average values of wave elements (Chapter 2, Section 3) or such of their values which have a given cumulative probability. When reference is had to shallow water regions, where the wave development is affected by the depth of the sea, then consideration is given also to:
  g) dimensionless sea depth

$$\tilde{H} = \frac{g\overline{H}}{w^2}, \tag{2.5.7}$$

where $\overline{H}$ is the average depth over the path of wave propagation, found by some method from the entire variety of depths.

For conditions of the deep, open sea, where the depth of the sea and shore outline can be neglected, the dependence of dimensionless wave elements on dimensionless $\tilde{x}$ and $\tilde{t}$ can be expressed by the following universal functions:

$$\tilde{h} = f_1(\tilde{x},\ \tilde{t}), \qquad (2.5.8)$$

$$\tilde{\tau} = f_2(\tilde{x},\ \tilde{t}), \qquad (2.5.9)$$

$$\tilde{\beta} = f_3(\tilde{x},\ \tilde{t}). \qquad (2.5.10)$$

These functions can be represented in another form, if we assume that:

a) the wind blows for such a long time that the wave dimensions practically cease depending on $\tilde{t}$, i. e., wave development has ceased; or

b) when waves develop at such a distance from the shore (or from the storm limit), that their dimensions do not depend on $\tilde{x}$.

In the first case all the dimensionless wave elements are functions of $\tilde{x}$ only:

$$\tilde{h} = f_4(\tilde{x}), \qquad (2.5.11)$$

$$\tilde{\tau} = f_5(\tilde{x}), \qquad (2.5.12)$$

$$\tilde{\beta} = f_6(\tilde{x}), \qquad (2.5.13)$$

and in the second of $\tilde{t}$ only:

$$\tilde{h} = f_7(\tilde{t}), \qquad (2.5.14)$$

$$\tilde{\tau} = f_8(\tilde{t}), \qquad (2.5.15)$$

$$\tilde{\beta} = f_9(\tilde{t}). \qquad (2.5.16)$$

The above functions $(2.5.8)-(2.5.16)$ should be supplemented by still another function, relating the dimensionless wave height to their dimensionless period:

$$\tilde{h} = f_{10}(\tilde{\tau}), \qquad (2.5.17)$$

since such a relationship does not follow from wave theory (Chapter 1, Section 6). Consequently, the determination of numerical relationships between wave elements and wave-generating factors reduces to finding the above functions in explicit form.

This problem is presently being solved using instrumental recordings of wind-wave elements. The wind-generating factors are estimated, as far as possible, by previously discussed methods (Chapter 2, Section 4).

In all cases the values being compared have an appreciable scatter. This is particularly noticeable when comparing $\tilde{h}$, $\tilde{\tau}$ and β with $\tilde{x}$ and $\tilde{t}$. This is attributable to both inevitable inaccuracies in the determination of the wind-wave elements, as well as primarily by the indeterminacy in estimating x and t. As a result the sought relationships $(2.5.11)-(2.5.16)$ take on different numerical values.

In general they have the form

$$\widetilde{h} = a_1 \widetilde{x}^m,$$
$$\widetilde{\tau} = a_2 \widetilde{x}^p,$$
$$\widetilde{\lambda} = a_3 \widetilde{x}^n,$$
$$\widetilde{\delta} = a_4 \widetilde{x}^{m-n},$$
$$\widetilde{t} = a_5 \widetilde{x}^q.$$

Here the oscillations in values of the dimensionless coefficients a_1, a_2, a_3, a_4 and a_5 are primarily attributable to the difference in the cumulative probability of the wave elements contained in the sought relationships. The non-uniformity in the exponents m, n, p and q is primarily due to difficulties in determining x and t and to errors in these determinations. Hence the values of the above exponents range within certain limits, depending on the author. On the average their values are close to the following: $m \approx 0.5$, $n \approx 0.6$, $p \approx 0.3$ and $q \approx 0.85$.

After the first waves appear (Section 1) the wind continues feeding energy into the waves, the dimensions of which increase. It is shown by numerous observations that initially both the height and length of waves increase; here the latter increases at a higher rate, for which reason the steepness decreases and the waves become shallower if the wind speed remains constant. If the latter increases, then it is the wave height which increases at a higher rate and the waves become steeper.

It follows from (2.2.15) that for $\overline{\beta} = 1$ the wind stops feeding energy to the waves. Observations under natural conditions show that actually the wave growth is slowed down appreciably at $\overline{\beta} \approx 0.8$, while for $\overline{\beta}$ equal to approximately unity it apparently ceases entirely. These experimental results show that the effectiveness of the wind's effect on the wave profile depends markedly on $\overline{\beta}$ and variation in the latter characterizes changes in the wind energy supply to the waves. Since the latter is proportional to the square of the wave height (Chapter 1, Section 7), the functional dimensionless relationship

$$\widetilde{h} = f_{11}(\overline{\beta}) \tag{2.5.18}$$

should describe variations in the height of developing waves.

Using recordings of wave height and period observations obtained recently by new recording apparatus and simultaneous instrumental measurements of the wind speed made by Soviet expedition vessels, expression (2.5.18) was expressed in the form (Titov, 1965):

$$\widetilde{h} = 0.146 \,\overline{\beta}^{1.5} \tag{2.5.19}$$

or

$$\overline{h} = \frac{0.146}{g}\, w^2 \overline{\beta}^{1.5} = 0.0152 w^2 \overline{\beta}^{1.5}, \tag{2.5.20}$$

and also

$$\overline{h} = \frac{0.146}{(2\pi)^{1.5}}\, g^{0.5} \overline{\tau}^{1.5} w^{0.5} = 0.029 \overline{\tau}^{1.5} w^{0.5} \tag{2.5.21}$$

In these expressions $\overline{h}$ is in meters, $\overline{\tau}$ in seconds, and $\overline{w}$ in m/sec.
From (2.5.19), using (1.3.4), we have

$$\overline{\delta} = \frac{0.023}{\overline{\beta}^{0.5}}.$$

(2.5.22)

Upon substitution of (2.5.20) we obtain

$$\overline{\lambda} \approx \frac{2\pi}{g} w^2 \overline{\beta}^2,$$

$$\overline{\tau} \approx \frac{2\pi}{g} w \overline{\beta}.$$

(2.5.23)

It is seen by comparing (2.5.20) and (2.5.23) that the wave height increases
over the range $\beta = 0.5$ to $\beta = 1$ by approximately 180%, while the wave length
increases 4-fold. Hence the wave steepness decreases with development of
wind waves. However, for a given value of β the wave is steeper the stronger
the wind.

It follows from analysis of the previously mentioned recordings of wave
elements (Titov, 1965) that (2.5.19) is valid for $\overline{\beta} \leqslant 1$; this agrees with
(2.2.15). Consequently, the limiting value of $\overline{\beta}$ is

$$\overline{\beta}_{lim} = 1:$$

(2.5.24)

for higher $\overline{\beta}$ the waves cease accepting the wind energy.

However, there should exist some minimum value of $\overline{\beta}$, i.e., $\overline{\beta}_{min}$, at
which wind waves start developing (Chapter 2, Section 1).

The highest possible steepness of waves in water, including also wind
waves, is approximately (Chapter 1, Section 5)

$$\delta = \frac{1}{7},$$

while the limiting wave height, with a cumulative probability close to 0.1%
(Chapter 2, Section 3), is equal to three times the average wave height.
As the most probable wave length one can take 0.8 of the average wave
height (Chapter 2, Section 3). Consequently (Krylov, 1958),

$$0.143 = \frac{3\overline{h}}{0.8\overline{\lambda}},$$

(2.5.25)

whence the limiting possible average wave steepness $(\overline{\delta}_{lim})$ is found to be

$$\overline{\delta}_{lim} = 0.038.$$

(2.5.26)

Substitution of (2.5.24) and (2.5.26) into (2.5.20), (2.5.21) and (2.5.22)
yields

$$\overline{h}_{max} = 0.0152 w^2,$$

(2.5.27)

$$\overline{h}_{min} = 0.0034 w^2,$$

(2.5.28)

$$\overline{\tau}_{max} = 0.64 w,$$

(2.5.29)

$$\overline{\tau}_{min} = 0.238 w,$$

(2.5.30)

$$\overline{\beta}_{min} = 0.372.$$

(2.5.31)

The dependence of the average wave height on the wind speed and the average period according to the above formulas is shown in Figure 2.5.1, while the dependence of the average wave steepness on β is given in Figure 2.5.2.

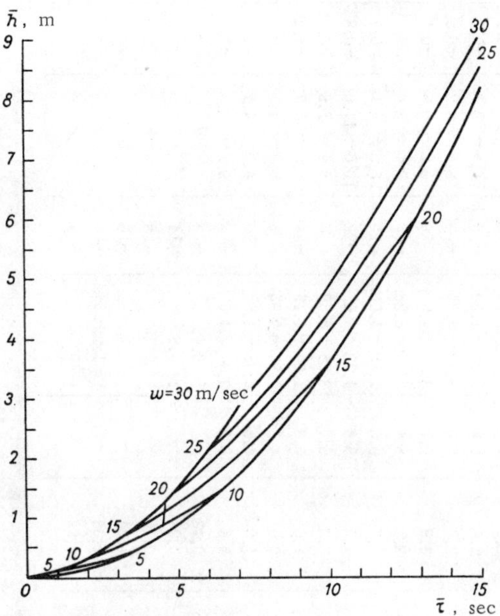

FIGURE 2.5.1. $h=f(\tau, w)$, according to (2.5.23).

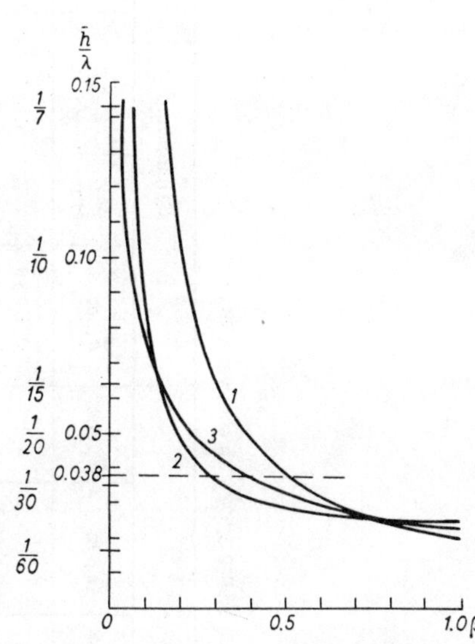

FIGURE 2.5.2. $\bar{\delta}=\dfrac{\bar{h}}{\bar{\lambda}}=f(\beta)$:

1 — from $\bar{\delta}=\dfrac{0.019}{\beta}$; 2 — according to (2.8.24); 3 — according to (2.5.22).

Expression (2.5.22) depends markedly on a similar relationship suggested by Krylov (1956):

$$\bar{\delta}=\frac{0.019}{\bar{\beta}},$$

depicted in Figure 2.5.2. The same figure also shows the relationship obtained by Shuleikin (expression (2.8.24)), which is in excellent agreement with (2.5.22). The latter expression yields

$$\bar{\tau}=10.5\,\bar{h}^{0.67}w^{-0.33},$$

which differs from a similar relationship obtained previously (Titov, 1951; Grushevskii, 1960):

$$\bar{\tau}=4.1\bar{h}^{0.6},$$

and from that used by Davidan (1964):

$$\bar{\tau}=4.8\bar{h}^{0.5}.$$

TABLE 2.5.1.   Comparison of wave elements calculated from formulas (2.5.21), (2.5.32) and (2.5.33) with observed values

Ordinal No.	Region of observations	Method of observations	Observations					Calculations				Kind of waves
			w, m/sec	x, km	t, hours	$\bar{h}$, m	$\bar{\bar{\tau}}$, sec	$\bar{h}$, m	$\Delta\bar{h}$, cm	$\bar{\bar{\tau}}$, sec	$\Delta\tau$, sec	
1	Norwegian Sea	Wave recorder	9 — 10	>1000	12	1.0	4.3	1.0 — 1.18	0	5.1 — 5.4	−0.8	Developing
2	North Sea	"	19 — 20	~ 175	> 6 — 8	2.1 — 2.2	7.0	2.35 — 2.50	−15	7.0 — 7.2	0	Fully developed
3	"	"	17 — 18	200	> 9 — 11	2.0 — 2.1	6.8 — 7.0	2.2 — 2.4	−10	7.0 — 7.2	0	"
4	"	"	17	225	>12 — 14	2.1	7.0	2.32	−22	7.2	−0.2	"
5	"	"	11 — 12	200	20	1.4	5.2	1.35 — 1.50	+ 5	5.9 — 6.1	−0.7	"
6	North Atlantic	"	15 — 17	500	20	2.7	8.0	2.72 — 3.30	− 2	8.5 — 9.2	−0.5	Developing
7	"	"	20 — 21	500	15	3.9	9.1	3.70 — 4.0	0	9.2 — 9.5	−0.1	"
8	"	"	18 — 19	350	20	2.7	8.4	3.03 — 3.28	−33	8.4 — 8.6	0	Fully developed
9	"	"	11 — 12	650	~12	1.7	6.8	1.32 — 1.50	+ 20	5.8 — 6.1	+ 0.7	Developing
10	"	"	18 — 19	900	27	4.0	9.8	4.28 — 4.6	−28	10.6 — 11.0	−0.8	"
11	"	"	21 — 22	350	10	3.3	9.0	3.25 — 3.50	+ 5	8.4 — 8.6	+ 0.4	"
12	"	"	11 — 12	>1000	60	1.6	5.8	1.84 — 2.19	−24	7.0 — 7.7	−1.2	Fully developed
13	Black Sea	"	14 — 15	40	>10	0.8	4.1	0.85 — 0.95	− 5	4.0 — 4.1	0	"
14	"	"	21 — 22	25	> 6	1.0	4.5	0.96 — 1.0	0	4.0 — 4.1	+ 0.4	"
15	White Sea depression	"	22 — 23	400	12	3.8	8.8	3.84 — 4.0	− 4	9.2 — 9.5	−0.4	Developing
16	Gulf of Finland	"	12	15	>48	0.43	—	0.45	− 2	2.9	—	Fully developed
17	"	"	15	32	>48	0.94	—	0.80	+14	3.7	—	"
18	"	"	13	28	>48	0.75	—	0.70	+ 5	3.5	—	"
19	"	"	16	52	>48	1.04	—	1.10	− 6	4.5	—	"
20	"	"	14	87	>48	1.08	—	1.22	−14	4.7	—	"
21	"	"	12	43	>48	0.62	—	0.76	−14	3.8	—	"

22	Lake Ladoga	Stereophoto-graphy	11	135 — 140	> 24	1.10	—	1.15	— 5	5.3	—	Fully developed
23	"	"	10	95 — 105	> 24	0.74	—	0.90	−16	4.6	—	"
24	"	"	9	25	> 24	0.40	—	0.42	− 2	2.8	—	"
25	"	"	9	60 — 75	> 24	0.53	—	0.67	−14	3.6	—	"
26	"	"	10	110	> 24	0.74	—	0.95	−21	4.7	—	"
27	"	"	11	125 — 130	> 24	0.97	—	1.12	−15	5.2	—	"
28	Kuibyshev Reservoir	Wave recorder	13	23	> 6	0.6	3.9	0.62	− 2	3.3	+ 0.6	"
29	"	"	8	16	12	0.4	2.7	0.30	+10	2.3	+ 0.4	"
30	"	"	10	20	10	0.4	2.7	0.45	− 5	2.8	−0.1	"
31	"	"	9	20	12	0.4	2.7 — 2.9	0.4	0	2.7	0	"
32	"	"	13	16	15 — 17	0.4	2.9 — 3.1	0.5	−10	2.9	0	"
33	"	"	12	16	19	0.4	2.7	0.42	− 2	2.7	0	"
34	"	"	12	18	> 2	0.4	2.6	0.47	− 7	2.9	−0.3	"
35	"	"	16	16	4	0.4	3.1	0.66	−26	2.7	+ 0.4	"
36	"	"	10	20	7	0.4	2.8	0.45	− 5	2.8	0	"
37	"	"	14	20	3	0.6	3.6	0.65	− 5	3.2	+ 0.4	"
38	"	"	9	10	24	0.2	2.3	0.30	−10	2.2	+ 0.1	"
39	"	"	12	23	10	0.5	3.3	0.53	− 3	3.2	+ 0.1	"
40	"	"	7	16	2	0.3	2.9	0.25	+ 5	2.2	+ 0.7	"
41	"	"	13	16	3	0.4	3.0	0.50	−10	2.9	+ 0.1	"
42	"	"	15	12	4	0.3	2.5	0.49	−19	2.7	−0.2	"
43	"	"	14	16	4	0.5	3.8	0.52	− 2	2.9	+ 0.9	"
44	Lake Onega	"	13	33	4	0.8	—	0.72	+ 8	3.8	—	"
45	"	"	15	120	5.5	1.2	—	1.55	−35	5.7	—	"
46	"	"	14	33	16	0.9	—	0.76	+14	3.9	—	"
47	"	"	13	56	30	1.0	—	0.93	+ 7	4.3	—	"
48	"	"	13	80	—	1.2	—	1.10	+10	4.8	—	"
49	"	"	11	60	9	1.0	—	0.80	+20	4.1	—	"
50	Krasnoe Lake	"	8	4.5	8	0.18	—	0.15	+ 3	1.7	—	"

Correlating wave-recorder observations of wave elements and instrumental measurements of the wind speed, fetch, and duration of wind from synoptic maps, we may express functions (2.5.12) and (2.5.15) in the form

$$\widetilde{\tau} = 2.26 \left(\widetilde{x} \right)^{0.30}, \tag{2.5.32}$$

$$\widetilde{\tau} = 2.26 \left(\widetilde{t} \right)^{0.35}, \tag{2.5.33}$$

from which it follows that

$$\overline{\tau} = \frac{2.26 w^{0.4} x^{0.3}}{g^{0.7}} = 0.457 x^{0.3} w^{0.4}, \tag{2.5.34}$$

$$\overline{\tau} = \frac{2.26 t^{0.35} w^{0.65}}{g^{0.65}} = 0.512 t^{0.35} w^{0.65}, \tag{2.5.35}$$

where $\overline{\tau}$ is in seconds, t in hours, w in m/sec, and x in kilometers.

Using (1.3.4) and (1.3.5) we derive the following from (2.5.32) and (2.5.33):

$$\widetilde{\beta} = 0.36 \left(\widetilde{x} \right)^{0.30}, \tag{2.5.36}$$

$$\widetilde{\beta} = 0.36 \left(\widetilde{t} \right)^{0.35}. \tag{2.5.37}$$

The latter two expressions give

$$x = 3.06 w^2 \overline{\beta}^{3.33}, \tag{2.5.38}$$

$$t = 1.89 \overline{\beta}^{2.86} w, \tag{2.5.39}$$

where x is in kilometers, w in m/sec, and t in hours.

In addition, it is found by equating (2.5.36) and (2.5.37) that

$$\widetilde{x} = \left(\widetilde{t} \right)^{1.17}. \tag{2.5.40}$$

When condition (2.5.24) is satisfied it follows from (2.5.38) and (2.5.39) that

$$x_{\mathrm{lim}} \approx 3.06 w^2 \approx 3.0 w^2, \tag{2.5.41}$$

$$t_{\mathrm{lim}} = 1.89 w \approx 1.9 w, \tag{2.5.42}$$

where x_{lim} is in kilometers, w in m/sec, and t_{lim} in hours.

These expressions define the limiting values of wave fetch and wind duration characterizing the generation of a fully developed sea.

The values of $\overline{h}, \overline{\tau}$ and t calculated from (2.5.21)—(2.5.31), (2.5.34) and (2.5.35) are listed in Table 3.3.1.

The rate at which the wave process propagates (Chapter 1, Section 8), according to the above relationships, is determined from (2.5.38) and (2.5.39):

$$\frac{\dfrac{dx}{d\overline{\beta}}}{\dfrac{dt}{d\overline{\beta}}} = \frac{dx}{dt} = u \approx 0.51 w \overline{\beta}^{0.5}. \tag{2.5.43}$$

The last expression defines the group velocity which (Chapter 1, Section 8) is equal to $0.5c$. It follows from (2.5.43) that

$$\frac{u}{c} = \frac{0.5}{\bar{\beta}^{0.5}} .$$
(2.5.44)

The above result does not satisfy the condition $u = \frac{1}{2}c$ due to an inaccuracy in the raw observational data and in the determination of wave-generating factors. These conditions are satisfied at $\bar{\beta} = 1$; however at $\bar{\beta} = 0.84$ $u = 0.52c$, and at $\bar{\beta} = 0.375$ (at the initial wave development stage) it is found that $u = 0.82c$.

The wave elements calculated from the above formulas (Table 3.3.1) are compared with observed data (Davidan, 1967; Davidan and Smirnova, 1969) in Table 2.5.1. The standard error in calculating the wave height is ± 12 cm, and in calculating the wave period it is ± 0.5 secs. These differences lie within the limits of experimental error.

The relative standard error in the wave height for waves 1 m or more high with a period of 4 sec or more is approximately $\pm 12\%$. For lower and shorter waves it will increase to $\sim 15 - 20\%$.

TABLE 2.5.2. Wind waves of limiting size

w, m/sec	For $\bar{\beta} = 1$						
	$h_{1\%}$, m	$\bar{\tau}$, sec	δ	t, hours	x, km	x^*, km	$F(k_x)$, %
5	0.92	3.2	1/18	9.5	75	500	86
10	3.7	6.4	1/18	18.9	300	500	55
15	8.3	9.6	1/18	28.4	675	332	13
20	14.7	12.8	1/18	37.8	1200	250	~ 1
25	23.0	15.9	1/18	47.5	1875	200	~ 0.01
30	33.0	19.2	1/18	56.7	2700	167	< 0.00001

w, m/sec	For $\bar{\beta} = 0.8$					
	$h_{1\%}$, m	$\bar{\tau}$, sec	δ	t, hours	x, km	$F(k_x)$, %
5	0.68	2.5	1/15	5.0	36	99
10	2.6	5.1	1/15	9.9	140	74
15	5.8	7.7	1/15	14.7	320	38
20	10.2	10.0	1/15	19.7	580	10
25	16.3	12.8	1/15	25.0	900	~ 1
30	23.2	15.4	1/15	30.2	1300	~ 0.04

Note. The calculations were carried out using formulas (2.5.20), (2.5.22), (2.5.23), (2.5.38) and (2.5.39) for $\bar{\beta} = 1$ and $\bar{\beta} = 0.8$; x^* and $F(k_x)$ were calculated from (2.4.3) and (2.4.2).

In observations under natural conditions the values of $\bar{\beta}$ calculated for wind waves of limiting height with wave velocities of 20 m/sec and more never reached 1.0. They usually lie within the limits of $0.5 - 0.8$. Near-unity values of $\bar{\beta}$ with lower wind velocities are encountered more frequently, the weaker the wind. It is presently assumed by the majority of investigators that $\bar{\beta} \approx 0.8$ for the entire wind-speed range. The above differences in estimates of $\bar{\beta}$ by various investigators can, it appears, be explained by considering the probability of the existence (Chapter 2, Section 4) of such limiting fetches at the end of which maximum-size waves develop.

Table 2.5.2 lists wave heights with cumulative probability of 1%, average period, steepness, duration of wave growth, limiting fetch needed for development of such waves and the cumulative probability of such fetches, calculated from previously mentioned formulas (Table 3.3.1) for $\bar{\beta} = 1$ and $\bar{\beta} = 0.8$.

It is easy to note that it is actually rather impossible to observe under natural conditions waves propagating at velocities equal to wind speeds of 25 m/sec and more. The probability of the existence of limiting fetches ($x_{\lim}$) needed for this is so small (a fraction of a percent), that the existence of such fetches is highly improbable.

It should also be taken into account that, in addition to the existence of such fetches, it is necessary that a wind of the given speed should prevail over the fetch for more than two days, the probability of which is naturally also very low. At $\bar{\beta} = 0.9$ the probability of fetches increases markedly — 100-fold for winds of 25 m/sec and 4000-fold for winds of 30 m/sec.

Consequently, expression (2.2.15) obtained from experimental studies is apparently correct. The contradictions in observations are a result of the low probability of existence in nature of conditions needed for the full development of waves which occurs with β close to unity.

6. Variety of wind-wave elements and the variability of wave-generating factors. Regime-climatic distribution functions of wave elements

The statistical regularities in the variety of wind-wave elements were considered in Section 3 of the present chapter. It was also pointed out that the appearance of different-size wind waves is apparently due to changes in wind direction and speed. It is shown in this section that actually there is a certain relationship between the variability of wave generation factors and the wave variety.

After the initial disturbances of the sea surface, produced by turbulent fluctuations of the wind pressure, wind waves develop and their future development is determined by the average wind speed (Chapter 2, Section 1). The latter undergoes continuous, time-varying oscillations at the point of measurement, and these are imposed on the principal velocity. This phenomenon is termed a wind gust; the wind speed in gusts may differ markedly from average. However, in general the stronger the gust the shorter it acts, usually for not more than several minutes.

It can hence be assumed that the above short-period inhomogeneity in the wind speed can have a substantial effect on the appearance of waves of such differing dimensions, as is found from observations. In addition, the distribution functions of relative wave elements (k_h, k_τ) are in satisfactory agreement, irrespective of wind speed (Chapter 2, Section 3). It is therefore highly doubtful that an assumption that short-time variations in wind speed for any value of the latter are governed by the same laws is valid. Hence the cause for the appearance of waves of markedly different dimensions should apparently be sought in the more permanent variations

in wind speed over the sea surface in time, as well as in space. Consequently, one should investigate the variability in the wind flows.

Obviously, the above short-duration fluctuations in the wind speed also
produce waves, but the amplitudes of these (Chapter 2, Section 1) are of
the order of $10^{-3} - 10^{-1}$ cm. These waves exist at the surface of the main
wave system. Their role in supplying the wind energy (Chapter 2,
Section 2) to the developing wind waves is appreciable; however, they are
not responsible for the variability in the main wave system.

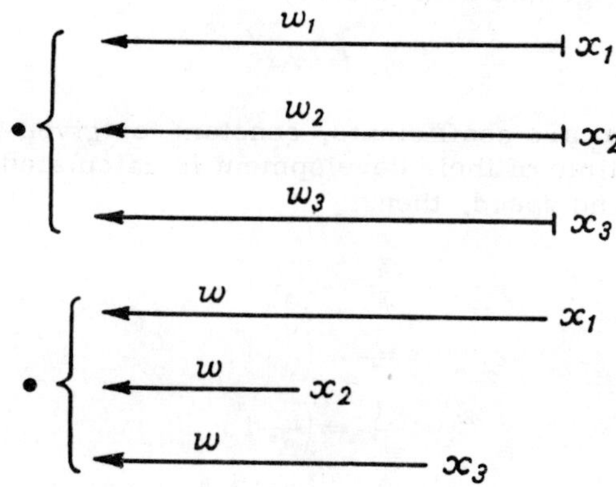

FIGURE 2.6.1.

Wind waves may arrive at a given point from the same distance but
due to winds of various speeds, or from different distances due to winds of
the same speeds (Figure 2.6.1). Naturally, a large variety of such distances,
i.e., in air flows, is possible. From this the wave dimensions may be
different. However, from our previous discussion (Chapter 2, Section 4)
the length of the air flow (provided its curvature does not exceed 25° and its
average velocity does not change by more than 2 m/sec) determines the
fetch. The variability of the latter under natural conditions is described
by expression (2.4.2):

$$F\left(\frac{x}{x^*}\right) = \exp\left[-\frac{x}{x^*}\right],$$

which can also be written in the form

$$F(k_x) = \exp[-k_x],$$

where $k_x = \dfrac{x}{x^*}$.

Consequently, the distribution function (2.4.2) of fetches and the
distribution functions of wave elements should, according to the above, be
interrelated in a certain manner. The first should define the second.

The dependence of wave elements on the fetch (x) for a given wind speed can be represented in the form (Chapter 2, Section 5)

$$\overline{h} = a_1 x^m,$$
$$\overline{\tau} = a_2 x^p,$$
$$\overline{\lambda} = a_3 x^n,$$
$$\overline{\delta} = a_4 x^{m-n}.$$

Using (2.5.14) together with (2.5.11), we have

$$t = a_5 x^q,$$

where a_1 through a_5 are coefficients, constant for given w. If the wave elements and the time of their development is calculated for any x and for x^* for the same wind speed, then

$$\frac{\overline{h}}{\overline{h}^*} = \left(\frac{x}{x^*}\right)^m,$$

$$\frac{\overline{\tau}}{\overline{\tau}^*} = \left(\frac{x}{x^*}\right)^p,$$

$$\frac{\overline{\lambda}}{\overline{\lambda}^*} = \left(\frac{x}{x^*}\right)^n,$$

$$\frac{\overline{\delta}}{\overline{\delta}^*} = \left(\frac{x}{x^*}\right)^{m-n},$$

$$\frac{t}{t^*} = \left(\frac{x}{x^*}\right)^q.$$

Here h^*, τ^*, etc., denote wave elements and the time of their growth (t), corresponding to average fetches x^* (Chapter 2, Section 2). All these relationships are more conveniently expressed in the form

$$\widetilde{k}_h = \left(\widetilde{k}_x\right)^m, \tag{2.6.1}$$

$$\widetilde{k}_\tau = \left(\widetilde{k}_x\right)^p, \tag{2.6.2}$$

$$\widetilde{k}_\lambda = \left(\widetilde{k}_x\right)^n, \tag{2.6.3}$$

$$\widetilde{k}_\delta = \left(\widetilde{k}_x\right)^{m-n}, \tag{2.6.4}$$

$$\widetilde{k}_t = \left(\widetilde{k}_x\right)^q. \tag{2.6.5}$$

The dimensionless ratios $\widetilde{k}_h$, $\widetilde{k}_\tau$, $\widetilde{k}_\lambda$, $\widetilde{k}_\delta$, $\widetilde{k}_t$ should contain elements of waves with the same cumulative probability, for example, their average values, corresponding to any value of the fetch or to its average value (x^*) $\widetilde{k}_h = \dfrac{\overline{h}}{\overline{h}^*}$. The wavy bar over $\widetilde{k}$ in (2.6.1)—(2.6.5) denotes its difference from (2.3.9) and (2.3.10). There this symbol denotes the ratio, for example, of a wave height of any cumulative probability to its average height $\overline{h}$, i.e., $k_h = \dfrac{h}{\overline{h}}$. It was noted (Chapter 2, Section 4) that the exponents of x in (2.6.1)—(2.6.5) vary in different relationships. Using some average values (Chapter 2, Section 3), we can express $(2.6.1)-(2.6.5)$ in the form

$$\widetilde{k}_x = \left(\widetilde{k}_h\right)^2,\tag{2.6.6}$$

$$\widetilde{k}_x = \left(\widetilde{k}_\tau\right)^{3,33},\tag{2.6.7}$$

$$\widetilde{k}_x = \left(\widetilde{k}_\lambda\right)^{1,7},\tag{2.6.8}$$

$$\widetilde{k}_x = \left(\widetilde{k}_\delta\right)^{-10},\tag{2.6.9}$$

$$\widetilde{k}_x = \left(\widetilde{k}_t\right)^{1,17}.\tag{2.6.10}$$

Using (2.3.20) one finds the distributions of the random variables $\widetilde{k}_h,\ \widetilde{k}_\tau,\ \widetilde{k}_\lambda,\ \widetilde{k}_\delta,\ \widetilde{k}_t$, which are functionally related to another random variable, in this case k_x, for which the distribution is known (expression (2.2.2)):

$$F\left(\widetilde{k}_h\right) = \int_0^{\left(\widetilde{k}_h\right)^2} -\exp\left[-\widetilde{k}_h\right]dk_x,$$

from which it follows that

$$F\left(\widetilde{k}_h\right) = 1 - \exp\left[-\left(\widetilde{k}_h\right)^2\right].$$

Making use of the fact that

$$F(X < x) = 1 - F(X > x),$$

we write

$$F\left(\widetilde{k}_h\right) = \exp\left[-\left(\widetilde{k}_h\right)^2\right].\tag{2.6.11}$$

Proceeding similarly, using $(2.6.7)-(2.6.10)$, one finds the distribution functions for $\widetilde{k}_\lambda,\ \widetilde{k}_\tau,\ \widetilde{k}_\delta,\ \widetilde{k}_t$ in the form

$$F\left(\widetilde{k}_\lambda\right) = \exp\left[-\left(\widetilde{k}_\lambda\right)^{1,7}\right],\tag{2.6.12}$$

$$F\left(\widetilde{k}_\tau\right) = \exp\left[-\left(\widetilde{k}_\tau\right)^{3,33}\right],\tag{2.6.13}$$

$$F\left(\widetilde{k}_t\right) = \exp\left[-\left(\widetilde{k}_t\right)^{1,17}\right],\tag{2.6.14}$$

$$F\left(\widetilde{k}_\delta\right) = \exp\left[-\left(\frac{1}{\left(\widetilde{k}_\delta\right)}\right)^{10}\right].\tag{2.6.15}$$

Distribution functions $F(\widetilde{k}_c)$ and $F(\widetilde{k}_\beta)$ will coincide with (2.6.13) from previous considerations (Chapter 2, Section 3). It should be noted that functions $(2.6.11)-(2.6.15)$ can be represented in the general form

$$F(k) = \exp\left[-\left(\widetilde{k}\right)^n\right].$$

Taking the logarithm of this expression twice, it is found that

$$\lg\left[-\lg F\left(\widetilde{k}\right)\right] = n\lg\left(\widetilde{k}\right) + \lg(\lg e).$$

Consequently, functions (2.6.11)—(2.6.15) will be plotted as straight lines if, for example, $\lg\lg F(k)$ is plotted on the ordinate and $\lg\widetilde{k}$ on the abscissa.

The corresponding probability density functions are obtained by differentiating (2.6.10)—(2.6.15), as a result of which

$$f\left(\widetilde{k}_h\right) = 2\widetilde{k}_h\exp\left[-\left(\widetilde{k}_h\right)^2\right], \tag{2.6.16}$$

$$f\left(\widetilde{k}_\lambda\right) = 1.7\left(\widetilde{k}_\lambda\right)^{0.7}\exp\left[-\left(\widetilde{k}_\lambda\right)^{1.7}\right], \tag{2.6.17}$$

$$f\left(\widetilde{k}_\tau\right) = 3.33\left(\widetilde{k}_\tau\right)^{2.33}\exp\left[-\left(\widetilde{k}_\tau\right)^{3.33}\right], \tag{2.6.18}$$

$$f\left(\widetilde{k}_t\right) = 1.17\left(\widetilde{k}_t\right)^{0.17}\exp\left[-\left(\widetilde{k}_t\right)^{1.17}\right], \tag{2.6.19}$$

$$f\left(\widetilde{k}_\delta\right) = 10\left(\frac{1}{\widetilde{k}_\delta}\right)^9\frac{1}{(\widetilde{k}_\delta)^2}\exp\left[-\left(\frac{1}{\widetilde{k}_\delta}\right)^{10}\right]. \tag{2.6.20}$$

These functions have their maxima at

$$\widetilde{k}_h \approx 0.72,$$
$$\widetilde{k}_\tau \approx 0.9,$$
$$\widetilde{k}_t \approx 0.3,$$
$$\widetilde{k}_\lambda \approx 0.6,$$
$$\widetilde{k}_\delta \approx 1.0.$$

The numerical values of $\tilde{k}_h$, $\tilde{k}_\lambda$, $\tilde{k}_\tau$, $\tilde{k}_t$, $\tilde{k}_\delta$ according to (2.6.11)—(2.6.15) for F in percents are listed in Table 2.6.1. Expressions (2.6.10)—(2.6.15) contain the wave elements and the times of their growth as a ratio to the same values, but corresponding to the average fetch, for example, h^*, τ^* and t^*. It is possible to replace ratios $\overline{h}$, $\overline{\tau}$ and t by their ratio to the values of the latter, corresponding not to x^* but to fetches of some other cumulative probability, for example, to the height, period and time of growth of waves corresponding to fetches of 46% ($x_{46\%}$) and 50% ($x_{50\%}$) cumulative probabilities. As a result

$$F\left(\widetilde{k}_{h46\%}\right) = \exp\left[-0.78\left(\widetilde{k}_{h46\%}\right)^2\right], \tag{2.6.21}$$

$$F\left(\widetilde{k}_{h50\%}\right) = \exp\left[-0.70\left(\widetilde{k}_{h50\%}\right)^2\right], \tag{2.6.22}$$

$$F\left(\widetilde{k}_{\tau50\%}\right) = \exp\left[-0.70\left(\widetilde{k}_{\tau50\%}\right)^{3.33}\right], \tag{2.6.23}$$

$$F\left(\widetilde{k}_{t50\%}\right) = \exp\left[-0.70\left(\widetilde{k}_{t50\%}\right)^{1.17}\right]. \tag{2.6.24}$$

Table 2.6.1 lists values obtained from (2.6.21)—(2.6.24), as well as values of k_x calculated from (2.4.2).

The distribution function of k_x in (2.4.2) was derived (Chapter 2, Section 4) from measured x and calculated x^* for long periods of time, i.e., for a storm and even longer periods (month, season), combined for different years. Consequently, to estimate the correctness of distribution functions (2.6.11)—(2.6.24) these should be compared with similar natural distribution of relative wave elements. For this purpose one may use applicable correlations of observations.

TABLE 2.6.1. Distribution of k_x, $\widetilde{k}_h$, $\widetilde{k}_\tau$, $\widetilde{k}_c$, $\widetilde{k}_\beta$, $\widetilde{k}_t$, $\widetilde{k}\lambda$

$F\%$	k_x	$\widetilde{k}_h$	$\widetilde{k}_{h_{46\%}}$	$\widetilde{k}_{h_{50\%}}$	$\widetilde{k}_\lambda$	$\widetilde{k}_t$ / $\widetilde{k}_c$ / $\widetilde{k}_\beta$	$\widetilde{k}_{\tau_{50\%}}$ / $\widetilde{k}_{c_{50\%}}$ / $\widetilde{k}_{\beta_{50\%}}$	$\widetilde{k}_\delta$	$\widetilde{k}_t$	$k_{t_{50\%}}$
1	2	3	4	5	6	7	8	9	10	11
0.01	~9.2	3.03	3.44	3.61	3.78	1.95	2.17	0.80	6.74	9.10
0.1	~6.9	2.62	2.98	3.12	3.20	1.78	1.98	0.82	5.27	7.12
1	4.6	2.14	2.44	2.55	2.49	1.58	1.76	0.86	3.71	5.00
2	3.91	1.98	2.25	2.36	2.26	1.50	1.67	0.87	3.22	4.35
3	3.50	1.87	2.12	2.23	2.13	1.46	1.62	0.88	2.94	3.97
4	3.21	1.79	2.04	2.13	2.03	1.42	1.58	0.89	2.72	3.67
5	3.00	1.73	1.97	2.05	1.93	1.39	1.54	0.90	2.57	3.47
10	2.30	1.52	1.73	1.81	1.66	1.28	1.42	0.92	2.05	2.77
15	1.90	1.38	1.57	1.64	1.47	1.21	1.34	0.94	1.74	2.35
20	1.60	1.27	1.44	1.51	1.34	1.15	1.28	0.96	1.50	2.03
25	1.38	1.18	1.34	1.40	1.22	1.10	1.22	0.97	1.33	1.80
30	1.20	1.10	1.25	1.31	1.12	1.06	1.18	0.98	1.17	1.58
35	1.05	1.03	1.17	1.23	1.03	1.01	1.12	0.99	1.04	1.40
37	1.00	1.00	1.14	1.19	1.00	1.00	1.11	1.00	1.00	1.35
40	0.91	0.95	1.08	1.13	0.97	0.97	1.08	1.01	0.92	1.24
45	0.80	0.89	1.01	1.06	0.87	0.94	1.04	1.02	0.83	1.12
46	0.78	0.88	1.00	1.05	0.86	0.93	1.03	1.03	0.81	1.09
50	0.70	0.84	0.95	1.00	0.81	0.90	1.00	1.04	0.74	1.00
55	0.60	0.77	0.87	0.92	0.74	0.86	0.96	1.05	0.64	0.86
60	0.51	0.71	0.81	0.85	0.67	0.82	0.91	1.07	0.56	0.75
65	0.43	0.66	0.75	0.79	0.60	0.78	0.87	1.08	0.48	0.65
70	0.36	0.60	0.68	0.71	0.54	0.74	0.82	1.10	0.42	0.57
75	0.29	0.54	0.61	0.64	0.48	0.69	0.77	1.13	0.35	0.47
80	0.22	0.47	0.53	0.56	0.40	0.63	0.70	1.16	0.27	0.36
85	0.16	0.40	0.45	0.48	0.33	0.58	0.65	1.20	0.21	0.28
90	0.10	0.33	0.37	0.39	0.27	0.52	0.58	1.24	0.15	0.20
95	0.05	0.22	0.25	0.26	0.17	0.41	0.46	1.35	0.08	0.13
99	0.01	0.10	0.14	0.15	0.06	0.25	0.22	1.59	0.02	0.03

Davidan (1967) used regular observations of the wave height at 12 weather stations and concurrent shipboard observations in the Atlantic and Pacific oceans. He used about 97,000 measurements of wave height and period for individual seasons of a number of years, pertaining to different regions in the above oceans, located between 13 and 62°N. Ratio $\dfrac{h}{h_{50\%}}$ was calculated for each region, and then the distribution of these dimensionless ratios was obtained. It was found that for all the regions and all seasons of the year, irrespective of wind speed, the distributions of $\dfrac{h}{h_{50\%}}$ and $\dfrac{\tau}{\tau_{50\%}}$ are sufficiently homogeneous, as can be seen from Figures 2.6.2 and 2.6.3. With this natural distribution one can compare directly distribution functions (2.6.22) and (2.6.23), since reference is had to the distribution of relative wave heights and periods. Naturally, natural distribution quantities $\dfrac{h}{h_{50\%}}$ and $\dfrac{\tau}{\tau_{50\%}}$ were calculated also for wave elements with the same cumulative probability.

127

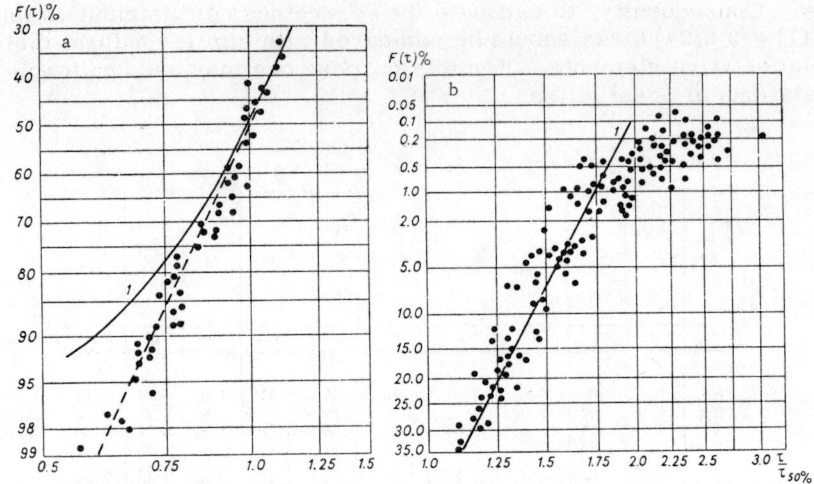

FIGURE 2.6.2. $F\left(\dfrac{\tau}{\tau_{50\%}}\right)$ by (2.6.23) (Table 2.6.1, column 8).

Dark circles represent observational data (Davidan, 1967).

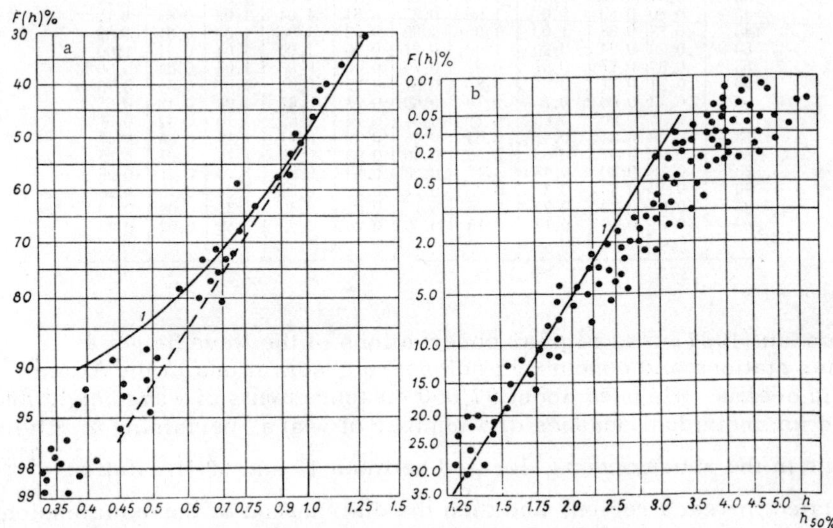

FIGURE 2.6.3. $F\left(k_{h_{50\%}}\right)$ in percents by (2.6.22) (Table 2.6.1, column 5).

Dark circles represent observational data (Davidan, 1967).

Figures 2.6.2 and 2.6.3 are plots of calculations from (2.6.22) and (2.6.23) (Table 2.6.1). The agreement between the calculated distribution of $\tilde{k}_{h_{50\%}}$ and $\tilde{k}_{\tau_{50\%}}$ with natural observations is quite satisfactory. Consequently, the validity of distribution laws (2.6.22) and (2.6.23) is verified.

Deviations are observed at $F = 0.01\%$ and less.   But it was pointed
out already that the distribution of k_x for $F \leqslant 0.01$ may contain errors
(Chapter 2, Section 4).   It should also be remembered that the values of
$\dfrac{h}{h_{50\%}}$ and $\dfrac{\tau}{\tau_{50\%}}$ obtained from observations may also be inexact, since
estimates of very high and very long waves are single and frequently
erroneous.   Deviations in the distribution of calculated values of $\tilde{k}_h$ and $\tilde{k}_\tau$
from the natural distribution also occur at $F \geqslant 70\%$.

The agreement of the distributions of $\tilde{k}_{h_{50\%}}$ and $\tilde{k}_{\tau_{50\%}}$ calculated from
(2.6.22) and (2.6.23) with their corresponding natural distribution allows
the assumption that the distribution described by (2.6.12), (2.6.14) and
(2.6.15) also applies for estimating the distributions of $\tilde{k}_\lambda$, $\tilde{k}_t$ and $\tilde{k}_\delta$ in
oceans over an extended period (month, season, year).

The probability of waves of different size in individual regions of seas
and oceans over a prolonged period, i.e., over individual months, seasons,
years, or for a long-term period, are termed regime or climatic-regime
distribution functions of wave elements.   However, the distributions of
relative wave elements, described by Table 2.6.1, are in no way regime
distributions of wave elements.

All these functions express the distribution of relative wave elements.
In order to convert them to absolute values of the latter, one must know
the values of these corresponding to average fetches, i.e., h^*, τ^*, etc.
But in each individual case these wave elements will be functions of not
only x^*, but also velocity (w), wind duration (t), and sea depth (H).
Consequently, one must have a relationship, for example, of the form

$$h^* = f_1(x^*,\ w,\ t,\ H),\tag{2.6.25}$$

or a similar expression for some other wave element.   Here x^* in (2.6.25)
can be expressed by (2.4.3).   Then (2.6.25) acquires the form

$$h^* = f_2(w,\ t,\ H).\tag{2.6.26}$$

Assuming that the wind always acts long enough for full development of
waves over any x, and setting $H = \infty$, function (2.6.26) becomes

$$h^* = f_3(w).\tag{2.6.27}$$

Upon determining from (2.6.27) the value of h^* (or of some other wave
element from a similar relationship), it is possible to calculate the
distribution of absolute values of wave heights, using Table 2.6.1.   Over
an extended period (month, season, year, etc.), the given sea region may
be subjected to winds of various speeds in many different combinations.
Consequently, the regime distribution function of wave elements for the
sea, ocean or individual region under study will be a result of the combination of all the above individual distributions.   By virtue of the fact that
the frequency of winds with different speeds in individual regions of the sea
with different free spaces can differ markedly, the regime distribution
functions will also differ from one another.   Consideration must also be
given to other factors affecting the development of wind waves, for
example, to the shallowness of a given region in the sea, as well as to the
existence of swell, arriving from another region.

Observations show that, in fact, regime functions of wave-element distributions are very different for different regions in the World Ocean, and also differ in each individual region from one season to another (Figure 2.6.4).

In order to avoid the previously mentioned complicated and prolonged calculations of the regime distribution function of wave elements, one may use another solution of this problem, due to Zubova (1969). The substance of this solution consists in the following. If we assume that the wave development process is steady for any x and the effect of shallow water is neglected, then one may write, for example,

$$h = f_4(w, x). \tag{2.6.28}$$

According to (2.4.3)

$$x = k_x \frac{50 \cdot 10^5}{w}, \tag{2.6.29}$$

where $k_x = \dfrac{x}{x^*}$.

Replacing x in (2.6.28) by its expression from (2.6.29), we derive the new function

$$h = f_5(k_x, w). \tag{2.6.30}$$

Solution for k_x yields

$$k_x = f_6(h, w). \tag{2.6.31}$$

The wind speed over a long period has a different probability which, by (3.2.2), is expressed as

$$F(w) = \exp\left[\left(-\frac{w}{\beta}\right)^{\gamma}\right].$$

The probability of k_x is described by (2.4.2) which, from the conditions under which it was obtained (Chapter 2, Section 4), does not depend on the wave speed or season and is valid, apparently, for any region in the World Ocean. Consequently $F(k_x)$ and $F(w)$ can be regarded as entirely independent quantities. On the other hand, wind-wave elements are functionally related to x and w (Chapter 2, Section 5). A method for determining the distribution of random variable h (or some other wave element) as a function of two random arguments is described in probability theory (Venttsel', 1962).

The required distribution function of h (or, for example, of τ) is obtained from the formula

$$\varphi(h) = F(k_x, w) \subset D = \iint\limits_{(D)} F(k_x, w)\, dk_x\, dw, \tag{2.6.32}$$

where

$$\subset D = [f_5(k_x, w)] \geqslant h_i.$$

Quantity h is contained in (2.6.32) implicitly, by way of integration limits.

Now $\varphi(h)$ is the required climate-regime distribution function of h (or, for example, of $\varphi(\tau)$). By virtue of the fact that (3.2.3) in explicit form differs markedly for individual ocean regions, expression (2.6.32) can be calculated approximately, but with an accuracy sufficient for practical purposes, by means of the following operations.

1. Function (2.6.31) is used in explicit form. This can be done by means of any of the available functions similar to (2.6.28) (Chapter 2, Sections 5 and 8).

2. Using the assumed function (2.6.31), values of h_i and w_i are specified in any intervals, k_x is found, and the latter, using (2.4.2), is employed for finding values of $F(k_x)$.

3. For each value of w its corresponding $f(w)$ is multiplied by $F(k_x)$ for the selected intervals $\geqslant h_i$.

4. For each level $\geqslant h_i$ the results of multiplying $f(w)$ by $F(k_x)$ are summed. As a result this yields $\varphi(h)$ for each level $\geqslant h_i$.

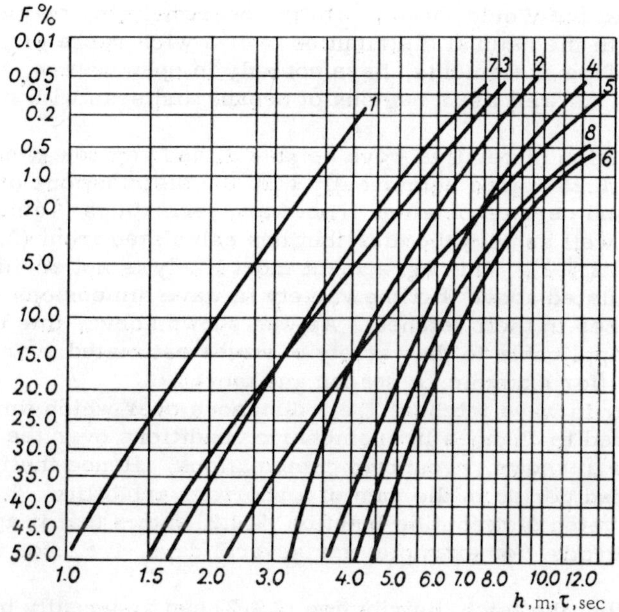

FIGURE 2.6.4. Distributions of the height and period of waves in different regions of the World Ocean from observations (Hogben, 1967):

Red Sea (average for a year), $n = 54{,}018$: 1 — wave height; 2 — wave period; North Sea (average for a year), $n = 7624$: 3 — wave height; 4 — wave period; North Atlantic, domain of weather ship B (average for a year), $n = 23{,}016$: 5 — wave height; 6 — wave period; North Atlantic, domain of weather ship B (summer, June — August), $n = 8635$: 7 — wave height; 8 — wave period.

The wave height is close to a cumulative probability of $3-5\%$, and the period is close to the average; n is the number of observations.

The same is done when determining $\varphi(\tau)$ or any other wave element. For example, using (2.5.21) and (2.5.34), one determines in explicit form functions (2.6.31), which take the form

$$k_x = \frac{7.03\bar{h}^{2.22}}{w^{1.44}}, \tag{2.6.33}$$

$$k_x = \frac{0.0027\bar{\tau}^{3.33}}{w^{0.33}}. \tag{2.6.34}$$

Then (2.3.2) is used to determine $F(k_x)$ for average heights and average periods. The former are then reduced to 3% cumulative probabilities (Tables 3.5.1 and 3.5.2). It is possible to determine k_x also for other wave elements and for the duration of their development, using appropriate expressions in Section 5 of Chapter 2. Examples of calculation of regime characteristics of wave elements are given in Section 5 of Chapter 3. A method for taking into account the existing swell is also considered there.

Numerous calculations of regime distribution functions by the above method yielded satisfactory agreement of calculations for a number of regions in the North Atlantic (Zubova, 1969), for individual seas and for other regions of the World Ocean. In the overwhelming majority of cases they agreed with the natural distribution over a wide range of cumulative probabilities of wave elements, here not only in open waters, but also at the coastal shoals, as well as in regions of oceans and seas with limited free surface.

The distribution of relative wave heights $\tilde{k}_h$ and periods $\tilde{k}_\tau$ according to (2.6.21) and (2.6.23) agree numerically with the distributions of relative wave heights and periods obtained from wave recordings (Tables 2.6.1 and 2.3.1), as well as with the distributions calculated from (2.3.18) and (2.3.21) (Table 2.3.2). This agreement most likely is not accidental.

It was postulated above that the variety in wave dimensions is a result of the differences in their fetches. As was shown above, this hypothesis is apparently applicable to the variety of waves estimated over a long period of time (for example, a season and more).

This variety in wave fetches, i.e., distances over which the wind has acted, is related to changes in anemobaric conditions over the ocean. The latter alternate in known, recurrent combinations. Hence their variety over an extended period is the sum of a shorter variability. It was noted above that the fetch distribution function (2.4.2) shows that it applies also for shorter periods, for example, for several days, i.e., for the duration of a storm.

Consequently, the fetch distribution (2.4.2) can apparently be extended to even shorter periods. Sometimes the variety of waves observed in the open sea by wave recordings will be expressed by the same distribution functions, i.e., (2.6.11)—(2.6.15) and (2.6.21)—(2.6.24), as for more extended periods. However, in making (2.4.2) universal, it cannot be extended to (2.4.3) or (2.4.4), which express the relationship between the average fetch and the wind speed. It will be different for short periods. But as the time interval over which averaging of an ever-increasing number of wave fetches is made larger, this relationship will approach that of the dependence of the average fetch on the wind speed, expressed by (2.4.3) or (2.4.4). This minimum time interval is possibly close to 5—10 days, i.e., to the natural synoptic period, which reflects with the required

completeness the variety of anemobaric conditions, and consequently also
the variety of fetches.

The difference between the previously discussed wave elements, i.e.,
between (2.6.21) on the one hand and (2.3.18) on the other, consists in the
fact that, for example, (2.3.18) contains the term

$$k_h = \frac{h}{\bar{h}},\qquad(2.6.35)$$

while (2.6.21) contains

$$\widetilde{k}_{h_{46\%}} = \frac{\bar{h}}{\bar{h}_{46\%}}.\qquad(2.6.36)$$

Here $\bar{h}_{46\%}$, as previously mentioned, corresponds for any wind velocity
to the relative fetch (k_x) having, from among the entire variety of fetches
during wave development and propagation, a cumulative probability of 46%.
For a given cumulative probability of h in (2.6.35) and of $\bar{h}$ for relative
fetches of the same cumulative probability in (2.6.36), the left-hand sides
of (2.6.35) and (2.6.36) will have identical numerical values (Tables 2.3.2
and 2.6.1). For example, the wave height with cumulative probability of
1% is 142% greater than the average wave height $\bar{h}$ (Table 2.3.2). The
average wave height for a relative fetch k_x with a 1% cumulative probability
is 144% (Table 2.6.1) greater than the wave height for a relative fetch of
46% cumulative probability, etc. Consequently, one may equate (2.6.35)
and (2.6.36) without much error. Hence

$$h = \widetilde{k}_{h_{46\%}}\,\bar{h}.\qquad(2.6.37)$$

However, according to Table 2.6.1

$$\widetilde{k}_{h_{46\%}} = 1.14\,\widetilde{k}_h.$$

Making use of (2.6.6), one writes

$$h = 1.14\,(k_x)^{0.5}\,\bar{h}_{\text{rec}},\qquad(2.6.38)$$

where $\bar{h}_{\text{rec}}$ is the average wave height obtained from recordings obtained
with a float-type wave recording instrument.

According to (2.6.38), the statistical characteristics of the wave heights,
determined from wave recordings, which were discussed above (Chapter 2,
Section 3), express the statistical characteristics of fetches which resulted
in the formation of waves of given dimensions. For example, a wave with
a cumulative probability of 1% is a wave which appeared at the end of a
fetch with 1% cumulative probability from their entire variety existing
during the given wind-wave process. The absolute value of such a fetch,
i.e., of a fetch with 1% cumulative probability, can be determined if x^*, i.e.,
the average value of all the wave fetches existing over the given time
interval, is known. However, the latter quantity cannot be calculated in
this case from (2.4.3) or (2.4.4), as mentioned above.

Proceeding in the same manner with respect to wave periods, one may write

$$\tau = 1.11 \, (k_x)^{0.3} \, \tilde{\tau}_{\text{rec}} , \qquad (2.6.39)$$

where τ_{rec} is the average wave period obtained from float-type wave recorders.

The average relative wave height, i.e., $k_h = 1$, obtained from processing of data obtained directly by float-type wave recorders, has a cumulative probability of 46% (Chapter 2, Section 3). According to distribution (2.6.11) the average relative height $(\tilde{k}_h)$ has a cumulative probability of 37% (Table 2.6.1). This disagreement apparently is due to the fact that float-type wave recorders do not record all the waves appearing at the point of observation. The maximum of the distribution of $\tilde{k}_h$, by (2.6.16), is $0.72 \, \tilde{k}_h$, i.e., is smaller than the maximum of the distribution obtained by processing wave recordings, which is close to 0.8 (Chapter 2, Section 3). The maximum of the distribution of $\tilde{k}_\tau$, which is 0.9 by (2.6.18), is also smaller than $k_\tau = 1.0$, obtained from wave recordings under natural conditions (Chapter 2, Section 3). It can hence be assumed that the probabilities of wave heights and periods calculated from wave recordings do not agree with their true values. For example, according to Table 2.3.2, wave heights with a cumulative probability of 5% will apparently have a cumulative probability of ~2% (Table 2.6.1). Similarly, a wave period with a cumulative probability of 1% (Table 2.3.2) will, according to Table 2.6.1, correspond to a period with cumulative probability of 0.1%, and so on. The higher the cumulative probability of the wave elements being compared, the smaller will be the differences. The average wave period and wave height obtained from float-type wave recorders most likely differ from their actual dimensions by, respectively, ~+11% and ~+14%.

It was noted (Chapter 2, Section 3) that the values of k_h and k_τ, determined under natural conditions, are highly nonuniform. In individual cases, particularly for low cumulative probabilities, they range over wide limits. The reason for these nonuniformities, in view of the above considerations on the causes of variability in wave elements, may be changes in the variety of fetches. With respect to individual anemobaric conditions (2.4.2) possibly should have, for example, the form

$$F(k_x) \approx \exp\left[-(k_x)^a\right], \qquad (2.6.40)$$

where a, the exponent of k_x, depends on specific weather conditions. For example, if we assume that a ranges from 0.9 to 1.1 (with an average value of unity, as in (2.4.2)) then $k_{h_{46\%}}$ and $k_{\tau_{50\%}}$ will, for $F(k_x) = 1\%$, range from 2.66 to 2.44 $(k_{h_{46\%}})$ and from 1.85 to 1.66 $(k_{\tau_{50\%}})$, which is in approximate agreement with the observed variations in k_h and k_τ (Chapter 2, Section 3). The above topic requires further study.

7. The energy balance equation for wind waves

Wind waves propagating under any condition, i.e., be it in deep or shallow waters, in different stages of their growth and decay are a form

of propagation of mechanical energy transmitted to them by the wind. The
balance of this energy (Makkaveev, 1937) in each elementary water column
participating in wave motion will be equal to the work of external forces
less the work of internal resistance forces. Using this principle, which is
valid for any mechanical system, it is possible to express the energy
balance of wave motion produced by the wind by examining some elementary
volume AA', BB', CC', DD' (Figure 2.7.1). Here its larger side is
aligned with the wind direction and the direction of wave propagation. The
thickness of layer AA' = BB' is unity, and the lower surface ABCD lies so
far down from the surface, that there is virtually no wave motion below it.
Distance $AD=dx$ has been selected so that within its limits the wave
elements of any given cumulative probability (Chapter 2, Section 3) vary
within negligible limits, i.e., some permanent wave system is continuously
in existence.

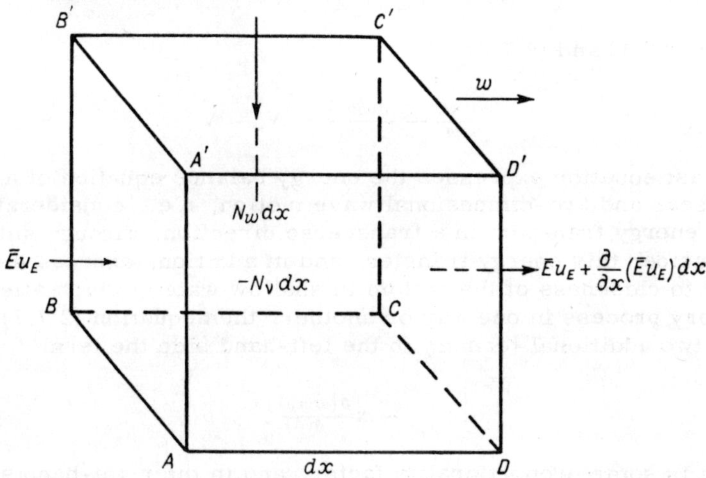

FIGURE 2.7.1.

The variation in the average wave energy $(\partial\bar{E})$ in the selected volume
per unit time is

$$\frac{\partial\bar{E}}{\partial t}\,dx, \tag{2.7.1}$$

where $\bar{E}$ is the average energy of the wave contained in a water column with
unit base and with height equal to that of the considered volume. The
energy arriving per unit time through surface AA'BB' is

$$\bar{E}u_E, \tag{2.7.2}$$

where u_E is the energy transfer rate. The energy leaving the volume
through side DD'CC' is

$$\bar{E}u_E + \frac{\partial}{\partial x}(\bar{E}u_E)\,dx. \tag{2.7.3}$$

The wind energy fed in through the upper side, i.e,, through the surface of the considered liquid volume, is

$$N_w \, dx. \tag{2.7.4}$$

Some of this energy is dissipated in the volume by viscosity forces:

$$-N_v \, dx. \tag{2.7.5}$$

Consequently, the total change in the average wave energy in the volume per unit time (with consideration of (2.7.2)—(2.7.5)) is

$$\overline{E}u_E - \left[\overline{E}u_E + \frac{\partial}{\partial x}(\overline{E}u_E)\, dx\right] + N_w \, dx - N_v \, dx =$$
$$= \left(-\frac{\partial}{\partial x}\overline{E}u_E + N_w - N_v\right) dx. \tag{2.7.6}$$

Equating (2.7.1) and (2.7.6),

$$\frac{\partial \overline{E}}{\partial t} + \frac{\partial(\overline{E}u_E)}{\partial x} = N_w - N_v. \tag{2.7.7}$$

The last equation expresses the energy balance equation of a wind wave in deep seas and two-dimensional wave motion, i.e., consideration is not given to energy transport in a transverse direction, through side BB'CC'. If we consider this energy transfer, and in addition, express the energy loss due to closeness of the bottom in shallow water, which affects the oscillatory process in one way or another, then equation (2.7.7) will acquire two additional terms: in the left-hand side the term

$$-m\frac{\partial(\partial \overline{E}u_E)}{\partial y}, \tag{2.7.8}$$

where m is some proportionality factor, and in the right-hand side the term

$$-N_H, \tag{2.7.9}$$

which accounts for energy losses due to shallowness of the water. Then equation (2.7.7) assumes the form

$$\frac{\partial \overline{E}}{\partial t} + \frac{\partial(\overline{E}u_E)}{\partial x} - m\frac{\partial(\partial \overline{E}u_E)}{\partial y} = N_w - N_v - N_H. \tag{2.7.10}$$

The first term in the left-hand side of equation (2.7.10) defines the variation in the wave energy in the considered volume per unit time, the second pertains to the change in energy transported through unit distance, and the third term represents the energy transmitted from the volume in the direction perpendicular to that of wave propagation. In the right-hand side of this equation the first term represents the energy transmitted by the wind, while the two other terms account for losses of this energy due to dissipation by turbulence and by the bottom effect, also referred to unit time per unit distance.

The rate of energy transport, i.e., the group velocity (Chapter 1, Section 8) is equal to half the phase velocity, while the wave energy (Chapter 1, Section 7) can be expressed by (1.7.6):

$$\overline{E} = \frac{1}{8}\, \rho g \overline{h}_0^2 .$$

The wave energy balance equation, for example, in the form of equation (2.7.7), can, with consideration of the above, be written as

$$\frac{\partial}{\partial t}\left(\frac{\rho g \overline{h}^2}{8}\right) + \frac{\partial}{\partial x}\left(\frac{\rho g h^2}{8}\,\frac{c}{2}\right) = N_w - N_v . \qquad (2.7.11)$$

Here N_w and N_v (Chapter 2, Section 2) also contain the wave elements $\overline{h}$ and $\overline{c}$ or $\overline{\tau}$, or $\overline{\lambda}$, and the wind speed contained in these relationships is known and constant.

By virtue of the same fact equation (2.7.11) contains two unknowns, $\overline{h}$ and $\overline{c}$. Hence in order to solve it one needs a second equation, relating the wave height to its phase velocity (or the period, or the length). Since the latter, according to the applicable formulas of wave theory (Chapter 1), can be expressed in terms of the wave length or period, the wave height can be expressed as a function of any of its three elements:

$$\left.\begin{aligned}
\overline{h} &= f_1(\overline{c}), \\
\overline{h} &= f_2(\overline{\tau}), \\
\overline{h} &= f_3(\overline{\lambda}),
\end{aligned}\right\} \qquad (2.7.12)$$

or, expressing the wave elements in dimensionless form (Chapter 2, Section 5), expressions (2.7.12) should be replaced by the expression

$$\overline{\delta} = f(\overline{\beta}). \qquad (2.7.13)$$

The specific form of these functions (Chapter 2, Section 5), as well as expressions for N_w and N_v (Chapter 2, Section 2), can be different. Obviously, the final solution of equation (2.7.11) may then also differ numerically.

The solution of the energy balance equation can be examined in general form (Grushevskii, 1960). Eliminating from equations (2.7.7) and (2.7.12) one of the unknowns, for example $\overline{c}$, and expressing the wave energy in terms of its height, this system can be reduced to a single equation with unknown wave height:

$$\frac{\partial F_1(\overline{h})}{\partial t} + \frac{\partial F_2(\overline{h})}{\partial x} = F(\overline{h}). \qquad (2.7.14)$$

The latter equation can be expressed in the form

$$\frac{dF_1(\overline{h})}{dh}\,\frac{\partial h}{\partial t} + \frac{dF_2(\overline{h})}{dh}\,\frac{\partial h}{\partial x} = F(\overline{h}). \qquad (2.7.15)$$

The coefficients of partial derivatives in equation (2.7.15) are some functions of $\bar{h}$. Division of equation (2.7.15) by $\dfrac{dF_1(\bar{h})}{dh}$ yields

$$\frac{\partial \bar{h}}{\partial t} + \Phi(\bar{h})\frac{\partial \bar{h}}{\partial x} = \psi(\bar{h}). \qquad (2.7.16)$$

Equation (2.7.16) is a particular case of a quasilinear partial differential equation which, as is known, has a discontinuous solution for a function of two independent variables, in our case x and t. This solution, in one part of the domain of existence, depends only on x (the fetch), i.e., in this case $\dfrac{\partial h}{\partial t} = 0$, while in the remaining part of this domain it depends only on the wind action time t, i.e., here $\dfrac{\partial h}{\partial x} = 0$.

Considering steady-state wave development, i.e., assuming that

$$\frac{\partial h}{\partial t} = 0,$$

equation (2.7.16) reduces to

$$x = \int_0^h \frac{\Phi(\bar{h})}{\psi(\bar{h})}\, d\bar{h} = \Phi_1(\bar{h}). \qquad (2.7.17)$$

Solution of this equation for $\bar{h}$ yields

$$\bar{h}_x = f_1(x). \qquad (2.7.18)$$

Consequently, in this case the wave height depends only on coordinate x (fetch), i.e., it is assumed that $t = t_\infty$.

Each wind speed has its own dependence on x. Consequently, there exists a family of curves $h(x)$ emanating from a common point $(0,\ 0)$ on the $(x,\ h)$ plane. Function (2.7.18) increases monotonically and asymptotically approaches some limiting value $h_{\lim}$, which varies for different wind speeds. It is also possible to construct the function

$$h_x = f(w), \qquad (2.7.19)$$

which is also an increasing function.

In distinction to the above, one may consider the case

$$\frac{\partial \bar{h}}{\partial x} = 0,$$

when the wave height depends only on the time of wind action, i.e., one may assume that coordinate x is sufficient for developing the given wave height ($x_{\lim}$). Equation (2.7.16) then yields

$$t = \int_0^{\bar{h}} \frac{d\bar{h}}{\psi(\bar{h})} = \Phi_2(\bar{h}) \qquad (2.7.20)$$

and, expressing $\bar{h}$ in terms of t, one obtains

$$\bar{h}_t = f_2(t). \tag{2.7.21}$$

In this case there exists a family of curves $\bar{h}(t)$ emanating from a common point $(0,0)$, but now on the (t, h) plane. Function $(2.7.21)$ increases monotonically and asymptotically tends to some limiting value h_{lim}, which varies for different wind speed. Hence it is possible, similarly to equation $(2.7.19)$, to find a function

$$h_t = f(w). \tag{2.7.22}$$

Functions such as $(2.7.18)$, $(2.7.19)$, $(2.7.21)$ and $(2.7.22)$ can be determined from observational data (Chapter 2, Section 5). In these relationships the values of h do not approach the values of $h_x = h_{\text{lim}}$ and $h_t = h_{\text{lim}}$ asymptotically, but rather with an inflection, which is due to the limited accuracy in obtaining the raw data.

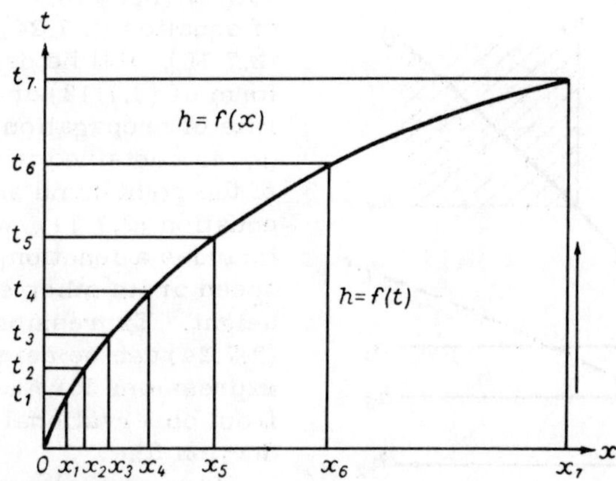

FIGURE 2.7.2.

Upon obtaining functions $(2.7.18)$ and $(2.7.21)$, i.e., upon establishing the relationship governing the wave development as a function of x and t, it is possible for each value of h_1, h_2, h_3, etc., to determine a corresponding value of x_1, x_2, x_3, etc., from equation $(2.7.18)$, and corresponding values of t_1, t_2, t_3, etc., from equation $(2.7.21)$, i.e., to find the expression

$$f_1(x) = f_2(t). \tag{2.7.23}$$

This (bounding) curve in the (x, t) coordinate system (Figure 2.7.2) will separate the half-plane $x \to 0$ and $t \to 0$ into two regions. In the region above the curve the wave height and its other elements are a function of x only, and this region is termed the domain of steady-state wave motion.

In fact (Figure 2.7.2), let a given t_i be equal to t_4, and let a given x_i be equal to x_1. In this case the wave height and its other elements will, for an increase in x_i, be determined only by the change in x_i from x_1 to x_4. Conversely, if $t_i = t_1$ while $x_i = x_4$, the wave elements will change only as a function of the increase in t_i from t_1 to t_4. Hence the region lying below the bounding curve is termed the domain of developing wave motion; in this region the change in wave height and its other elements is determined solely by changes in time, i.e., is a function of t. Consequently, by plotting specified values of x and t on the (x, t) plane (Figure 2.7.2), one may determine the domain which contains these values and also the relationship to be used in calculating the wave elements, i.e., function (2.7.18) or (2.7.21).

The bounding curve equation can be found from (2.7.18) and (2.7.20), namely,

$$\frac{dx}{dt} = u_E = u_{b.c.} = \Phi(h),\qquad(2.7.24)$$

where $u_{b.c.}$ is the rate of propagation of the bounding curve, i.e., of the boundary between domains of steady and unsteady wave motion. This is the rate of propagation of the disturbance front. The former increases in size at the expense of the latter with changes in x and t. The form of equation (2.7.24), i.e., of $\Phi(h)$ in (2.7.16), will be determined by the form of (2.7.12) or (2.7.13). The rate of propagation of wave energy (u_E) is contained in the second term of the right-hand side of the energy equation (2.7.11), which hence includes a function relating the wave speed or its other element to its height. Expressions (2.7.23) and (2.7.24) can be determined also from expressions for $h(x)$ and $h(t)$ obtained from observational data (Chapter 2, Section 5).

If, over each point on the (x, t) plane, we construct a z-axis segment, corresponding to the value of h for the given values of x_1 and t_1, then we obtain the surface $h[f(x, t)]$, which will have an edge above the bounding curve (Shuleikin, 1956).

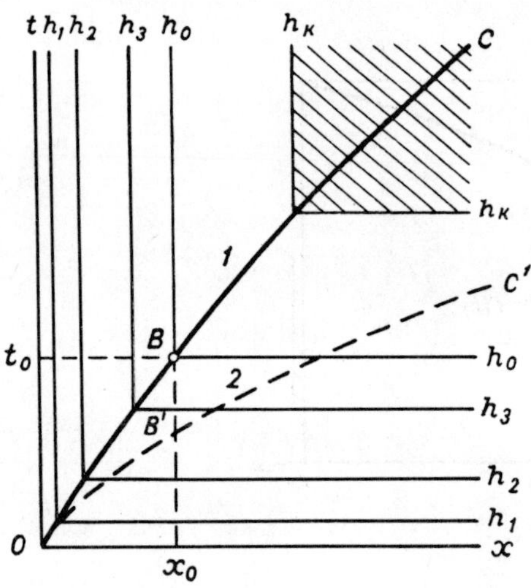

FIGURE 2.7.3. Bounding curve for wind velocities of w_1(1) and w_2(2) (after Grushevskii).

This edge will divide the plane into two parts, each of which consists of a cylindrical surface with generatrices parallel to one of the axes (x, or t). Sections produced in surface $h = f(x, t)$ by horizontal planes can be expressed by isolines in the (x, t) plane. These isolines will correspond to $h = \text{const}$. They consist (Figure 2.7.3) of broken lines consisting each of two branches: straight lines, parallel to the coordinate axes, and lines intersecting at a right angle. If these isolines are drawn through equal, small intervals of variations in wave height (Δh), then the distance between these isolines will increase with increasing wave height. This is due to the gradual reduction in the rate of wave growth as a function of x as well as of t. For some value of h close to limiting $(h_{\lim})$, for a given wind speed, this interval increases to infinity, since the following value of h will exceed $h_{\lim}$. For example, let $h_k < h_{\lim}$, but $h_{k+1} > h_{\lim}$; then it is possible to assume approximately that $h_k < h_{\lim}$, and hence it is permissible to extend the isoline $h = h_{\lim}$ over the entire part of the plane to the right and upward of this isoline. This region is crosshatched in Figure 2.7.3. The coordinates x_k and t_k of the corner point of this line can be regarded as virtually corresponding to an infinite x and t and can be denoted by $x_{\lim|}$ and $t_{\lim|}$.

The dashed line in Figure 2.7.3 shows the curve for $w_2 > w_1$. This bounding curve will pass below curve 1 for lower wind speeds. This is due to the fact that the growth of wave elements, in particular of its height, occurs more rapidly and the front of steady-state wave motion moves at a higher speed. The increase in $\dfrac{dx}{dt}$ corresponds to a reduction in the slope of the curve with respect to the abscissa.

Thus, at $x_{\lim|}$ and $t_{\lim|}$ the wave height, as well as its other elements, attain their limiting values. The latter is denoted by some authors by the symbol "∞". Each wind velocity will have its values of $x_{\lim}$ and $t_{\lim}$, as well as the limiting values of wave elements $(h_{\lim},\ \lambda_{\lim},$ etc.$)$. It is, for example, possible to determine the relationship for the limiting wave height for a given wind speed and for corresponding $x_{\lim}$ and $t_{\lim|}$:

$$h_{\lim} = f(w).$$

Relationships of this kind are also obtained from observational data (Chapter 2, Section 5).

As was previously noted, the solution of equation (2.7.7) or (2.7.10) consists of the use in explicit form of functions (2.7.12) or (2.7.13), as well as of substitution into the right-hand side of equation (2.7.7) or (2.7.10) of applicable expressions for N_w and N_v pertaining to the deep sea or, additionally, of N_H for shallow-sea conditions (2.7.10). Naturally this requires determining the numerical values of constants contained in the expression for the wind energy fed into the waves and dissipated in them (Chapter 2, Section 2). The subsequent, purely mathematical, operations yield expressions for the wave elements as a function of x and t in accordance with those conditions which were presented above in considering the general case of solving wind-wave energy balance equations (2.7.14) and (2.7.24).

8. Solutions of the energy balance equation for wind waves

There exist many solutions of the wind-wave energy balance equations, implemented by a number of investigators. Here the expression of the wave energy in terms of the square of its velocity, and the definition of the wave group velocity as half of its phase velocity in accordance with expressions (1.8.6) and (1.7.6), is used by all the investigators. Differences, and sometime quite appreciable, exist in using the second of equations (2.7.12) and of expressions for N_w and N_v (Chapter 2, Section 2). This results in a situation whereby the results of individual solutions sometimes differ markedly from one another. In addition, many solutions lack directions as to the cumulative probability of wave elements contained in equation (2.7.12) as well as of those substituted into the expressions for N_w, N_v and N_H and in the left-hand side of energy balance equation (2.7.11). These circumstances make it impossible in individual cases to objectively compare the results thus obtained with observed facts, as well as with one another, and to assess the validity of the different coefficients used. Solutions of this kind are, for example, those by Makkaveev (1937, 1951), Brovikov (1954), Ivanov (1955), Khvan (1958) and others. Studies which, in addition to sufficiently substantiated solutions of the energy balance equation, contain indications of the cumulative probability wave elements, are of definite interest. These solutions include those by Braslavskii (1952), Krylov (1956, 1958), Shuleikin (1956, 1963), and Zubova (1968).

Shuleikin (1956) uses as the second equation (equation (2.7.12)) for solving equation (2.7.11) the expression

$$\frac{h}{h_0} = 0.278 \frac{\lambda}{\lambda_0} + 0.722 \left(\frac{\lambda}{\lambda_0}\right)^{\frac{1}{3}} \qquad (2.8.1)$$

or

$$\delta = 0.04 + 0.103 \left(\frac{\lambda_0}{\lambda}\right)^{\frac{2}{3}}, \qquad (2.8.2)$$

where λ_0 and h_0 are the wave length and height at the stage with highest steepness, corresponding to the start of wave generation. These expressions were obtained by Shuleikin by employing the angular momentum theorem to water particles moving in waves along circular orbits.

The energy expended for wave development (Shuleikin, 1956) is expressed by equation (2.2.17):

$$N_w = \frac{\gamma \rho' g^2}{4\pi} h^2 (w - c)^2 c^{-3},$$

which is written as

$$N_w = \frac{2\gamma \rho' r^2 (w - c)^2}{Rt}. \qquad (2.8.3)$$

The value of N_v is estimated by means of (2.2.19); here the value of μ is replaced by ν_τ according to (2.2.45) for the surface:

$$\nu_\tau = \frac{\varkappa^2 \pi}{18} \frac{h^2}{\tau}. \qquad (2.8.4)$$

Expression (2.2.19) thus assumes the form

$$N_v = \frac{4\pi}{9} \varkappa^2 \rho g \frac{r^4}{R^2 \tau}, \qquad (2.8.5)$$

where $\varkappa = 0.1$ (Chapter 2, Section 2). Expression (2.8.5) can also be written as

$$N_v = \frac{2\pi}{9}\,\varkappa^2 \rho g\,\frac{r^4}{R^2\tau}\,.\qquad (2.8.6)$$

The energy needed for producing fully developed wave motion is equal to the energy expended for the internal turbulent process. Consequently, in this case

$$N_w = N_v\,.\qquad (2.8.7)$$

Equating (2.8.3) to (2.8.6), ascribing subscript "∞" to quantities r and R and solving (2.8.7) for r_∞, we find that

$$r_\infty = \frac{9}{\pi}\,\frac{\gamma}{\varkappa^2}\,\frac{\rho'}{\rho}\left(\frac{R_\infty}{r_\infty}\right)\frac{(w-c)^2}{g}\,.\qquad (2.8.8)$$

Using (2.8.3) and (2.8.5), equation (2.7.11) is rewritten in the form

$$\rho g r\,\frac{\partial r}{\partial t} + \frac{5}{8}\,\rho g^{\frac{3}{2}}\left(\frac{R}{r}\right)^{\frac{1}{2}} r^{\frac{3}{2}}\,\frac{\partial r}{\partial x} = \frac{2\gamma\rho' r^2 (w-c)^2}{R\tau} - \frac{2\pi}{9}\,\varkappa^2\rho g\,\frac{r^4}{R^2\tau}\,.\qquad (2.8.9)$$

First both sides of equation (2.8.9) are divided by $\rho g r$, and then by (2.8.8). Then it is assumed that at every wave development stage

$$\gamma\left(\frac{R}{r}\right)(w-c)^2 \approx \mathrm{const.}$$

As a result, one derives the following dimensionless equation instead of equation (2.8.9):

$$\frac{\partial\eta}{\partial\tau} = 1 - \eta - \eta^{\frac{1}{2}}\,\frac{\partial\eta}{\partial\zeta}\,,\qquad (2.8.10)$$

where η is the dimensionless wave height, $\eta = \dfrac{h}{h_\infty} = \dfrac{r}{r_\infty}$, τ is the dimensionless time of wave growth, $\tau = \dfrac{t}{\tau_\infty}$, with t measured in hours and the period of fully developed waves (τ_∞) in seconds; ζ is the dimensionless distance $\zeta = \dfrac{x}{w t_\infty}$, where x is the distance in kilometers, w in meters/sec and t_∞ in seconds.

In equation (2.8.10)

$$dx = 0.895 g^{\frac{1}{2}}\,\frac{\tau}{\varkappa^2}\left(\frac{R}{r}\right)^{\frac{5}{2}} r_\infty^{\frac{1}{2}}\,d\zeta,\qquad (2.8.11)$$

$$dt = \frac{9}{2\pi}\,\frac{\tau}{\varkappa^2}\left(\frac{R}{r}\right)^2 d\tau,\qquad (2.8.12)$$

$$\eta = \frac{r}{r_\infty}\,.\qquad (2.8.13)$$

It is assumed on the basis of cumulative observational data (Shuleikin, 1963) that

$$h_\infty = 0.0186 w^2 \tag{2.8.14}$$

or

$$\frac{r_\infty}{w^2} = 0.093.$$

However, it is assumed (Shuleikin, 1963) that the constant coefficient in (2.8.14) should be increased for waves of limiting size by 10%, i.e.,

$$h_\infty = 0.0205 w^2. \tag{2.8.15}$$

Then

$$\frac{r_\infty}{w^2} = 0.103. \tag{2.8.16}$$

According to (1.6.27)

$$c^2 = \frac{g\lambda}{2\pi},$$

which can be represented in the form

$$c^2 = gR,$$

or

$$c^2 = g\left(\frac{R}{r}\right)r.$$

Dividing both sides of the last expression by w^2, one obtains for waves of limiting size

$$\beta_\infty^2 = \frac{r_\infty}{w^2}\frac{R_\infty}{r_\infty}g.$$

Using (2.8.16) and assuming (Shuleikin, 1956) that $\beta = 0.82$, one derives from the last expression

$$\left(\frac{\lambda}{h}\right)_\infty = 21.$$

With (2.8.2) one obtains

$$\frac{1}{21} = 0.04 + 0.103\left(\frac{\lambda_0}{\lambda_\infty}\right)^{\frac{2}{3}},$$

whence

$$\frac{\lambda_\infty}{\lambda_0} = 50; \quad \frac{h_\infty}{h_0} = 16.7.$$

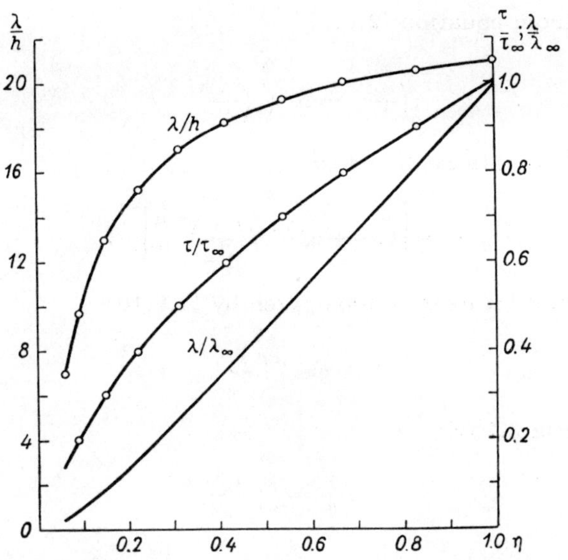

FIGURE 2.8.1. Curves of λ/h, λ/λ_∞ and $\dfrac{\tau}{\tau_\infty}$ as a function of η according to (2.8.19) and (2.8.21) (after Shuleikin).

Expression (2.8.1) is transformed on the basis of obvious identities

$$\frac{h}{h_0} = \frac{h_\infty}{h_0}\,\eta = 16.7\eta, \tag{2.8.17}$$

$$\frac{\lambda}{\lambda_0} = \frac{\lambda_\infty}{\lambda_0}\frac{\lambda}{\lambda_\infty} = 50\frac{\lambda}{\lambda_\infty}. \tag{2.8.18}$$

From equations (2.8.1), (2.8.17) and (2.8.18) one finds

$$\eta = 0.833\,\frac{\lambda}{\lambda_\infty} + 0.167\left(\frac{\lambda}{\lambda_\infty}\right)^{\frac{1}{3}}. \tag{2.8.19}$$

Figure 2.8.1 shows the curve of $\dfrac{\lambda}{\lambda_\infty}$ as a function of η according to (2.8.19). This figure also displays the curve $\dfrac{\tau}{\tau_\infty}$, which was obtained from the expression

$$\frac{\tau}{\tau_\infty} = \left(\frac{\lambda}{\lambda_\infty}\right)^{\frac{1}{2}}.$$

Using the previously given expression

$$\frac{\lambda_\infty}{\lambda_0} = 50,$$

we can derive from equation (2.8.2)

$$\left(\frac{\lambda_0}{\lambda}\right)^{\frac{2}{3}} = 50^{-\frac{2}{3}} \left(\frac{\lambda}{\lambda_\infty}\right)^{-\frac{2}{3}},$$

in which case (2.8.2) is replaced by

$$\frac{\lambda}{h} = \left[0.04 + 0.00757 \left(\frac{\lambda}{\lambda_\infty}\right)^{-\frac{2}{3}} \right]^{-1}. \qquad (2.8.20)$$

Having the explicit expressions given by (2.8.19):

$$\eta = f\left(\frac{\lambda}{\lambda_\infty}\right)$$

and by equation (2.8.20):

$$\frac{\lambda}{h} = f\left(\frac{\lambda}{\lambda_\infty}\right),$$

one finds

$$\frac{\lambda}{h} = f(\eta), \qquad (2.8.21)$$

shown in Figure 2.8.1.

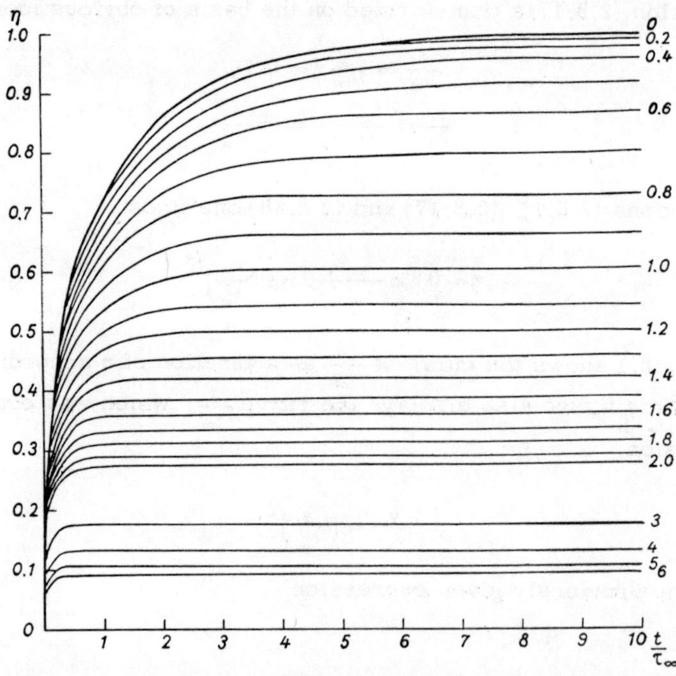

FIGURE 2.8.2. Curves of $\eta = f\left(\dfrac{t}{\tau_\infty}\right)$ according to (2.8.22) (after Shuleikin).

Relationships (2.8.11) and (2.8.12) are rewritten in a form more convenient for numerical integration:

$$\frac{t}{\tau_\infty} = 4 \cdot 10^{-3} \int \left(\frac{\tau}{\tau_\infty}\right)\left(\frac{\lambda}{h}\right)^2 \frac{d\eta}{1-\eta} \qquad (2.8.22)$$

and

$$\frac{x}{w\tau_\infty} = 1.54 \cdot 10^{-3} \int \left(\frac{\tau}{\tau_\infty}\right)\left(\frac{\lambda}{h}\right)^{\frac{5}{2}} \frac{\eta^{\frac{1}{2}} \, d\eta}{1-\eta} . \qquad (2.8.23)$$

The results of numerical integration of these expressions are shown (Shuleikin, 1963) in Figures 2.8.2 and 2.8.3 as the uppermost solid curves. Figure 2.8.4 makes it possible to decide whether the wave motion is developing or already established. The dimensionless distances $\frac{x}{w\tau_\infty}$ are plotted on the abscissa and the dimensionless times of wind action $\frac{t}{\tau_\infty}$ on the ordinate. Consequently, the curve shown in this figure is a bounding curve (Chapter 2, Section 7).

The cumulative probability of wave elements calculated from the above relationships can be estimated by using (2.5.22):

$$\bar{h} = 0.0152 w^2 \beta^{1.5} .$$

Equating the above expression to (2.8.15) and making use of the fact that $\beta = 0.82$, it is found that the cumulative probability of wave heights in (2.8.15) is $\sim 5\%$. As to the cumulative probability of the wave length and period contained in the above expressions, these for deep-water conditions are apparently equal to their average values.

The above estimate of wave-element probability makes it possible to express (2.8.20) for average wave-element magnitudes, for which purpose this expression is written in the form (Titov, 1965)

$$\bar{\delta} = 0.021 + \frac{0.03}{\beta^{1.33}} . \qquad (2.8.24)$$

Imposing the limitations of (2.5.25) on (2.8.24), it is found that 0.038 of the highest waves have their limiting steepness at

$$\beta_{min} \cong 0.27.$$

According to our previous assumption, the limiting wave height appears at

$$\beta_\infty = 0.82.$$

It follows from (2.8.24) that

$$\bar{h} = 0.0335\bar{\tau}^2 + 0.00259\bar{\tau}^{0.67} w^{1.33} . \qquad (2.8.25)$$

Figure 2.5.2 shows a plot of (2.8.24), which is in satisfactory agreement with (2.5.22) obtained directly from observational data.

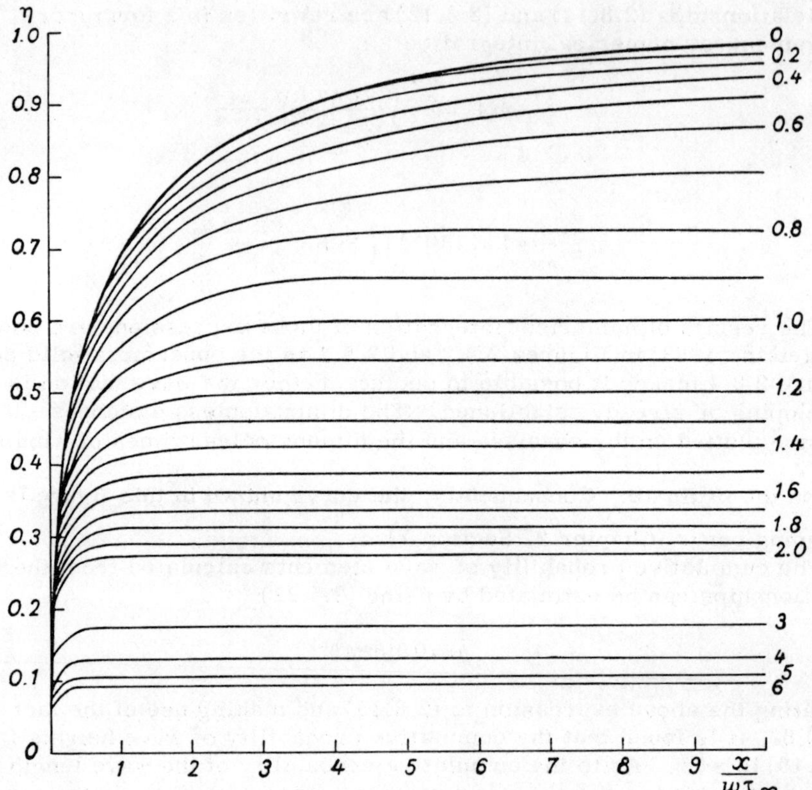

FIGURE 2.8.3. Curves of $\eta = f\left(\dfrac{x}{w\tau_\infty}\right)$ according to (2.8.23) (after Shuleikin).

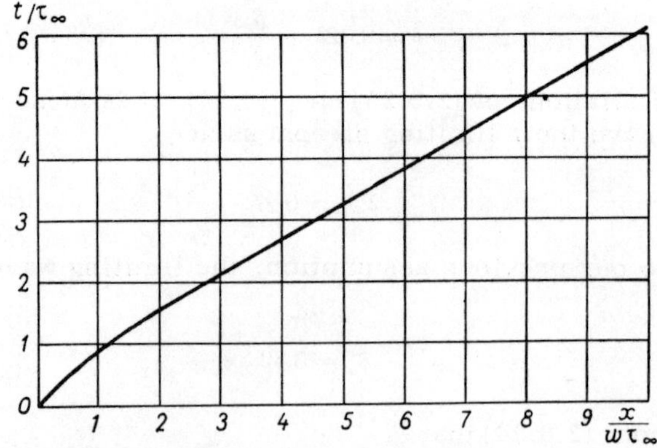

FIGURE 2.8.4. Bounding curve (after Shuleikin).

Equation (2.8.25) is plotted in Figure 2.8.5. Figure 2.8.6 shows a curve corresponding to

$$\tau_\infty = f(w).$$

One may put

$$\tau_\infty = \frac{2\pi}{g}\,\frac{c_\infty}{w}\,w$$

or

$$\tau_\infty = \frac{2\pi}{g}\,(\beta_\infty)\,w^2.$$

Assuming $\overline{\beta} = 0.82$, we obtain

$$\tau_\infty = 0.525w.$$

This can also be written in the form

$$\tau_\infty w = 0.525w^2.$$

This relationship is also given in Figure 2.8.6. Finally, this figure also shows the plot of (2.8.15).

In shallow seas, even at a large distance from the windward shore, waves can never attain the dimensions which would have been possible in deep sea. The breaking of waves due to proximity of the bottom (Chapter 2, Section 11) limits the wave growth. The average energy lost by the waves per unit surface due to proximity of the bottom, is expressed by the formula

$$N_H = \frac{\pi\varepsilon}{2}\,\frac{\rho g}{\tau H_1}\,r^3, \tag{2.8.26}$$

where ε is the fraction of the total energy lost due to partial spillover of the crest; r, as before, is half the wave height; τ is the wave period; H_1 is the depth in shallow seas, which is expressed as

$$H_1 = H\,\frac{\operatorname{sh}2\dfrac{H}{R}}{2\dfrac{H}{R}},$$

where

$$R = \frac{\lambda}{2\pi}.$$

Expression (2.8.26) is added as a new term to the energy balance equation (2.7.10), i. e., it is introduced into equation (2.8.9). After transformations similar to those carried out when solving equation (2.8.9), one derives

$$\frac{dr}{dt} = a - br - er^2, \qquad (2.8.27)$$

where

$$a = \frac{2\varkappa}{\tau} \left(\frac{r}{R} \right) \frac{\rho}{\rho'} \frac{(w - c)^2}{g},$$

$$b = \frac{2\pi}{9} \frac{\varkappa^2}{\tau} \left(\frac{r}{R} \right)^2,$$

$$e = \frac{\pi\varepsilon}{2\tau H_1}.$$

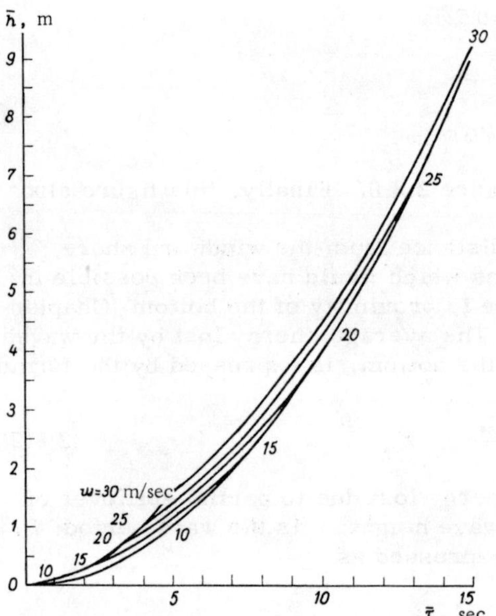

FIGURE 2.8.5. Curves of $h = f(\tau, w)$ according to (2.8.25).

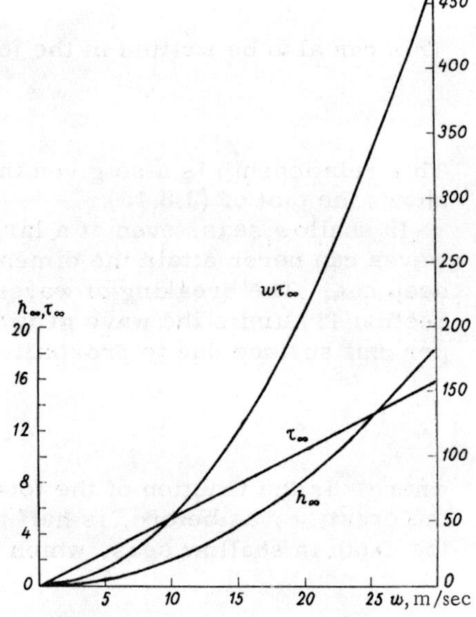

FIGURE 2.8.6. Curves of $h_\infty = f(w)$ and of $\tau_\infty = f(w)$ (after Shuleikin).

It follows from differential equation (2.8.27) that the waves stop growing in height and attain the limiting value of $2r$, which is possible in shallow waters with a given wind, when its right-hand side vanishes. By virtue of the same fact, the limiting value of r is given by

$$r = \frac{1}{2\varepsilon} \left[-b + \sqrt{4ae + b^2} \right].$$

Using, according to the previously cited example, the dimensionless variables η and τ and transforming to them from the variables τ and t, it becomes possible to express the law governing wave growth in shallow seas in the form

$$\eta = \operatorname{th} \tau, \qquad (2.8.28)$$

where

$$\frac{t}{\tau} = \frac{2}{\sqrt{4ae - b}},$$

$$\frac{r}{\eta} = r_2.$$

As can be seen, equation (2.8.28) differs from (2.8.10) obtained for deep seas.

In extremely shallow water, i.e., when the depth of the sea is always less than half the wave length at the surface, the energy loss in waves due to turbulent friction can be neglected as compared with energy losses produced by partial spillover of the wave crests due to proximity of the bottom. By virtue of the same fact the value of br in the expression for $\dfrac{dr}{dt}$ is neglected as compared with er^2, and quantity b is omitted in the expression for r. In addition, in the case under study $H_1 = H$. Then, taking into account the values of a, b and e, contained in the expression for $\dfrac{dr}{dt}$, one replaces the previous expression for r by (Khvan, 1967)

$$r_M = 2(w - c)\left[\left(\frac{\varkappa}{\pi\varepsilon}\right)\left(\frac{r}{R}\right)_M\left(\frac{\rho'}{\rho}\right)\left(\frac{H}{g}\right)\right]^{\frac{1}{2}}, \qquad (2.8.29)$$

where the subscript "M" pertains to the wave height in shallow waters.

It can be noted by comparing (2.8.29) with (2.8.8), that in shallow waters the wave height depends on the wind speed to the first power (as compared to the square for deep water) and on the square root of the sea depth. The value of ε in (2.8.29) is taken as approximately 0.02 (Khvan, 1958), while $\varkappa$, as before, equals 0.1. The value of $\dfrac{r}{R}$, i.e., the wave steepness in (2.8.29), is expressed (Khvan, 1958), as for deep water, by (2.8.2).

The solution obtained by Kao Wen-Hsiu and Wang Hsiao-Nan (1961) made it possible to establish that the limiting height of waves in a sea with arbitrary depth H_1 can be expressed as

$$h_M = Fw^2,$$

where (Shuleikin, 1963)

$$F = \frac{-1 + \left[1 + 0.019\chi_2^2 \dfrac{w^2}{H_1}\right]^{\frac{1}{2}}}{0.91\chi_2^2 \dfrac{w^2}{H_1}}. \qquad (2.8.30)$$

Here

$$\chi_2 = \frac{\lambda_M}{h_M}.$$

Parameter $\frac{w^2}{H_1}$ in (2.8.30) is replaced by $\frac{w}{(gH)^{1/2}}$, which makes possible graphical determination of η as a function of not only wave-generating factors, as was done above, but also of $\frac{w}{(gH)^{1/2}}$, i.e., for a sea of any depth. Figures 2.8.2 and 2.8.3 show the variation in η as a function of $\frac{w}{(gH)^{1/2}}$, marked on the right-hand scales of these figures.

9. Decay of wind waves. Swell

When no longer subjected to the wind, wind waves start decaying. As the criterion for determining the decaying stages one can use the condition that for swell $\overline{\beta} = \frac{\overline{c}}{w} \geqslant 1$ (Chapter 2, Section 5). In the entire variety of wind waves (Chapter 2, Section 3) prevailing on the sea surface, a known proportion of them satisfying $\overline{\beta} > 0.5$ also satisfy $\overline{\beta} > 1$. Consequently, even among wind waves there exist such which, on the basis of the above, can be treated as well. However, these waves should be classified as "wind swell" (Rzheplinskii, 1965), i.e., as swell which appears also during development of wind waves, particularly on short-duration reduction in the wind speed, which occurs frequently when wind blows in gusts. Hence the term swell proper, so to speak "pure swell," should be applied to these waves when they move with a markedly calm wind, i.e., when $\overline{\beta} = \frac{\overline{c}}{w} \approx 1$. In this case all the waves have a phase velocity exceeding the wind speed. A marked reduction in the wind speed is usually due to the fact that the wind waves move out from the region with strong winds and propagate unhindered into that part of the ocean where weak winds and even calm prevail. These conditions are usually produced at the outskirts of deep atmospheric lows (Chapter 2, Section 4). From here they can travel hundreds of miles over open free water spaces. The latter is to some extent limited in seas, but sufficiently extensive in oceans. This circumstance is responsible for a major feature in the propagation of wind waves in oceans.

Figures 2.9.1 and 2.9.2 show the distribution of the frequency (in percents) of wind waves and swell 3 and more meters high for the North Atlantic, separately for winter and summer. In addition, arrows in these figures show the prevailing direction of wind waves and of swell. It is seen by comparing these figures that regions where large wind waves are frequently encountered are limited to areas with a high frequency of storm winds, while regions with heavy swell extend over much larger ocean areas. Naturally, the direction of propagation of wind waves is that of prevailing winds. The direction of travel of swell coincides only with that of winds with either high speed or stable direction. Wind waves which move from under the wind and become swell retain heights of 3 m and more for

hundreds of miles, particularly due to strong and prolonged storms
which generate this swell.

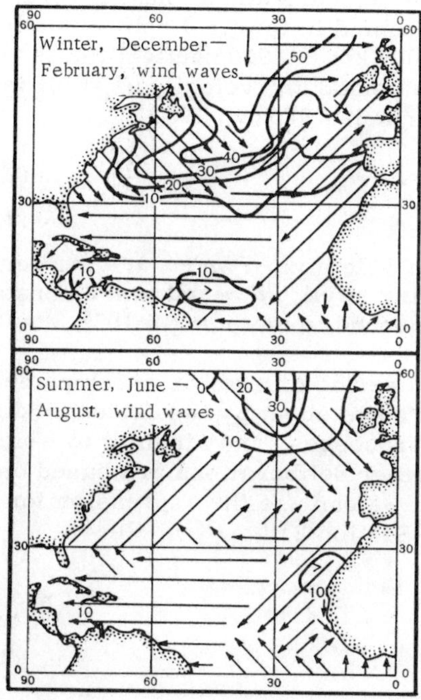

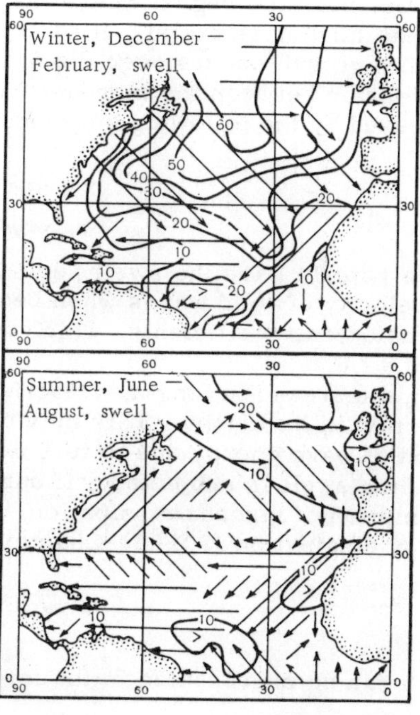

FIGURE 2.9.1. Frequency of wind waves 3 m
and higher.

The numbers on the isolines show the frequency, in percent, per season. The arrows
denote the prevailing direction of wind-wave
propagation during the given season.

FIGURE 2.9.2. Frequency of swell 3 m and
higher.

The numbers on the isolines show the frequency, in percent, per season. The arrows
denote the prevailing direction of swell propagation during the given season.

It is easily seen by comparing Figure 2.9.1 with Figure 2.9.2 that the
frequency of swell 3 m and higher in the equatorial Atlantic clearly exceeds
the frequency of wind waves of the same height both in winter and summer.
This is a result of the fact that trade winds quite regularly generate high
swell. The prevalence of swell in winter is attributable to the higher
speeds of northeasterly and southeasterly tradewinds in winter than in
summer.

The above distribution of wind waves and swell for the North Atlantic
is typical of all the oceans. Consequently, heavy swell may be encountered
in the oceans over much larger spaces than heavy wind waves. This is
the most characteristic feature in the propagation of wind waves in the
ocean.

At the same time, the swell which travels through the previously
mentioned large distances should, it would seem, propagate at constant
velocity as free gravity waves, if molecular viscosity forces are
neglected, retaining their shape and dimensions as long as the water depth

exceeds half the wave length. However, it was found from numerous observations that the speed, length and period of swell increase by approximately 70% as they move from the generation zone into the storm area. This growth is not uniform over the entire path. It occurs at a higher rate initially, near the region where the swell appears and at quite low rates at areas farther removed. It can be assumed in the first approximation that this growth occurs over the first 2000 km ($\sim$ 1000 nautical miles). The relationship between the increase in the period and velocity of swell is expressed by the following empirical formulas (Titov, 1951):

$$\tau_{sw} = \tau_0 m (1 - ne^{-pt}), \qquad (2.9.1)$$

$$c_{sw} = c_0 m (1 - ne^{-pt}), \qquad (2.9.2)$$

where τ_{sw} and c_{sw} are the swell period and velocity; τ_0 and c_0 are the period and velocity of wind waves when becoming swell; p, n and m are constant dimensionless coefficients, which were found to be: $p = 2.5 \cdot 10^{-5}$, $n = 0.375$, and $m = 1.6$.

This increase in length, velocity and period of swell is accompanied by reduction in its height. This may be brought about by two factors: dissipation of wave energy due to turbulent processes, and air drag to which traveling swell is subjected. If our further deliberations are based on the assumption of viscous dissipation, then we can use the expression for energy dissipation (2.2.22) (Chapter 2, Section 2):

$$N_v = \frac{1}{2} \rho \nu_\tau g \left(\frac{2\pi}{\lambda} \right)^3 h^2 c^2,$$

which can be represented in the form

$$N_v = \frac{1}{2} \nu_\tau \rho g^3 c^{-4} \overline{h^2}, \qquad (2.9.3)$$

where ν_τ is replaced (Chapter 2, Section 2) by the expression for the eddy viscosity coefficient given by (2.2.46) (Table 2.2.3):

$$\nu_\tau = 6 \cdot 10^{-9} \beta^2 w^3.$$

When considering swell, quantity ν_τ in the above expression should not be related to the wind speed, but to the phase velocity of the waves. Hence, using the expression

$$w = \frac{c}{\beta}$$

and assuming that the majority of wind waves (Chapter 2, Section 5) attain their maximum height at $\beta \approx 0.8$, we can write the following expression for the kinematic viscosity coefficient for swell:

$$\nu_\tau = kc^3 = 8 \cdot 10^{-9} c^3. \qquad (2.9.4)$$

Substitution of (2.9.4) into (2.9.3) yields

$$N_v = \frac{1}{2} \rho k g^3 \overline{h^2} c^{-1}, \qquad (2.9.5)$$

or

$$N_\nu = \frac{d\overline{E}}{dt} = -\frac{1}{2}\rho k g^3 \overline{h}^2 c^{-1}. \qquad (2.9.6)$$

Since

$$\frac{d\overline{E}}{dt} = \frac{d\left(\frac{1}{8}\rho g \overline{h}^2\right)}{dt} = \frac{1}{4} g\rho\overline{h}\,\frac{d\overline{h}}{dt},$$

consequently

$$\frac{d\overline{h}}{dt} = -\frac{2g^2 k\overline{h}}{c}. \qquad (2.9.7)$$

Using (2.9.2), we obtain

$$\frac{dh}{dt} = -\frac{2g^2 k\overline{h}}{c_0 m\left(1-ne^{-pt}\right)}. \qquad (2.9.8)$$

Now the variation in swell height is given by the expression

$$\int_{h_0}^{h}\frac{d\overline{h}}{h} = -\left[\frac{2g^2 k}{mc_0}\right]\int_0^t\frac{dt}{1-ne^{-pt}}.$$

which yields

$$\ln\frac{\overline{h}}{h_0} = -\frac{2kg}{c_0 m}\left[t + \frac{1}{p}\ln\left(\frac{1-ne^{-pt}}{1-n}\right)\right]. \qquad (2.9.9)$$

Introducing the previously mentioned constants, equation (2.9.9) becomes

$$\ln\frac{\overline{h}}{\overline{h_0}} = -\frac{0.01}{c_0}\left[t + 4\cdot 10^4\ln\left(\frac{1-0.375e^{-2.5\cdot 10^{-5}t}}{0.625}\right)\right]. \qquad (2.9.10)$$

During a time t the swell will travel a distance x_{sw}, which is given by

$$x_{sw} = \int_0^t u\,dt, \qquad (2.9.11)$$

where u is the group velocity. Consequently, (2.9.2) yields

$$x_{sw} = 0.8c_0\left[t - 1.52\cdot 10^{-4}\left(1-e^{-2.5\cdot 10^{-5}t}\right)\right]. \qquad (2.9.12)$$

Using (2.9.1), (2.9.10) and (2.9.12) one can determine the size of swell during its travel. The results calculated with these formulas are listed in Table 3.3.2.

Variation in elements of swell waves can be calculated by reference to the drag they experience when moving at a speed above that of the presently prevailing wind, and neglecting the effect of turbulence. In this case the energy loss due to this drag can be expressed by (2.2.8) (Sverdrup and Munk, 1947):

$$N_w = \frac{s\rho' g^2}{8} h^2(w-c)^2 c^{-3}.$$

However, w in this expression for swell is either zero, or very small as compared with c. Then (2.2.8) reduces to

$$R = -\frac{1}{8}\, s\rho' g^2 h^2 c^{-1},\qquad (2.9.13)$$

where R denotes the energy loss of the swell due to wind drag. These energy losses are divided, largely artificially, into energy losses due to reduction in height (R_h) and increase in velocity (R_c), as follows:

$$R_h = -\left(1 + \frac{r}{\alpha}\right) R,\qquad (2.9.14)$$

$$R_c = -\frac{r}{\alpha} R,\qquad (2.9.15)$$

where r and α are constants: $r = 0.580$ and $\alpha = 2.500$.

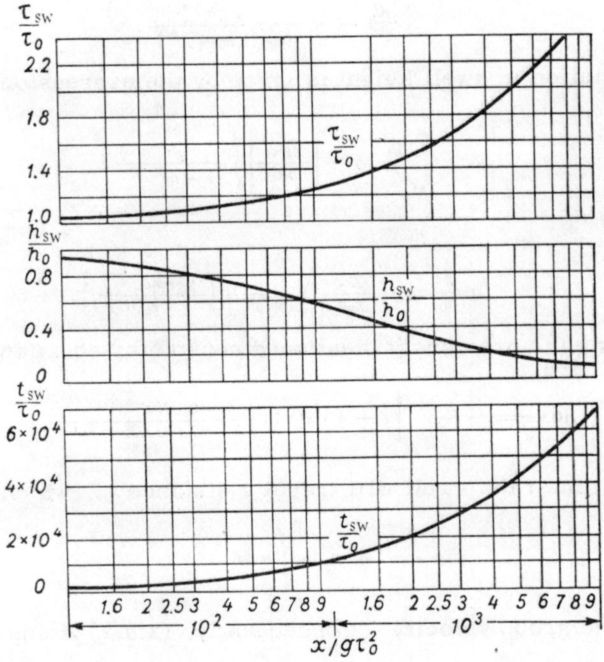

FIGURE 2.9.3. Variations in $\dfrac{\tau_{sw}}{\tau_0}$, $\dfrac{t_{sw}}{\tau_0}$ and $\dfrac{h_{sw}}{h_0}$ according to expressions (2.9.18)—(2.9.20).

The latter expressions assume the form

$$\frac{1}{h}\frac{dh}{dx} = -(r - \alpha)\,A\,\frac{4\pi^2}{g}\,\tau^{-2}\qquad (2.9.16)$$

and

$$\frac{d\tau}{dx} = \frac{8\pi^2}{g} Ar\tau^{-1},$$ (2.9.17)

where

$$A = 5.12 \cdot 10^{-3} \frac{\rho'}{\rho}.$$

Integration of (2.9.16) and (2.9.17) yields an expression for the variation in periods of swell at the start (τ_0) and termination (τ_{sw}) of the distance traveled by the swell:

$$\frac{\tau_{sw}}{\tau_0} = \left[1 + 16\pi^2 Ar \left(\frac{x}{g\tau_0^2}\right)\right]^{\frac{1}{2}}.$$ (2.9.18)

The time (t_{sw}) during which the swell travels from the start (x_{sw}) to the end, is given by the expression

$$\frac{t_{sw}}{\tau_0} = \frac{1}{2\pi Ar}\left(\frac{\tau_{sw}}{\tau_0} - 1\right).$$ (2.9.19)

The variation in the height of swell is calculated from

$$\frac{h_{sw}}{h_0} = \left(\frac{\tau_{sw}}{\tau_0}\right)^{-(r+\alpha)\,2r}.$$ (2.9.20)

The results of calculations using (2.9.18)—(2.9.20) are plotted in dimensionless form in Figure 2.9.3.

The differences in numerical results when using (2.9.10)—(2.9.12) on the one hand and (2.9.18)—(2.9.20) on the other, are not too high. According to the latter the height of swell decreases slower during the first few hours of propagation. Over sufficiently large distances the results are virtually identical.

10. The spectral method of wind-wave study

The underlying idea in all the previous deliberations was consideration of an individual two-dimensional sinusoidal or trochoidal wave, i.e., a monochromatic wave. The use of statistical characteristics of the variety of waves (Chapter 2, Section 3) allows one to convert to values of wave elements of any cumulative probability. However, this approach did not take into account, for example, the problem of the distribution of the energy of the entire wave field among the individual waves and its other features.

The wind-wave motion can be approached from the standpoint of spectral analysis. The latter is used in various branches of science and technology. With respect to sea waves it is based on the principle that the complicated structure of the wind-wave field in the form seen by an observer is expressed as the final results of a complex combination of an infinite number of elementary components, which are taken to be two-dimensional,

157

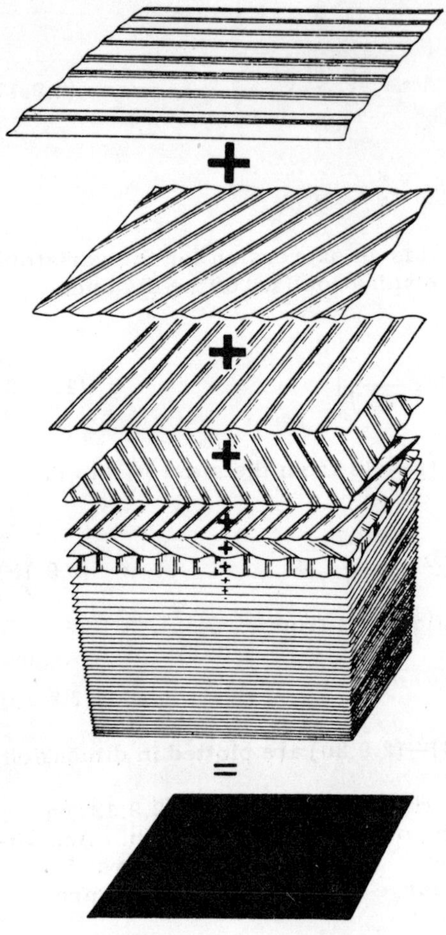

FIGURE 2.10.1. Schematic diagram representing a real undulated surface as a sum of simple sinusoidal waves (after Pierson).

sinusoidal waves of different amplitudes, phases and directions of propagation (Longuet—Higgins, 1957). The question as to whether such elementary waves actually exist, or whether they serve only as a convenient means for quantitative description of the wind-wave process, is open. Figure 2.10.1 represents graphically this approach to the analysis of wind-wave covered sea surfaces. The superimposition of an infinite number of systems of sinusoidal waves, shown in the upper part of Figure 2.10.1, results in the final count in the observed surface, which is shown at the bottom of this figure, beneath the equality sign. Each elementary wave carries with it a certain amount of energy, which is referred to unit surface and depends on the wave period (frequency). The distribution of wave energies by frequencies and directions of propagation is termed the energy spectrum of wind waves. It can be determined from the following considerations. It was mentioned above (Chapter 2, Section 3) that the sea wave field can be represented as a stochastic (probability) process. The wave field in its turn is expressed as some quasistationary function of time and horizontal coordinates, possessing the property of ergodicity. This makes it possible to regard some elementary area of the sea surface with wind waves traveling over it as an area at which the average wave characteristics will change little, but at the same time an infinite variety of wave forms will exist at this area.

An orthogonal coordinate system xoy is set up on the still-water level (Figure 2.10.2) (Krylov, 1966). The x axis is aligned with the wind direction. It is assumed that a two-dimensional wave propagates on the sea surface and that the ray of this wave makes an angle α with the x axis (Figure 2.10.2). In this case the shape of the sea surface is described by the equation

$$z(x,\ y,\ t) = a\cos[\omega t + \theta - k(x\cos\alpha + y\sin\alpha)], \qquad (2.10.1)$$

where θ is a random phase with a uniform distribution between 0 and 2π. In deep sea ω and k are related by (1.3.2):

$$\omega^2 = kg.$$

The undulated surface can be represented as the sum of a large number of these elementary waves (Longuet—Higgins, 1957). Each wave is expressed as

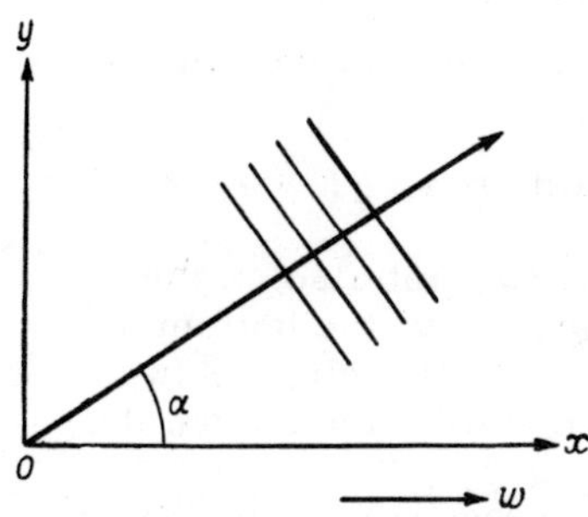

FIGURE 2.10.2.

$$z_{ij}(x,\ y,\ t) =$$
$$= a(\omega_i \alpha_j) \sqrt{\Delta\omega\,\Delta\alpha}\,\cos\left[\omega_i t + \theta_{ij} - k_i(x\cos\alpha_j + y\sin\alpha_j)\right]. \qquad (2.10.2)$$

Now the undulated surface is represented as a sum of such elementary waves:

$$z(x,\ y,\ t) = \sum_{i=1}^{n} \sum_{j=1}^{m} z_{ij}.$$

It is assumed that the phase θ_{ij} of each of the components in equation (2.10.2) is a random variable, uniformly distributed between 0 and 2π. Hence $z(x, y, t)$ is also a random variable.

With consideration of (1.7.6), the total energy of each component is

$$\frac{1}{2} g\rho a^2(\omega_i \alpha_j)\,d\omega\,d\alpha; \qquad (2.10.3)$$

or

$$e(\omega_i \alpha_j) = \frac{1}{2} g\rho a^2(\omega_i \alpha_j). \qquad (2.10.4)$$

Equation (2.10.4) is termed the two-dimensional energy spectrum. Its physical meaning consists in the fact that quantity $e(\omega, \alpha)\,d\omega\,d\alpha$ in (2.10.3) equals the amount of wave energy of the spectral components with frequencies from ω to $\omega + d\omega$ and directed over angles from α to $\alpha + d\alpha$. Summing the spectral components over angle α, one obtains the spectrum as a function of only the frequency ω. Hence the two-dimensional spectrum is related to its one-dimensional counterpart by the expression

$$e(\omega) = \int_{-\pi}^{+\pi} e(\omega,\ \alpha)\,d\alpha.$$

The one-dimensional (frequency) energy spectrum is expressed in the form

$$e(\omega) = \frac{1}{2} g\rho a^2(\omega_i)\,d\omega \qquad (2.10.5)$$

or

$$a^2 = \frac{2e(\omega)}{g\rho}\,d\omega. \qquad (2.10.6)$$

The latter expressions are frequently represented in the form

$$a^2 = A^2(\omega)\, d(\omega), \qquad\qquad (2.10.7)$$

where

$$A(\omega) = \sqrt{\frac{2\varepsilon(\omega)}{g\rho}}\,.$$

If the two- or one-dimensional spectrum is known, it is possible to calculate not only the quantity $a = \dfrac{h}{2}$, but also a number of other important characteristics of the undulated surface.

The problem of calculating various wave parameters reduces, in spectral theory, to finding the function $e(\omega, \alpha)$.

The energy spectrum depends not only on ω and α, but also on the characteristics of the wind field and on the topography of the bottom and shores of the body of water. In the elementary cases it includes, in the form of parameters, the wind speed, its duration and distance from the leeward shore.

Empirical studies of the frequency spectrum at present are based on correlation analysis of automatic wave recordings. The correlation function $z(t)$ of a stationary random process, possessing the ergodicity property (Krylov, 1966), is defined as

$$k(\tau) \approx \frac{1}{T-\tau} \int\limits_0^{T-\tau} z(t)\, z(t+\tau)\, dt. \qquad\qquad (2.10.8)$$

This formula is represented in the form of a finite series. For this, segment $(0, t)$ is subdivided into N segments by points $t_0 = 0$; $t_1 = \Delta t$; $t_2 = 2\Delta t$; ...; $t_k = k\Delta t$; ...; $t_n = N\Delta t$. Hence the correlation function

$$k(m\,\Delta t) = \frac{1}{N-m} \sum_{k=0}^{N-m} z(k\,\Delta t)\, z(k\,\Delta t + m\,\Delta t). \qquad\qquad (2.10.9)$$

For $m = 0$

$$k(0) = \frac{1}{N} \sum_{k=0}^{N} z^2(k\,\Delta t).$$

For $m = 1$

$$k(\Delta t) = \frac{1}{N-1} \sum_{k=0}^{N-1} z(k\,\Delta t)\, z(k\,\Delta t + \Delta t) = \frac{z(t_0)\, z(t_2) + z(t_1)\, z(t_3)\, \dots}{N-1}\,.$$

For $m = 2$

$$k(2\Delta t) = \frac{z(t_0)\, z(t_2) + z(t_1)\, z(t_3) + \dots}{N-2}$$

and so on.

Thus formula (2.10.9) can be employed for calculating successive values of the correlation function, and hence, if necessary, also for constructing a graph of this function.

In spite of their simplicity, these calculations are still very labor-consuming. Consequently they are implemented by means of specially constructed computing instruments, called correlators. These instruments directly convert the recording from the recorder tape into the corresponding correlation function.

The energy spectrum is uniquely expressed by the corresponding correlation function:

$$\frac{e(\omega)}{g\rho} = \frac{2}{\pi} \int_0^\infty k(\tau)\cos\omega\tau\,d\tau. \tag{2.10.10}$$

Substitution of function $k(\tau)$ into this formula makes it possible to determine the corresponding energy spectrum by numerical integration. This method involves errors, because the record length is always finite. However, these errors can be estimated and partially eliminated (Rozkhov, 1967).

The most promising approach involves determining the relationship between the wind-flux characteristics and the energy spectrum of wind waves. But such theoretical solutions are difficult to obtain, because the wind does not blow past individual elementary waves, but past a complex wave surface, consisting of a sum of all the components. Determination of the relationship between the spectrum of turbulent fluctuations and the wave spectrum is a problem which has not as yet been solved. Hence at present a different approach is generally used. It consists in determining the relationship between the energy spectrum of the waves and the directly observed wave statistic, for example, the average height, average period, etc. The relationship between the latter and the wind-field characteristics, i.e., w, x and t, has already been established to a certain approximation in solving the energy balance equation of wind waves (Chapter 2, Sections 7 and 8) or empirically (Chapter 2, Section 5). This makes it possible, as a result, to determine the numerical dependence of the wave spectrum on the wave-generating factors. However, such solutions were so far obtained only to the most elementary cases of wind-wave development, i.e., to cases of steady wind waves corresponding to the one-dimensional spectrum.

On the basis of correlation analysis of wave measurements many formulas have been obtained up to present for describing the structure of the energy spectrum over a frequency range. The latter can be represented in generalized form (Burling, 1957) as

$$\frac{e(\omega)}{g\rho} = \frac{c_1}{\omega^p} \exp\left(-\frac{c_2}{\omega^r}\right); \tag{2.10.11}$$

here the dimensions of the spectrum are $\dfrac{L^2}{T}$. In this equation $e(\omega)$ is the specific energy of elementary waves for an infinitesimal spectral interval, g is the acceleration of gravity, ρ is the density of sea water, $\omega = \dfrac{2\pi}{\tau}$ is the circular frequency, c_1 and c_2 are dimensional parameters, and p and r are dimensionless parameters, which may be functions of x, w and t. For a fully developed sea (Chapter 2, Sections 7 and 8) the energy spectrum is defined as

$$\frac{e(\omega)}{g\rho} = f(\omega,\ w,\ g). \tag{2.10.12}$$

With consideration of (2.10.12), expression (2.10.11) can in this case
be written in the form (Strekalov, 1961)

$$\frac{e(\omega)}{g\rho} = \frac{A}{\omega^n}\,\frac{g^{n-3}}{w^{n-5}}\,\exp\left[-\left(\frac{Bg}{\omega w}\right)^m\right],\qquad (2.10.13)$$

where A, B, n and m are unknown dimensionless constants.

In general these can be treated as functions of the dimensionless time of
wind duration, $\tilde{t}$, and dimensionless distance (fetch) $\tilde{x}$ (Chapter 2, Section 4).
In its most elementary form the spectrum represented by (2.10.13)
contains as a parameter only one value of the wind speed (w). Hence it can
be represented graphically as a family of curves $[A(\omega)]^2$, corresponding to
different wind speeds (Figure 2.10.3). As the wind speed increases, the
spectral density maximum (ω_{max}) shifts toward smaller frequencies, i.e.,
longer periods, which is natural, since the wave periods increase with
increasing wind speeds. The area beneath the spectrum then increases.

The spectrum in the form of (2.10.13) has been determined by a number
of investigators. Strekalov (1961) used wave-recorder measurements of
wave elements in the Atlantic for determining the constants contained in
(2.10.13). These he found to be: $n = 6$, $m = 2$, $A = 1.2 \cdot 10^{-2}$, and $B = 0.88$.

It was also assumed that $\dfrac{\bar{\tau}}{\tau_{max}} \cong$ const during the entire wave development
and $\omega = \dfrac{2\pi}{\bar{\tau}}$.

This made it possible to represent the energy spectrum for the fully
developed sea in the form

$$\frac{e(\omega)}{g\rho} = \frac{Ag^3}{\omega^6 w}\,\exp\left[-B\left(\frac{g}{\omega w}\right)^2\right].\qquad (2.10.14)$$

Figure 2.10.4 shows a family of energy spectra obtained empirically and
calculated for corresponding wind speeds from (2.10.14). As can be seen,
the latter reflects quite precisely the location of the spectral maximum
(ω_{max}), and also gives a satisfactory approximation with respect to area.

Another expression for the energy spectrum was obtained by Davidan
(1967), who approximated the frequency spectrum obtained from automatic
wave recordings in the open parts of the Barents Sea and in the Atlantic
by the expression

$$\frac{1}{g\rho}\,e(\omega) = Ag^2\omega^{-5}\exp\left[-m\left(\frac{\omega_{max}}{\omega}\right)^4\right],\qquad (2.10.15)$$

where $A = 1.56 \cdot 10^{-2}$, $m = 1.1$, $\omega_{max} = 0.8\dfrac{2\pi}{\bar{\tau}}$, whence it follows that $\tau_{max} = 1.1\bar{\tau}$.

Since $\bar{\tau}$ increases with the developing waves, ω_{max} should shift with the
growth of waves in the direction of smaller frequencies.

The above formulas pertaining to the frequency spectrum, as well as
those suggested by others, differ quantitatively, but also have a number of
common features (Krylov, 1966). All these relationships represent
structurally the product of two functions (2.10.13): a power-law function
ω^{-n} and an exponential of another power-law function, $\exp[-B\omega^{-m}]$, where
$n \geqslant 5$ and $m \geqslant 2$. For low frequencies function $\omega^{-n}\exp[-B\omega^{-m}]$ tends very
rapidly to zero, due to the vanishing of $\exp[-B\omega^{-m}]$, but at higher frequencies

it decreases much less rapidly. This property of the spectrum is attributable to the fact that long waves do not receive energy from the wind, but conversely, encounter aerodynamic drag of the air and decay. Average and short waves are continuously supplied by wind energy, and hence the spectrum is sufficiently "enriched" by these components. It is also characteristic of all the frequency spectra that the spectral period τ_{max} corresponding to the energy maximum is always greater than τ, the average spectrum of visible waves, and this relationship differs little for different wave motion stages (Krylov, 1966): $\tau_{max} \approx 1.2\bar{\tau}$.

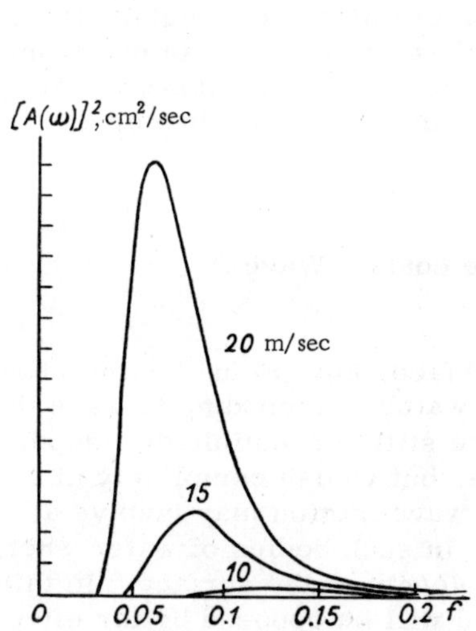

FIGURE 2.10.3. Wave spectra for different wind speeds.

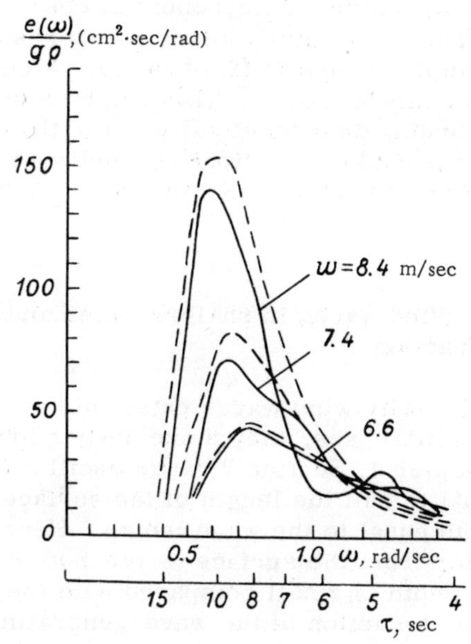

FIGURE 2.10.4. Family of empirical curves (solid lines) of a fully developed sea and of spectra calculated from equation (2.10.14) (after Strekalov) (dashed curves).

The two-dimensional spectrum $e(\omega, \alpha)$ has been much less studied for a simple reason. In order to study it, it is necessary to record the undulated surface over a known, sufficiently large area, as a function of the x and y coordinates (for example, by stereophotography). After this, the data thus obtained are processed in a manner similar to that of automatic wave recordings. The main difficulty consists in determining the topography of the wave field over the [relatively] large area of several square kilometers. This area should contain a large number of waves. Otherwise, spectral analysis yields very large errors.

In view of the difficulty of covering such a large area, one employs an artificial device, consisting in repeated photographing of a relatively small segment of the sea surface (Matushevskii and Strekalov, 1963).

In addition, one considers a new characteristic of the two-dimensional spectrum — the function of the angular distribution of spectral energy, or the one-dimensional angular spectrum (Matushevskii, 1964). This function is obtained from the two-dimensional spectrum by integrating over all the frequencies:

$$e(\alpha) = c_1 \int_0^\infty e(\omega, \alpha)\, d\omega. \tag{2.10.16}$$

It was found that in the open sea this function is close to $\cos\alpha^2$, while in coastal waters the exponent of $\cos\alpha$ becomes larger.

The use of given formulas expressing the energy spectrum, for example, equation (2.10.14) for calculating the latter, is contingent upon determination of τ. This can be found, for example, from automatic wave recording or theoretically. The theoretical value of τ is obtained from corresponding relationships between wave elements and wave-generating factors (Chapter 2, Section 8) by techniques discussed in Chapter 3.

11. Wind waves in shallow water and at the coast. Wave refraction

Initially wind waves exist only at the surface, but gradually, as their size increases, they move deeper into the water. According to wave theory (Chapter 1, Section 2) wave oscillations are still perceptible at a depth equal to half the length of the surface wave, but vanish completely at a depth equal to the wave length. Hence the wave motion may involve all the water from the surface to the bottom only in such bodies of water where the depth is small compared with the wave length at the surface. In this case the action of the wave-generating wind will be opposed by the effect of the bottom, which is expressed by deforming the wave profile (Shuleikin, 1956). Hence waves in shallow seas will never reach the limiting height determined by the wind force, its duration and the fetch (Chapter 2, Section 5). For such a sea the difference in the distance from the surface of the wave crest to the bottom and from its trough to the bottom will already be perceptible.

Hence, using (1.2.12), we may write for a point on the wave crest

$$c^2_{\mathrm{cr}} = \frac{g\pi}{2\pi}\,\mathrm{th}\,\frac{2\pi}{\lambda}\left(H + \frac{h}{2}\right),$$

for the median line

$$c^2_{\mathrm{med}} = \frac{g\lambda}{2\pi}\,\mathrm{th}\,\frac{2\pi}{\lambda}\,H$$

and for a point at the wave trough

$$c^2_{\mathrm{tr}} = \frac{g\lambda}{2\pi}\,\mathrm{th}\,\frac{2\pi}{\lambda}\left(H - \frac{h}{2}\right).$$

Consequently, the phase velocity of a point at the crest will be higher than that for a point at the median line, and the latter will be higher than the phase velocity of a point at the trough; in other words

$$c_{cr} > c_{med} > c_{tr} \ .$$

The wave crest at the surface of a shallow sea moves with a higher speed than the trough of the same wave. By virtue of the same fact the leeward slope becomes increasingly steeper, the crest becomes unstable and spills over. This process involves a perceptible loss in wave energy and limits the growth of wind waves developing on the surface of a shallow sea.

During time t a point on the crest travels a distance

$$x_{cr} = c_{cr} t.$$

During this same time a point on the median line travels a distance

$$x_{med} = c_{med} t.$$

Moving in the direction of wind propagation, a point on the crest will overtake a point on the median line. When the difference between x_{cr} and x_{med} becomes $1/4\lambda$ the wave inevitably breaks, since the crest will "overhang" the wave trough. The time which passes until the wave breaks should satisfy

$$t \leqslant \frac{\bar{\lambda}}{4\,(c_{cr} - c_{med})} \ ,$$

and the corresponding distance should satisfy

$$x \leqslant \frac{\bar{\lambda} c_{cr}}{4\,(c_{cr} - c_{med})} \leqslant \frac{\bar{\lambda}}{4n} \ . \tag{2.11.1}$$

where

$$n = 1 - \frac{c_{med}}{c_{cr}}.$$

Inequality (2.11.1) implies that when the extent of the shallow section of the sea exceeds manyfold the average wave length ($\bar{\lambda}$) in deep water, then (2.11.1) is satisfied at a very low value of n. For example, if

$$x \leqslant 1000\bar{\lambda},$$

then

$$n \leqslant 0.00025.$$

Consequently, over the entire shallow-water area, $\bar{\lambda}$ will be close to approximately three times the depth (Chapter 1, Section 4), since the bottom starts exerting a perceptible effect on the phase velocity of waves at this depth. After the wave crest spills over, the waves again start

growing in height, but a new spillover stops the growth of the wave height
and of its period.

Expression (1.6.26), namely

$$\overline{\tau} = \left(\frac{2\pi\overline{\lambda}}{g}\right)^{\frac{1}{2}},$$

can be rewritten (assuming that $\overline{\lambda} = 3.33\,H$) in the form

$$\overline{\tau} = \left(\frac{6.66\pi\overline{H}}{g}\right)^{\frac{1}{2}},$$

where $\overline{H}$ is the mean depth of the shallow sea.

Expressing the wave period in seconds and the sea depth $(\overline{H})$ in meters,
the above expression becomes

$$\overline{\tau}_{max} = 1.47\left(\overline{H}\right)^{\frac{1}{2}}, \tag{2.11.2}$$

i.e., the average wave period $(\overline{\tau})$ in a shallow sea with average depth $\overline{H}$
cannot exceed the period defined by (2.11.2). For example, for a depth of
80 m the period cannot exceed ~ 13 sec (Figure 2.11.1).

Expression (2.11.2) can be used for estimating that limiting depth (H_{lim})
at which the waves can fully develop, i.e., it is assumed that in this case
$\overline{\beta} \approx 1$ (Chapter 2, Section 5).

Replacing $\overline{\tau}$ in (2.11.2) by $\overline{c}$ according to (1.6.26) and (1.6.27), and
dividing both sides of the resulting expression by w^2, one obtains

$$H = 0.18\overline{\beta}^2 w^2.$$

For a fully developed sea $\beta = 1$; hence

$$\overline{H}_{lim} = 0.18w^2. \tag{2.11.3}$$

It is seen from Figure 2.11.1 that, for example, for a wind of 10 m/sec
the sea will be "shallow" for a depth of 18 m and less, while for a wind of
30 m/sec it will become "shallow" for a depth of 162 m. Consequently, the
sea becomes "shallower" with increasing wind speed.

Wind waves in a shallow sea first develop in accordance with the wind
speed, its duration and the fetch, as in an infinite-depth sea. When the
average wave period attains the value given by (2.11.2), future growth of
wind waves ceases and they retain their dimensions over the remainder of
the shallow sea. If the manner in which the wave height depends on wind
speed and wave period is known, it is possible to determine the possible
wave height for the given wind speed at the surface of a shallow sea with
average depth $\overline{H}$. For example, using (2.5.21) and employing (2.11.2) it
can be derived that

$$\overline{h} = 0.052\overline{H}^{\frac{3}{4}} w^{\frac{1}{2}}, \tag{2.11.4}$$

where $\overline{H}$ and $\overline{h}$ are in meters and w in meters per second.

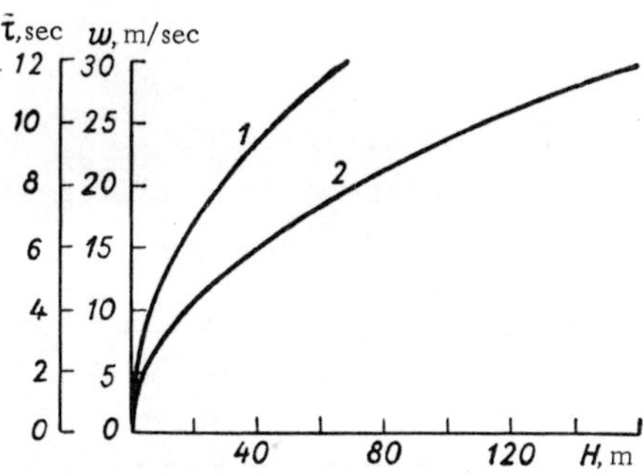

FIGURE 2.11.1:

1 — according to (2.11.2); 2 — according to (2.11.3).

As was mentioned previously, all the above pertains to shallow-water bodies with relatively level underwater topography. Typical examples of the latter should be sought not only in seas, but in inland bodies of water, i. e., among many natural and water-storage lakes. In addition, the Sea of Azov, with a mean depth of $\sim 14\,\mathrm{m}$, can serve as an example of a shallow sea. Certain closed shallow gulfs of individual seas, for example, the Gulf of Riga in the Baltic, can also be counted among such natural shallow bodies of water.

Table 2.11.1 lists the possible average wave height calculated from (2.11.4) with data from Table 3.3.3, as well as $\bar{\tau}$ according to (2.11.2), for an average shallow-sea of depth from 5 to 30 m and for wind speeds from 5 to 30 m/sec. For example, the highest possible wave height for the Sea of Azov $(H \approx 14\,\mathrm{m})$ is $\bar{h} = 2.00\,\mathrm{m}$ and the highest possible average period is $\bar{\tau} = 5.7$ sec for a wind speed of 30 m/sec. Here, $x = 50$ km and $t = 1.8$ hours. Consequently, the limiting height of wind waves in the Sea of Azov (defining the latter as the height of waves with a 1% cumulative probability) is, using data from Table 2.3.2, $\sim 4.8\,\mathrm{m}$.

TABLE 2.11.1. Limiting wave dimensions in shallow seas

w, m/sec	Average sea depth, m											
	5				10				15			
	$\bar{h}$, m	$\bar{\tau}$, sec	x, km	t, hours	$\bar{h}$, m	$\bar{\tau}$, sec	x, km	t, hours	$\bar{h}$, m	$\bar{\tau}$, sec	x, km	t, hours
4	0.23	2.5	48	7.6	0.23	2.5	48	7.6	0.23	2.5	48	7.6
10	0.54	3.3	40	3.3	0.95	4.6	100	7.3	1.25	5.7	200	13.6
15	0.65	3.3	20	1.3	1.15	4.6	60	3.4	1.55	5.7	120	6.3
20	0.65	3.3	15	0.8	1.25	4.6	40	1.9	1.80	5.7	80	3.6
25	0.65	3.3	10	0.5	1.45	4.6	30	1.3	2.00	5.7	60	2.4
30	0.65	3.3	8	0.3	1.50	4.6	25	1.2	2.00	5.7	50	1.8

w, m/sec	Average sea depth, m							
	20				30			
	$\bar{h}$, m	$\bar{\tau}$, sec	x, km	t, hours	$\bar{h}$, m	$\bar{\tau}$, sec	x, km	t, hours
4	0.23	2.5	48	7.6	0.23	2.5	48	7.6
10	1.52	6.4	300	18.9	1.52	6.4	300	18.9
15	1.85	6.6	170	9.1	2.65	8.2	400	18.0
20	2.25	6.6	140	5.8	3.00	8.2	260	9.9
25	2.45	6.6	100	3.7	3.40	8.2	200	6.8
30	2.65	6.6	80	2.7	3.70	8.2	160	4.8

Table 2.11.2 compares wind-wave elements in shallow waters calculated from Table 3.3.1 with consideration of Table 2.11.1, with their observed values. The standard error in the wave height is ± 13 cm and in the period,

TABLE 2.11.2. Comparison of observed and calculated wind-wave elements for shallow seas

Ordinal number	Region of observations	Method of observations	H, m	w, m/sec	x, km	Observed		Calculated			
						$\bar{h}$, m	$\bar{\tau}$, sec	$\bar{h}$, m	$\Delta\bar{h}$, cm	$\bar{\tau}$, sec	$\Delta\bar{\tau}$, sec
1	Aral Sea	Automatic wave recorder	5	9	118—230	0.43	3.3	0.53	—10	3.3	0
2	"	"	23	14	100	1.34	5.6	1.30	+4	5.8	—0.2
3	"	"	23	9.3	50	0.84	4.3	0.60	+24	3.6	+0.7
4	"	"	6	4	200	0.23	2.9	0.23	0	2.5	+0.4
5	Kakhovka Reservoir	Step gage	6.5	9—12	34	0.65	3.5	0.52—0.70	0	3.3—3.7	0
6	"	"	6.5	9—10	15	0.46	3.2	0.35—0.38	+8	2.4—2.5	+0.7
7	"	"	6.5	8—11	54	0.46	3.4	0.50—0.70	—4	3.4—3.8	0
8	"	"	6.5	8—10	15	0.42	3.1	0.30—0.38	+4	2.4—2.7	+0.4
9	"	"	6.5	10—13	11	0.44	3.5	0.30—0.42	+2	2.3—2.7	+0.8
10	"	"	7.5	16—20	20	1.10	3.5	0.70—0.85	+25	3.3—3.7	0
11	"	"	6.5	14—16	31	0.71—0.91	3.3—3.4	0.75—0.95	0	3.6—3.7	—0.2

12	Kakhovka Reservoir	Step gage	6.5	7—10	14	0.42—0.54	2.7—3.4	0.25—0.38	+4	2.2—2.5	+0.2
13	"	"	6.5	10—12	15	0.49	3.1	0.38—0.42	+7	2.5—3.2	0
14	"	"	6.5	12—14	11	0.56	3.2	0.35—0.40	+16	2.5—3.7	—0.5
15	Northern part of the Caspian Sea	Automatic wave recorder	5	13	30	0.56	3.3	0.65	—10	3.3	0
16	"	"	5	11	63	0.35	3.4	0.55	—20	3.3	+0.1
17	"	"	5	12	63	0.43	3.4	0.55	—12	3.3	+0.1
18	"	"	5	9	63	0.31	2.9	0.50	—19	3.3	—0.4
19	"	"	4	15	94	0.56	2.9	0.55	+1	2.9	0
20	"	"	3	13	94	0.56	2.9	0.40	+16	2.5	+0.4
21	"	"	4	18	63	0.67	3.1	0.50	+17	2.9	+0.2
22	"	"	4	17	63	0.63	3.0	0.50	+13	2.9	+0.1
23	"	"	4	13	63	0.57	2.7	0.50	+7	2,9	—0.2

±0.3 sec. These errors lie within the limits of observational accuracy. The question of wind-wave development at the surface of shallow seas can be examined by considering the energy balance of wind waves, discussed in Section 8 of this chapter.

In seas and oceans the bottom effect usually becomes perceptible as the waves approach the bank shoal (the shelf). First the shelf usually slopes gently up to a depth of several tens of meters, and then becomes steeper (the continental slope). The effect of the bottom on waves moving onto the shelf is at first slight and almost imperceptible. Gradually, as the waves move into ever-decreasing depths, this effect is felt increasingly more and, in spite of the most varied local features of the bottom topography, has a number of common features.

First, the waves approaching the shore change direction. Irrespective of the direction outside the shoals, the waves approach the shore in a near-normal direction. This phenomenon is called wave refraction. It is usually accompanied by a reconstruction of the wave system, i.e., three-dimensional waves become two-dimensional. This process is the more intensive, the shallower the water at the shelf. The kinematic and energetic reconstruction of waves is also appreciable and has common features. The cause of this is that the wave-energy density, i.e., the amount of energy passing through unit area in unit time, increases due to decreasing sea depth. As the waves approach the edge of the water, the energy flux density continuously increases and attains a critical value at which the wave breaks. Depending on the bottom topography, this break can occur several times before the wave disintegrates entirely, when the entire wave energy is converted to its kinetic form. Each such successive spill damps out a part of the wave energy, the wave decreases in size, its crest first becomes shallow, then becomes steeper, and finally breaks. Complete breakup usually occurs at a critical depth (H_{cr}) of about $1.3-2.0\,\overline{h}$.

The shore belt of the sea is usually subdivided into the following zones (Figure 2.11.2): the first zone $(H \geqslant 0.5\lambda)$ is the deep-water zone, which virtually does not affect the shape and dimensions of the waves; the second zone $(0.5\lambda > H > H_{\mathrm{cr}})$ is the shallow-water zone, located closer to the shore; the bottom effect in this zone affects perceptibly the wave form and dimensions; in the third $(H < H_{\mathrm{cr}})$, tidal zone, the wave crests break, i.e., tidal waves appear and then break, accompanied by ejection of foam and spray; the fourth zone is that at the water edge, in which the water from fully broken waves periodically rises up (is splashed or rolled up) onto the shore or escarpment.

The length of these zones from the shore into the sea depends, in addition to the depth of each of them, also on the size of waves, particularly on their height and length. Hence each of these zones may be either narrow, or conversely, may extend very far from the shore into the sea.

The theory of waves in finite-depth water (Chapter 1, Section 2) does not reflect all the features and details of wave propagation in shallow waters and, naturally, is to some degree approximate. Hence the different aspects of this complicated natural phenomenon and its quantitative estimates should be supplemented by observations under actual conditions and in the laboratory. At the same time the imperfection of the former and the inevitable schematization of the reproduced phenomena in the latter result in some degree of approximation and sometimes also in contradiction of results.

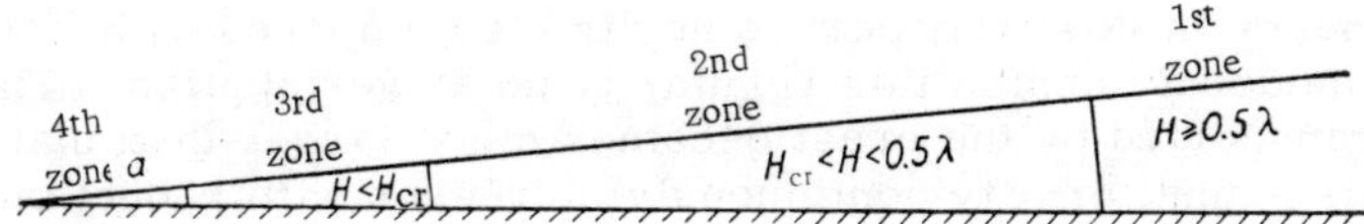

Figure 2.11.2. Subdivision of the coastal region of the sea into zones (after specifications SN 92-60).

Numerous observations and experimental laboratory studies show that the wave length decreases with decreasing depth, and that the phase velocity of the waves decreases also. The wave propagation velocity, calculated from (1.2.12), is in satisfactory agreement with that measured. The difference between calculations and observations up to $\dfrac{H}{\lambda} = 0.10$ is on the average about +10%. This inaccuracy is not so significant as to make this formula unsuitable for practical calculations. The period of an individual wave does not change markedly (Vilenskii and Glukhovskii, 1955). The reduction in wave length occurs primarily due to reduction in crest width (L_{cr}), measured along the static horizon line (Introduction, Section 2). The trough width (L_n) measured along the same line varies much less. The ratio $L_{cr}:L_n$, which is close to unity in deep water, gradually decreases in shallow water and lies within the limits of $\sim 0.35-0.45$ at the point where the wave is close to breaking. The crest height (h_{cr}) measured from the static horizon increases in the majority of cases as the waves approach the shore. While in deep water the ratio of h_{cr} to the wave height (h) is close to 0.52; at the time of breaking it is about $0.8-0.9$. The wave height as they approach the surf zone first decreases and then, particularly when breaking, increases.

As a result of these changes in wave elements, the wave profile loses its initial trochoidal or sinusoidal shape. The crests become short and steep, separated by shallow troughs. The kinematics of the wave in shallow waters for a relative depth of $\dfrac{H}{\lambda} \geqslant 0.35$, in general, according to observations in nature and laboratory experiments (Kondrat'ev, 1953), is described satisfactorily by the theory of waves in finite-depth seas (Chapter 1, Section 2). The particle orbits have an almost regular elliptical shape. However, when the waves travel into the region of smaller relative depths, i.e., when $\dfrac{H}{\lambda} < 0.35$, the particle orbits deform and are no longer elliptical (Figure 2.11.3). The upper part of the orbit, above the static horizon, becomes much larger than the bottom half. The horizontal axis becomes markedly larger, and the part of the vertical axis above the static horizon becomes larger than its lower half. The velocity of particles in the upper part of the orbit increases and decreases in the lower part. Hence, instead of the smooth, uniform particle motion observed at great depth, the particles move rapidly toward the shore at the trough and return slowly into the initial position in the trough. Variations in the orbital motion of the particles and changes in the shape of the orbits also determine the energy reconstruction of waves approaching the shore, i.e., the breaking zone. While in deep water the energy is transported (Chapter 1, Section 8) in almost equal amounts by the passage of crests and troughs, and the portion

of kinetic energy in this transport is negligible compared with the potential
energy; at moderate depths this regularity no longer applies. The amount
of energy transported by the crest becomes much larger than that by the
trough. This nonuniformity continuously increases with reduction in the
relative depth. At the same time the portion of kinetic energy carried by
the wave increases. Upon each spill of the crest an increasing amount of
energy is converted into the kinetic form, and finally, all of it dissipates
when the wave finally breaks. This explains the large destructive forces
of shallow-water waves, which is the greater, the closer such a wave is
to its complete destruction.

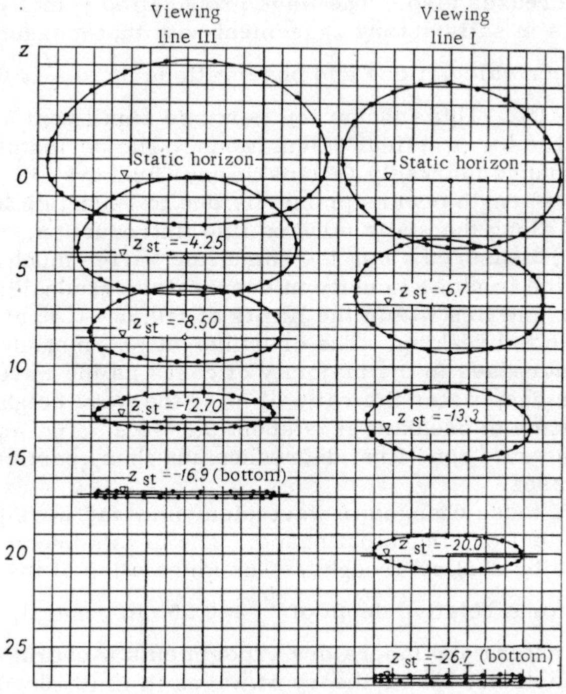

FIGURE 2.11.3. Averaged orbits of liquid mass in a shallow-water
wave at the start of transformation (viewing line I) and on substantial transformation (viewing line III) (after Kondrat'ev).

The continuous destruction of waves in the shore belt results in the
accumulation of water masses between the shore line and the surf line.
At individual points this water breaks through the surf belt and moves at
a high speed into the sea, producing rip currents. These currents may
be quite dangerous to persons in water or on light vessels. This danger
is increased by the fact that rip currents usually form suddenly and
irregularly at different points of the shore shoals. In the shallow-water
shore strip one should also take into account the wave current (Chapter 1,
Section 5), which produces, in addition to some increase in water level
in the shore zone (particularly when the water approaches the shore line
in the normal direction), a water transport along the shore line.

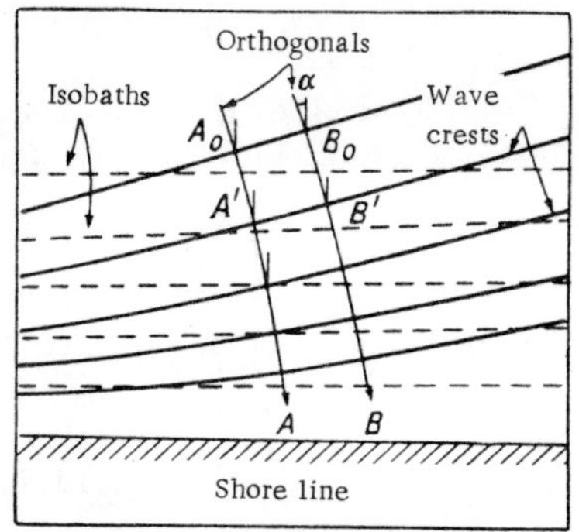

FIGURE 2.11.4. Refraction of waves along a beach.

As was previously mentioned, waves approaching the shore change direction. The angle between the wave travel direction and the "orthogonal" (see Figure 2.11.4) normal to the shore or to any isobath decreases. If it is assumed that the isobaths are rectilinear and parallel, then, of two points A_0 and B_0 on the wave crest, the former travels faster than the latter, since it corresponds to a location above a greater depth than point A_0. By virtue of this the wave front curves and tends to occupy a position parallel to the shore line. This phenomenon is termed "refraction" and from the physical point of view is similar to the curving of light rays in optical systems. When moving from a less to a more dense medium, the light ray bends. Waves in water also propagate from the less "dense" medium (deep sea) to the more "dense" medium (shallow water). Their rate of propagation decreases and their direction changes.

The general law of refraction for sea waves moving from the deep sea into the shallow, shore belt assumes the form (Shuleikin, 1935)

$$\frac{\sin \alpha_2}{\sin \alpha_1} = \frac{c_2}{c_1} = n, \qquad (2.11.5)$$

where α is the angle between the orthogonal and normal to the given isobath (Figure 2.11.4), c is the phase velocity of the waves, and n is the index of refraction. Subscripts 1 and 2 pertain to points where waves pass over the given isobath. Here point 2 is farther away from the shore than point 1. Since the relationship

$$\lambda = c\tau$$

is applicable for waves in both deep and shallow water, i.e., at any depth, then

$$\frac{\lambda_{sh}}{\lambda_d} = \frac{c_{sh}}{c_d} \frac{\tau_{sh}}{\tau_d},$$

where λ_{sh}, c_{sh} and τ_{sh} are the length, phase velocity and period of shallow-water waves, and the corresponding quantities with subscript d pertain to wave elements in deep water. Assuming that the period of an individual wave does not change, or changes little as it moves in shallow water, and neglecting this change, we have

$$\frac{\lambda_{sh}}{\lambda_d} = \frac{c_{sh}}{c_d}. \qquad (2.11.6)$$

Using $(1.2.12)$ and $(1.2.13)$ this becomes

$$\frac{\lambda_{sh}}{\lambda_d} = \frac{\left(\frac{g\lambda_{sh}}{2\pi}\right)^{\frac{1}{2}}\left(\operatorname{th} 2\pi\,\frac{H}{\lambda_{sh}}\right)^{\frac{1}{2}}}{\left(\frac{g\lambda_d}{2\pi}\right)^{\frac{1}{2}}},$$

which yields

$$\frac{\lambda_{sh}}{\lambda_d} = \operatorname{th} 2\pi\,\frac{H}{\lambda_{sh}}. \tag{2.11.7}$$

This expression is made inconvenient by the fact that its right-hand side includes the quantity λ_{sh}, which itself requires determination. Hence $(2.11.7)$ is expressed as

$$\frac{\lambda_{sh}}{\lambda_d} = \operatorname{th}\left[2\pi\,\frac{H}{\lambda_d}\left(\frac{\lambda_{sh}}{\lambda_d}\right)^{-1}\right],$$

which represents an expression of the form

$$\frac{\lambda_{sh}}{\lambda_d} = f\left(\frac{H}{\lambda_d}\right). \tag{2.11.8}$$

Using $(2.11.8)$, relationship $(2.11.7)$ assumes the form

$$\frac{\lambda_{sh}}{\lambda_d} = \operatorname{th} 2\pi\,\frac{H}{\lambda_d}\left(\frac{\lambda_{sh}}{\lambda_d}\right)^{-1}. \tag{2.11.9}$$

Recalling that the use of $(1.2.12)$ is limited to the range $\dfrac{H}{\lambda_d} < 0.5$ by accuracy requirements (Chapter 1, Section 4), one obtains values of $\dfrac{\lambda_{sh}}{\lambda_d}$, which should be multiplied by λ_d in order to obtain the wave length in shallow water (Table 2.11.3).

TABLE 2.11.3. Ratio $\dfrac{\lambda_{sh}}{\lambda_d}$ according to formula $(2.11.9)$

$\dfrac{H}{\lambda_d}$	0.05	0.01	0.02	0.03	0.04	0.05	0.06	0.07
$\dfrac{\lambda_{sh}}{\lambda_d}$	0.17	0.25	0.35	0.41	0.49	0.54	0.58	0.61
$\dfrac{H}{\lambda_d}$	0.08	0.09	0.1	0.15	0.2	0.3	0.4	0.5
$\dfrac{\lambda_{sh}}{\lambda_d}$	0.63	0.67	0.71	0.82	0.89	0.97	0.99	1.00

If it is assumed that c_2 in $(2.11.5)$ corresponds to the phase velocity of waves propagating over a sea with depth $H \geqslant \frac{1}{2}\lambda$, while c_1 is the corresponding quantity for depth $H < \frac{1}{2}\lambda$, then ascribing subscript "d" to c_2 and c_1 subscript "sh" to c_1, expression $(2.11.5)$ becomes

$$\frac{\sin\alpha_d}{\sin\alpha_{sh}} = \frac{c_d}{c_{sh}}. \tag{2.11.10}$$

Using (2.11.6) and (2.11.9), the latter expression is

$$\frac{\sin \alpha_d}{\sin \alpha_{sh}} = \frac{1}{\operatorname{th} 2\pi \dfrac{H}{\lambda_d}\left(\dfrac{\lambda_{sh}}{\lambda_d}\right)^{-1}},$$

whence

$$\sin \alpha_{sh} = \sin \alpha_d \ \operatorname{th} 2\pi \frac{H}{\lambda_d}\left(\frac{\lambda_{sh}}{\lambda_d}\right)^{-1}. \qquad (2.11.11)$$

This expression is employed to obtain Table 2.11.4, which is applicable to all cases when the isobaths are rectilinear and parallel to one another. It follows from Table 2.11.4 that the waves tend to approach the shore line in a normal direction, even if in deep waters they propagate almost parallel to the shore $(\alpha = 85°)$.

TABLE 2.11.4.   Angles of approach of waves (degrees) as a function of the angle of their approach to the shallow-water line (α_d), calculated from (2.11.1)

$\dfrac{H}{\lambda_d}$	α_d						
	0°	15°	30°	45°	60°	75°	85°
0.01	0	4	7	10	13	14	16
0.02	0	4	9	13	17	19	20
0.03	0	6	12	17	21	23	24
0.04	0	7	13	19	23	26	27
0.05	0	8	15	22	28	31	32
0.10	0	10	21	30	38	43	45
0.15	0	12	23	34	45	.51	54
0.20	0	13	26	38	50	60	63
0.30	0	14	28	43	58	70	75
0.40	0	15	29	44	59	73	81

If it is assumed that one need not consider loss of wave energy due to the bottom, i.e., the friction effect, and postulating also that the energy is transported only along orthogonals, and finally, that no energy is supplied by the wind or that there is no energy supply at all (swell), or that very little energy is supplied due to the limited width of the shallow-water shore belt, then one can write (Munk and Taylor, 1947)

$$E_d\, n_d\, c_d\, m_d = E_{sh} n_{sh} c_{sh} m_{sh}, \qquad (2.11.12)$$

where, as before, subscript d pertains to deep and sh to shallow water; E is the wave energy, c is its phase velocity, n is the friction of energy transported by the wave at group velocity u, while m is the distance between neighboring orthogonals.

In deep water (see (1.8.6))

$$n_d = \frac{1}{2}, \qquad (2.11.13)$$

and in shallow water (see (1.8.11))

$$n_{sh} = \frac{1}{2}\left[1 + \frac{2kH}{\operatorname{sh} 2kH}\right].$$
(2.11.14)

Using (1.7.9), i.e., referring the wave energy to the entire length, assuming that the wave crest length is unity, and utilizing the fact that (1.7.9) applies also to shallow-water conditions, it can be assumed that relationship (2.11.12) may be rewritten in the form

$$\frac{h_{sh}^2}{h_d^2} = \frac{n_d}{n_{sh}}\,\frac{c_d}{c_{sh}}\,\frac{m_d}{m_{sh}}.$$
(2.11.15)

If one considers "long waves" (Chapter 1, Section 4), with which one can identify, for example, swell which flows onto a very shallow shore belt, then for these waves (see (1.2.14))

$$c = (gH)^{\frac{1}{2}}$$

and, consequently, according to (2.11.12),

$$n_d = n_t = 1,$$

where subscript t pertains to tidal waves.

Neglecting refraction, and utilizing the above assumptions, (2.11.15) yields

$$\frac{h_{sh}}{h_d} = \left(\frac{H_d}{H_{sh}}\right)^{\frac{1}{4}}.$$
(2.11.16)

This equation, known as Green's formula, shows that when the sea depth along the path of propagation of a long wave decreases by a factor of two, the wave height increases by 20%.

With consideration of (2.11.6), expression (2.11.15) can be rewritten in the form

$$\frac{h_{sh}^2}{h_d^2} = \frac{n_d}{n_{sh}}\,\frac{\lambda_d}{\lambda_{sh}}\,\frac{m_d}{m_{sh}},$$
(2.11.17)

where n_d and n_{sh} are given by (2.11.13) and (2.11.14), while $\dfrac{\lambda_d}{\lambda_{sh}}$ is given by (2.11.19). Quantity $\dfrac{m_d}{m_{sh}}$ becomes (using (2.11.5))

$$\frac{m_d}{m_{sh}} = \frac{\cos \alpha_{sh}}{\cos \alpha_d}.$$

As a result of (2.11.9) and (2.11.11), this latter expression takes the form

$$\frac{m_d}{m_{sh}} = \left[\frac{\cos^2 \alpha_d}{1 - \left(\frac{\lambda_{sh}}{\lambda_d}\sin \alpha_d\right)^2}\right]^{\frac{1}{2}}.$$
(2.11.18)

Consequently, expression (2.11.17) may be replaced by

$$\frac{h_{sh}}{h_d} = \left\{ \frac{1}{1 + \dfrac{4\pi \dfrac{H}{\lambda_d}\left(\dfrac{\lambda_{sh}}{\lambda_d}\right)^{-1}}{\operatorname{sh} 4\pi \dfrac{H}{\lambda_d}\left(\dfrac{\lambda_{sh}}{\lambda_d}\right)^{-1}}} \cdot \frac{1}{\operatorname{th} \dfrac{2\pi H}{\lambda_d}\left(\dfrac{\lambda_{sh}}{\lambda_d}\right)^{-1}} \right\}^{\frac{1}{2}} \left[\frac{\cos^2 \alpha_d}{1 - \left(\dfrac{\lambda_{sh}}{\lambda_d}\sin \alpha_d\right)^2} \right]^{\frac{1}{4}} . \qquad (2.11.19)$$

The first two factors in this expression are functions only of the relative depth $\left(\dfrac{H}{\lambda_d}\right)$. Consequently, the ratio $\dfrac{h_{sh}}{h_d}$ does not change about any assumed isobaths. The last factor is the index of refraction, which defines the change in wave height as a function of the angle of approach of the wave to the start of the shallow-water belt (α_d), i. e., to that isobath for which $\dfrac{H}{\lambda_d} = 0.3 - 0.5$. The results of calculations using (2.11.19) are listed in Table 2.11.5.

TABLE 2.11.5. Values of $\dfrac{h_{sh}}{h_d}$ according to (2.11.19)

α_d°	$\dfrac{H}{\lambda_d}$							
	0.005	0,01	0.02	0.03	0.04	0.05	0,06	0,07
0	1.71	1.41	1.20	1.14	1.05	1.02	1.00	0.97
15	1.67	1.38	1.18	1.12	1.03	1.00	0.98	0.95
30	1.59	1.32	1.11	1.06	0.99	0.96	0.94	0.92
45	1.44	1.10	1.02	0.98	0.91	0.89	0.88	0.85
60	1.21	1.02	0.86	0.83	0.78	0.77	0.76	0.75
75	0.84	0.71	0.62	0.58	0.57	0.55	0.54	0.53

α_d°	$\dfrac{H}{\lambda_d}$							
	0.08	0.09	0.1	0.15	0.2	0.3	0.4	0.5
0	0.96	0.95	0.93	0,90	0.92	0.95	0.98	1.00
15	0.95	0.94	0.92	0.89	0.91	0.94	0.98	1.00
30	0.91	0.90	0.89	0.87	0.90	0.93	0.97	1.00
45	0.85	0.84	0.83	0.82	0.87	0.92	0.96	1.00
60	0.75	0.75	0.74	0.76	0.81	0.90	0.95	1.00
75	0.53	0.54	0.55	0.58	0.64	0.75	0.90	1.00

When the waves approach the start of the shallow-water belt at the shore in the normal direction (assuming the isobaths lie parallel to the shore line and the reduction in depth is smooth) the wave height first decreases (Table 2.11.5 and Figure 2.11.5), but when $\dfrac{H}{\lambda_d} < 0.15$ it starts increasing with reduction in $\left(\dfrac{H}{\lambda_d}\right)$. Ratio $\dfrac{h_{sh}}{h_d}$ changes in the same manner also when the wave approaches the shallow-water belt at different angles, the only difference being that the maximum reduction in wave height occurs with ever-decreasing $\dfrac{H}{\lambda_d}$ (Figure 2.11.5 and Table 2.11.5).

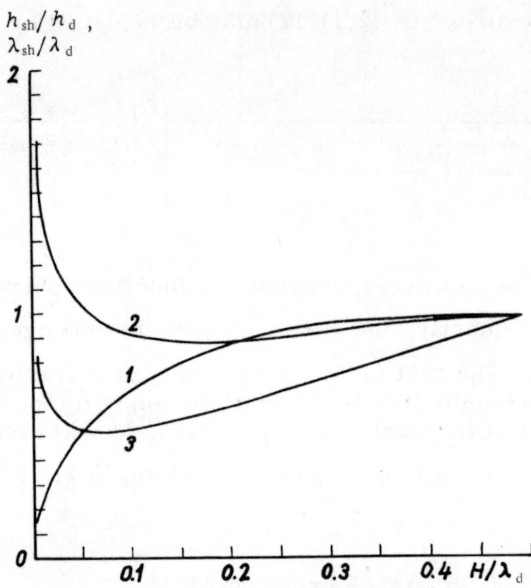

FIGURE 2.11.5. Variation in wave elements in the shallow shore belt:

1 — according to (2.11.9); 2 — according to (2.11.19) when $\alpha = 0°$; 3 — according to (2.11.19) when $\alpha = 75°$.

To estimate the variation in steepness (δ) of waves moving from the shallow-water limit $\left(\dfrac{H}{\lambda_d} < 0.5\right)$ one should use (2.11.19) and (2.11.9). Denoting the right-hand sides of these expressions by k_1 and k_2, it is found that

$$\frac{h_{sh}}{\lambda_{sh}} = \delta_{sh} = \frac{k_1}{k_2}\,\delta_d \; , \tag{2.11.20}$$

where, as before, sh denotes wave steepness in shallow water and d in deep water.

Calculations by (2.11.20) in the absence of refraction are shown in Figure 2.11.6. As the relative depth $\left(\dfrac{H}{\lambda_d}\right)$ (along the abscissa) decreases, the steepness of deep-water waves (ordinate) decreases also. This reduction would have been continuous if the shallow sea bottom did not cause the wave to break. Disintegration of waves, as shown by observations and laboratory experiments, occurs at depth H_{cr}, equal to ~ 1.3 of the wave height:

$$H_{cr} = 1.3 h_{sh}, \tag{2.11.21}$$

or

$$H_{cr} = 1.3 k_1 h_d \; ,$$

which can be represented in the form

$$\frac{H_{cr}}{\lambda_d} = 1.3k_1\,\frac{h_d}{\lambda_d}\,.$$

Now the left-hand side of the last expression, i.e., the relative critical depth, can be expressed as

$$\frac{H_{cr}}{\lambda_d} = f\left(\frac{H}{\lambda_d}\,;\ \ \frac{h_d}{\lambda_d}\right),\tag{2.11.22}$$

since

$$k_1 = f\left(\frac{H}{\lambda_d}\right).$$

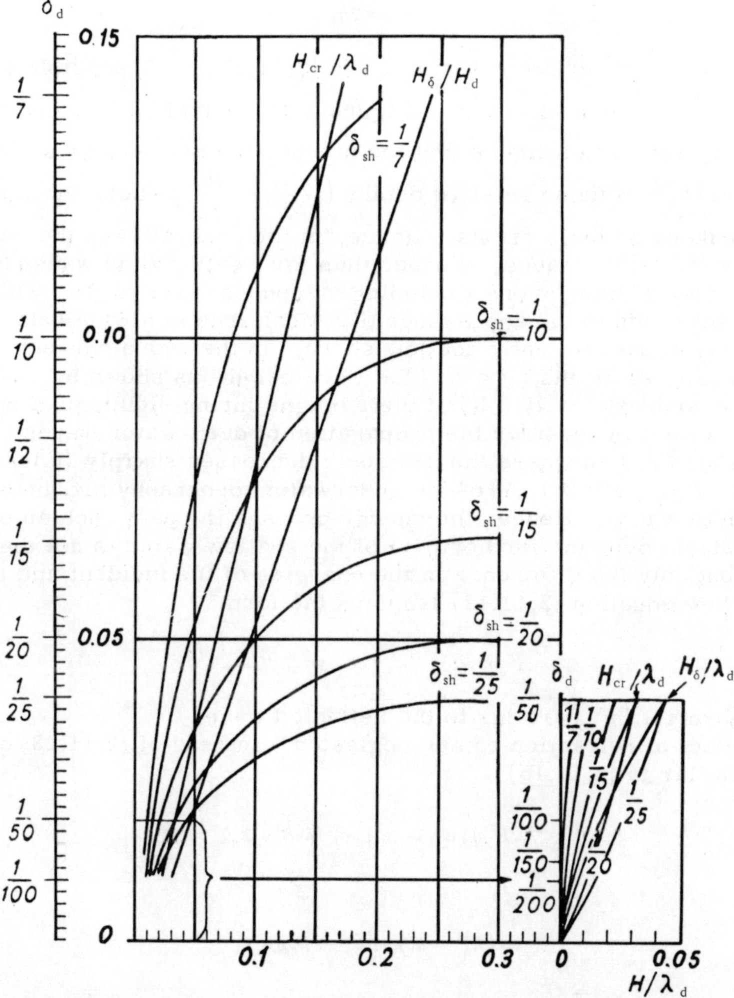

FIGURE 2.11.6. Variation in wave steepness in the shallow-shore region according to (2.11.20).

Calculations using equation (2.11.22) are shown in Figure 2.11.6. For example, waves with $\delta_d = {}^1\!/_{25}$ do not attain a curvature of $\delta_{sh} = {}^1\!/_{10}$ at relative depth $\dfrac{H}{\lambda_d} = 0.03$, since they break earlier, at a depth of $\dfrac{H}{\lambda_d} \cong 0.06$, when they have a curvature $\delta_{sh} \cong {}^1\!/_{15}$. Their height (h_{sh}) at this instant, according to Table 2.11.5 (Figure 2.11.5), will be close to $1.0\,h_d$. The smaller the curvature of the deep-water waves (Figure 2.11.6), the smaller the relative depth at which they disintegrate, but also the more intensively does their height increase (Figure 2.11.5). Consequently, for example, long swell will produce a stronger tide than steep wind waves of the same length.

Before the waves break due to the bottom effect (i.e., at $H_{cr} \approx 1.3\,h$), their crests start spilling. This process becomes particularly perceptible (when the wind blows from the sea) as early as at depth

$$H_\delta = 2\overline{h}_{sh}.$$

Making the appropriate substitution in (2.11.22), it is possible to determine $\dfrac{H_\delta}{\lambda_d}$, which is shown in Figure 2.11.6 (straight line marked $\dfrac{H_\delta}{\lambda_d}$). It is easily seen in this figure that the steeper the deep-water wave, the wider the range of those relative depths $\left(\dfrac{H_{cr}}{\lambda_d} - \dfrac{H_\delta}{\lambda_d}\right)$ where the waves carry breakers on their crests. Hence, in the general case the surf belt in wind waves (steep waves) is wider than for swell. Wind waves have a larger variety of dimensions (including steepness) than swell. This circumstance aids in the appearance of a tidal zone in a wide belt, particularly at a sufficiently shallow shore. In the shore current with less than critical depths $(H < H_{cr})$ the wave height (as shown by observations) amounts to approximately 0.85 of their height during disintegration.

It is possible to consider the propagation of deep-water waves in a shallow shore region where the sea depth decreases sharply (Lappo, 1956). This feature (Figure 2.11.7) of the underwater topography produces a partial reflection of waves. Hence, the energy propagating past such an underwater obstacle over the further part of the shallow water is not the total energy, but only the difference in the energies of the incident and reflected waves. Now equation (2.11.12) assumes the form

$$E_d\, n_d\, c_d - E_r\, n_d\, c_d = E_{sh} n_{sh} c_{sh}, \qquad (2.11.23)$$

where subscript "r" pertains to the reflected wave.

The effect of refraction can be neglected. Instead of (2.11.23) one may write, similar to (2.11.15),

$$h_d^2\, n_d\, c_d - h_r^2\, n_d\, c_d = h_{sh}^2 n_{sh} c_{sh},$$

or

$$\left(h_d^2 - h_r^2\right) n_d\, c_d = h_{sh}^2 n_{sh} c_{sh}. \qquad (2.11.24)$$

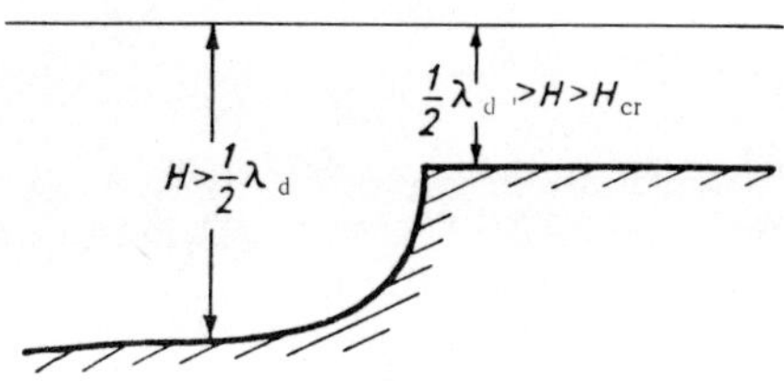

FIGURE 2.11.7. Feature of underwater topography
producing wave reflection (after Lappo).

The amplitudes of wave oscillations pertaining to the direct and
reflected waves are approximately equal at the boundary between the deep-
and shallow-water zones:

$$h_d + h_r = h_{sh}. \tag{2.11.25}$$

Solving the last two expressions for h_{sh}/h_d ,

$$\frac{h_{sh}}{h_d} = \frac{2c_d\,n_d}{c_d\,n_d + c_{sh}n_{sh}}.$$

Using (2.11.6), (2.11.9), (2.11.13) and (2.11.14), the last expression
yields

$$\frac{h_{sh}}{h_d} = \frac{2}{1 + \operatorname{th} 2\pi\,\frac{H}{\lambda_d}\left(\frac{\lambda_{sh}}{\lambda_d}\right)^{-1}\left[1 + \dfrac{4\pi\,\dfrac{H}{\lambda_d}\left(\dfrac{\lambda_{sh}}{\lambda_d}\right)^{-1}}{\operatorname{sh} 4\pi\,\dfrac{H}{\lambda_d}\left(\dfrac{\lambda_{sh}}{\lambda_d}\right)^{-1}}\right]}, \tag{2.11.26}$$

from which it follows that if the sea is very shallow ($H \to 0$), the wave height
in shallow water is approximately twice the wave height in deep water.
Consequently, it can be assumed that the height of waves approaching steep
or vertical underwater obstacles will double as a result of interference
between the incident and reflected waves, obviously, when the periods of
such waves are close to another. For values of $\dfrac{H}{\lambda_d}$ equal to 0.05 and
above, the results of calculations with (2.11.19) and (2.11.26) are virtually
identical. For $\dfrac{H}{\lambda_d} \leqslant 0.05$ the values of $\dfrac{h_{sh}}{h_d}$ are approximately as follows:

$$\frac{H}{\lambda_d} \; \ldots \ldots \ldots \quad 0.01 \quad 0.02$$

$$\frac{h_{sh}}{h_d} \; \ldots \ldots \ldots \quad {\sim}1.7 \quad {\sim}1.5$$

Chapter 3

FUNDAMENTALS OF TECHNIQUES OF COMPUTING WIND-WAVE ELEMENTS

1. Purposes and substance of computations of wind-wave elements. Reliability considerations

Knowledge of wind-wave elements is required to satisfy the needs of a number of branches of modern technology (Introduction, Section 1). In some geographic regions these elements are not observed at all, or the observations do not suffice for the intended purpose. These data can be supplied by computation. Such computation is made for oceans, seas and natural and water-storage lakes. The calculations differ depending on the wave-generating conditions, i.e., whether the waves propagate over deep or shallow waters, and whether they reach the shore regions. Computation will also differ depending on whether one wishes to determine elements of developing wind waves or their dimensions under steady conditions. Finally, it is also significant whether the wave elements are calculated for specific weather conditions, or estimated for prolonged time intervals, for example, for the duration of a storm, month, year or a long-term period. The latter are termed climatic-regime wave characteristics.

In all the above cases one usually calculates the height, length and period of waves and the time of their growth. Computation is made for one point in the coastal strip of the sea, or in its open part, or for a number of such points. In the latter case the results of calculations are presented in the form of maps, on which the distribution of computed wave elements in space is shown in an appropriate manner. The climatic-regime wave-motion characteristics can be represented similarly. The latter are usually supplemented by an estimate of their frequency over the period under study.

The calculations consist primarily in determining the numerical values of the wave-generating factors (Chapter 3, Section 2), and then in carrying out direct calculations, the techniques of which depend on the intended purpose. These techniques can be in general subdivided into: 1) method of wind-wave element computation for different weather conditions in the open sea (Chapter 3, Section 3); 2) method of computation in shallow-water and shore regions (Chapter 3, Section 4); and 3) method of computing climatic-regime characteristics of wave motion (Chapter 3, Section 5).

The reliability of these calculations can be determined approximately and in the most general form on the assumption that the main wave-generating factors (Chapter 2, Section 4) are the wind speed (w), its duration (t) and the fetch (x), and hence the results of subsequent

computation of wind-wave elements depend on the accuracy in determining
the three above quantities. The dependence of wind elements on the aforementioned quantities can be expressed approximately in the general form
(Chapter 2, Section 5)

$$h \approx m_1 w x^{0.5}, \qquad (3.1.1)$$

$$h \approx m_2 w^{1.5} t^{0.5}, \qquad (3.1.2)$$

$$\tau = m_3 w^{0.5} x^{0.3}, \qquad (3.1.3)$$

$$\tau \approx m_4 w^{0.7} t^{0.4}, \qquad (3.1.4)$$

where m_1, m_2, m_3 and m_4 are constants.

The relative error $(P\%)$ in determining x, w and t can in no way be less
than 10%. In this case, using $(3.1.1)-(3.1.4)$ and employing standard
methods of determining errors, it will be found that the relative error $(P\%)$
in calculating the wind elements is approximately

$$P(h_x) \approx 10\% + 5\% \approx 15\%,$$
$$P(h_t) \approx 15\% + 5\% \approx 20\%,$$
$$P(\tau_x) \approx 5\% + 3\% \approx 8\%,$$
$$P(\tau_t) \approx 7\% + 4\% \approx 11\%.$$

It follows that the error in estimating the wind speed affects more
appreciably the accuracy of computational results than that in determining
the fetch. The accuracy in calculating the wave period is higher than that
in determining the wave height.

The error in calculating wind elements as a function of wind duration
is greater than as a function of fetch. A relative error of 10% in determining the duration, wind speed and fetch is apparently the highest
permissible, if it is desired that the error in the final results should not
exceed a relative error of $\pm 15-20\%$. The latter apparently can be regarded
as the highest permissible error in calculating wind-wave elements for
various applied purposes. The criterion of reliability of these calculations
is their agreement with observational data. Determination of wind-wave
elements, even by means of instruments, involves on the average a relative
error of the order of $\pm 10-15\%$. At the same time an error arises in
determining wave-generation factors responsible for the waves being
measured. Frequently the estimate of computational reliability by comparison with experimental data is essentially indeterminate; sometimes the
solution of this problem becomes an independent and rather extensive study.
At the same time agreement within $\pm 15-20\%$ between observed and computed
wind-wave elements is a sufficiently satisfactory result, showing the
practical reliability of calculations made from given formulas.

2. Determination of wave-generating factors

In the practice of shipboard observations, wind speed is determined at
some height above sea level, which depends on the location of the wind-speed
recorders. This height ranges on the average from 6 to 10 m above sea

level. In observations carried out from the shore the elevation of instruments above sea level very frequently differs. Conversion from wind speed
measured from ships in the open sea at some elevation (H) above sea level
to wind speed at 6 m above sea level or to any other wind speed can be
carried out by means of factors listed in Table 3.2.1 (Sorkina, 1958).

TABLE 3.2.1. Ratio of wind speed measured at elevation H to that measured at an elevation of 6 m above
sea level (after Sorkina)

Elevation above sea level, m	State of the atmosphere		
	stable	marginally stable	in equilibrium and slightly unstable
30	1.38	1.19	1.12
20	1.27	1.14	1.09
10	1.10	1.06	1.04
8	1.07	1.03	1.02
6	1.00	1.00	1.00
4	0.92	0.96	0.97
2	0.80	0.89	0.92
1	0.70	0.82	0.86

The state of the atmosphere (in Table 3.2.1) is estimated in conjunction
with the difference between the water temperature (t°_w) and the air temperature (t°_a) from Table 3.2.2 (Sorkina, 1958).

TABLE 3.2.2. Relationship between the state of the atmosphere and the water-air temperature difference
(after Sorkina)

State of the atmosphere	$t^\circ_w - t^\circ_a$
Stable	-0.5
Marginally stable	from -0.5 to -0.1
In equilibrium or slightly unstable	from 0.0 to 2
Unstable	> 2.0

The problem of converting wind speed determined at points on the shore
to that at some given elevation is more complicated. If one takes 10 m
above sea level as the given elevation, then this conversion can be carried
out from expression (Standards SN 92-60, 1960):

$$w_{10} = k w_H ,$$

(3.2.1)

where w_{10} is the wind speed at the 10 m elevation, w_H is the wind speed
measured at elevation H, and k is a conversion factor, which is taken,
depending on H, from Table 3.2.3.

TABLE 3.2.3. Values of conversion factor k

H, m	2	6.5	8	10	12	17	28
k	1.25	1.05	1.03	1.0	0.98	0.94	0.89

The wind speed and its direction do not remain constant at the point of measurement. They sometimes vary within wide limits, particularly when the wind blows in gusts. Hence the measured (usually over 100 secs) wind speed and direction reflect some average direction and speed over the measurement interval. However, these characteristics can change also over longer time intervals due to variations in weather conditions. If these variations do not exceed 2 m/sec with respect to speed and 2 compass points with respect to direction, then wind speed and direction are regarded as constant.

An estimate of wind speed and direction only at individual points of the shore strip or at individual points in the open sea on the basis of observational data is frequently insufficient for wave calculations. It may be substantially different at intermediate points. This circumstance makes it impossible to objectively estimate the wind speed which determines the wave development. Hence maps of atmospheric pressure are employed for obtaining a more complete estimate of the wind speed. These data are used for computing wind speed and direction at any number of points. This makes it possible to obtain maps of wind fields (Sorkina, 1958). Such maps can be constructed from the distribution of atmospheric pressure for individual synoptic durations, or for a more extended period. Finally, such maps can serve as "typical maps of wind fields," i.e., as maps corresponding to specific types of atmospheric processes occurring over the given body of water. An ensemble of such typical maps can be used for describing all the principal wind conditions above the body of water. By virtue of the same fact, data on wind speed can also be obtained with the greatest completeness for any region or point of the body of water with consideration of different weather conditions.

The above methods of calculating wind speed are used primarily for computing wave elements under specific weather conditions. As was previously noted (Chapter 3, Section 1), wave elements may be calculated in order to obtain some idea about their values over an extended period of time, i.e., for determining the climatic-regime characteristics. In this case the wind speed used in calculations should be estimated from the point of view of its frequency (or cumulative probability) over the period for which the wave elements are to be calculated. Information on the frequency (cumulative probability) of winds of different speeds is available in appropriate climatic handbooks. Similar data are also provided on maps of climatic atlases.

It is often necessary to estimate the maximum possible wind speed and its probability. This is done by means of distribution functions of the integral frequency (cumulative probability) of wind speed in the form (Anapol'skaya and Gandin, 1960)

$$F(w) = \exp\left[-\left(\frac{w}{\beta}\right)^{\gamma}\right], \tag{3.2.2}$$

where β and γ are parameters defined by

$$\beta = \frac{\overline{w}}{\Gamma\left(1 + \dfrac{1}{\gamma}\right)}, \tag{3.2.3}$$

$$\frac{\sigma}{\overline{w}} = \left(\frac{\Gamma\left(1 + \dfrac{2}{\gamma}\right)}{\Gamma^2\left(1 + \dfrac{1}{\gamma}\right)}\right)^{\frac{1}{2}}; \tag{3.2.4}$$

here $\overline{w}$ is the mean velocity, σ is the standard deviation of the velocity from the mean, and Γ is the gamma function. Since $\Gamma\left(1 + \dfrac{1}{\gamma}\right)$ does not vary significantly in the range of usually encountered values of γ (approximately $1.0 < \gamma < 2.5$), β is proportional to the mean wind speed ($\overline{w}$) with an almost constant proportionality factor. The value of γ is uniquely related to the relative variance, decreasing with an increase in the latter. Instead of determining β and γ it is possible to solve (3.2.2) graphically. [For this purpose] it is written in the form

$$\lg\left[-\lg F(w)\right] = \gamma \lg w - \gamma \lg \beta + \lg \lg e. \tag{3.2.5}$$

It is clear from this expression that, if expression (3.2.2) is satisfied, then the observational points plotted in the $\lg w$, $\lg[-\lg F(w)]$ system should lie on the same straight line. Using appropriate graph paper (Figure 3.2.1) and plotting available data on w and $F(w)$, it is possible to draw, with some accuracy, a straight line representing (3.2.2). Hence it is not necessary to compute values of β and γ. This graphical relationship makes it possible, when necessary, to extrapolate the straight line to wind speeds with low cumulative probability and to determine the value of w corresponding to some given cumulative probability [$F(w)$]. The latter quantity will then be defined by

$$F(w) = \frac{m}{tN \cdot 365n}\, 100\%, \tag{3.2.6}$$

where m is the wind duration in hours, needed for producing steady wave motion; n is the number of years during which the wind speed with given cumulative probability is encountered once; N is the number of wind speed observations over a 24 hour period; t is the time in hours between observations (N). For example, if $m = 24$ hours, observations are made once daily; hence $t = 24$ hours and $N = 1$. Then (3.2.6) may be expressed in the form

$$F(w) = \frac{1}{365n}\, 100\%. \tag{3.2.7}$$

If $n = 10$ years, then $F(w) \approx 0.03\%$. If it is possible to plot the cumulative probability of a wind speed (obtained from wind observations) in the manner shown in Figure 3.2.1, then by extrapolating the straight line to the value $F(w) = 0.03\%$ it is possible to find the wind speed corresponding to the computed cumulative probability. This speed should be approximately 34 m/sec (Figure 3.2.1). Expression (3.2.6) can also be used for solving the inverse problem, i.e., of determining n from given $F(w)$.

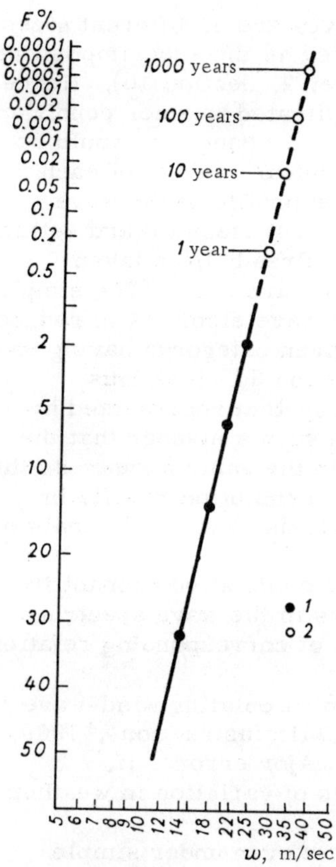

FIGURE 3.2.1. Determination of the cumulative probability of a wind with different speeds:

1 — observational data; 2 — derived by extrapolation.

Unit fetch (x) in the simplest case can be defined as the extent (in kilometers) of open and free water surface in a direction opposite to that of the wind. In this case the initial point of the fetch is the shore line or the edge of the ice. However, this definition of fetch is very imprecise, particularly when the extent of open water exceeds 100 km. In this case the wind direction may be highly nonlinear as the wind speed over a fetch measured in this manner may differ from its real distribution. This method of fetch measurement, even if valid, can be used for bodies of water for which free-water extents are less than 100 km. Hence fetches for large bodies of water, seas and especially oceans should be determined from synoptic maps or from maps of wind fields (see above). The fetches can then be determined in the same manner as they are measured (Chapter 2, Section 4).

The duration of wind action can be taken as the time interval between the times of composition of synoptic maps, from which the wind speed and fetch are determined, or as the difference between the dates of wind observation.

If it is necessary to determine the fetch (x) for computing wave elements in order to estimate climatic-regime wave-motion characteristics (Chapter 3, Section 2), then it is possible to use the distribution function of the integral frequency (cumulative probability) of relative fetches $\left(\dfrac{x}{x^*}\right)$ in the form of (2.4.2), or of their absolute values (2.4.3).

3. Computation of wind-wave elements for deep-water seas

Underlying the formulas expressing the dependence of wind-wave elements of wind-generating factors (Chapter 2, Sections 5 and 8) is the assumption that the latter remain constant during the entire wave development. Actually, wind speed and direction change under natural conditions. These changes produce changes in the fetch. Consequently, waves develop under variable weather conditions. Waves may develop, decay, and again increase in size. Hence in calculating wave elements consideration must be given to changes in wind-generating factors.

It should also be noted that wind-generated waves are of different sizes
(Chapter 2, Section 3). This variety can be treated as the superimposition
of an infinite number of elementary waves (Chapter 2, Section 10). Hence,
in order to estimate wave dimensions under complicated weather conditions
during which the wave-generating factors themselves change, it would be
sensible to estimate the development of each individual wave or of each
wave system, and then to sum the results, find the profile of the wave
surface, and obtain its quantitative estimates. First steps toward setting
up such or similar computational techniques have already been taken
(Krylov, 1966). As yet other, approximate methods are used. The simplification consists in the fact that the calculation of wave elements is reduced
to estimating the dimensions of waves of only a given category, having one
or another cumulative probability (Chapter 2, Section 3). It is thus
assumed that the development of the complicated system represented by a
wave field consisting of different waves occurs in such a manner that the
elements of waves of the selected category vary in the same manner as the
elements of all the other waves. Naturally, this assumption results in
inaccuracies and in the best case describes variations in wave elements of
only that category selected for calculation.

Conversion to dimensions of waves of different cumulative probability
(Chapter 2, Section 3) or an estimate of variations in the wave spectrum
(Chapter 2, Section 10) are carried out by means of corresponding relation-
ships.

At the same time, it is shown by experience in calculating wind-wave
elements based on the above assumption (Titov, 1951; Instructions, 1960)
that apparently these calculations do not involve major errors, if,
obviously, these are applied to the simplest cases of variation in weather
conditions.

Wind speed and direction do not remain constant even under simple
weather conditions. If the wind speed varies within 2 m/sec, while the
direction is constant within 2 compass points ($\sim 24°$), then (Chapter 2,
Section 4) these variations are disregarded. If, however, these variations
overstep these limits, they must be taken into account by using certain
techniques of computing wind-wave elements (Titov, 1951). These techniques,
in their most general form, pertain to changes in wave-generating factors
when: 1) the wind speed and its direction remain unchanged during the
entire wind action; 2) only the wind speed changes, while its direction
remains the same; 3) the wind speed does not change, and only its
direction changes.

The terms "wind speed variation" and "wind direction variations" mean
that these changes occur not only at a given point, but over the entire sea
area surrounding it, thus also producing changes in the fetch.

The examples of computing wind-wave elements presented below can be
solved using all those existing formulas in which the wave elements are
expressed as a function of velocity, fetch and duration of wind action
(Chapter 2, Sections 5 and 8).

Formulas (2.5.21)—(2.5.31) and (2.5.34)—(2.5.35) were used for generating
Table 3.3.1. Calculations of elementary swell encountered in the examples
were made with reference to Table 3.3.2 (Chapter 2, Section 11). It should
be remembered that a more complete description of techniques for calculating wind-wave elements, as well as recommendations on the use of a given
set of formulas, can be found in special publications (Titov, 1951;
Instructions, 1960; Selyuk, 1961; Directions, 1968).

TABLE 3.3.1. Wind-wave elements

x, km	$\bar{h}$, m	$\bar{\tau}$, sec	t, hours
$w = 4$ m/sec			
5	0.08	1.3	1.1
10	0.10	1.6	1.9
20	0.15	2.0	3.7
40	0.20	2.4	6.3
48	0.23	2.5	7.6
$w = 5$ m/sec			
5	0.10	1.4	0.9
10	0.15	1.8	1.8
20	0.20	2.1	2.8
40	0.30	2.7	5.8
60	0.35	3.0	7.9
75	0.38	3.2	9.5
$w = 6$ m/sec			
5	0.15	1.5	0.8
10	0.20	1.9	1.5
20	0.25	2.3	2.6
40	0.35	2.9	5.2
60	0.40	3.2	6.9
80	0.45	3.5	8.8
100	0.55	3.7	10.5
108	0.55	3.8	11.3
$w = 7$ m/sec			
5	0.15	1.6	0.7
10	0.20	2.0	1.3
20	0.30	2.5	2.5
40	0.40	3.0	4.2
60	0.50	3.4	6.0
80	0.55	3.7	7.7
100	0.60	4.0	9.6
120	0.65	4.2	11.1
140	0.70	4.4	12.7
147	0.75	4.5	13.2
$w = 8$ m/sec			
5	0.15	1.7	0.6
10	0.25	2.1	1.2
20	0.35	2.6	2.1
40	0.45	3.2	3.9
60	0.55	3.6	5.5

x, km	$\bar{h}$, m	$\bar{\tau}$, sec	t, hours
80	0.65	3.9	6.9
100	0.70	4.2	8.6
120	0.80	4.5	10.5
140	0.85	4.7	11.9
160	0.90	4.9	13.4
180	0.95	5.0	14.3
192	0.97	5.1	15.1
$w = 9$ m/sec			
5	0.20	1.8	0.6
10	0.30	2.2	1.0
20	0.40	2.7	1.9
40	0.55	3.4	3.6
60	0.65	3.8	5.1
80	0.70	4.1	6.2
100	0.80	4.4	7.7
120	0.90	4.7	9.3
140	0.95	4.9	10.5
160	1.00	5.1	11.8
180	1.05	5.3	13.2
200	1.10	5.4	13.9
220	1.15	5.6	15.5
240	1.21	5.8	17.0
243	1.23	5.8	17.0
$w = 10$ m/sec			
5	0.20	1.9	0.6
10	0.30	2.3	1.0
20	0.45	2.8	1.8
40	0.60	3.4	3.3
60	0.70	3.9	4.5
80	0.80	4.3	6.0
100	0.90	4.6	7.3
120	1.00	4.9	8.8
140	1.05	5.1	9.9
160	1.10	5.3	11.1
180	1.20	5.5	12.3
200	1.25	5.7	13.6
220	1.30	5.8	14.4
240	1.35	6.0	15.8
260	1.40	6.1	16.6
280	1.45	6.3	18.2
300	1.52	6.4	18.9

TABLE 3.3.1 (Continued)

x, km	$\bar{h}$, m	$\bar{\tau}$, sec	t, hours	x, km	$\bar{h}$, m	$\bar{\tau}$, sec	t, hours
				140	1.35	5.7	8.3
	$w=11$ m/sec			160	1.50	5.9	9.2
5	0.20	1.9	0.5	180	1.60	6.1	10.1
10	0.35	2.4	0.9	200	1.65	6.3	11.1
20	0.45	2.9	1.6	220	1.75	6.5	12.1
40	0.65	3.6	3.0	240	1.80	6.7	13.2
60	0.80	4.1	4.3	260	1.85	6.8	13.8
80	0.90	4.5	5.7	280	1.95	7.0	15.1
100	1.00	4.8	6.9	300	2.00	7.1	15.7
120	1.10	5.1	8.2	400	2.25	7.7	19.8
140	1.15	5.3	9.2	500	2.50	8.3	24.6
160	1.25	5.5	10.3	507	2.57	8.3	24.6
180	1.30	5.7	11.4				
200	1.35	5.9	12.6				
220	1.45	6.1	13.7		$w=14$ m/sec		
240	1.50	6.2	14.5	5	0.25	2.1	0.4
260	1.55	6.3	15.2	10	0.40	2.7	0.8
280	1.60	6.5	16.7	20	0.65	3.2	1.3
300	1.65	6.6	17.4	40	0.85	4.0	2.6
363	1.84	7.0	20.8	60	1.00	4.5	3.6
				80	1.15	4.9	4.6
				100	1.30	5.3	5.8
	$w=12$ m/sec			120	1.45	5.6	6.8
5	0.25	2.0	0.5	140	1.50	5.8	7.6
10	0.35	2.5	0.9	160	1.65	6.1	8.8
20	0.50	3.1	1.6	180	1.70	6.3	9.6
40	0.75	3.8	2.9	200	1.80	6.5	10.5
60	0.90	4.3	4.2	220	1.85	6.7	11.5
80	1.00	4.6	5.2	240	1.95	6.9	12.5
100	1.10	5.0	6.6	260	2.00	7.0	13.0
120	1.20	5.2	7.3	280	2.10	7.2	14.1
140	1.30	5.5	8.6	300	2.15	7.3	14.8
160	1.35	5.7	9.6	400	2.45	7.9	18.6
180	1.45	5.9	10.6	500	2.70	8.5	22.9
200	1.50	6.1	11.8	588	2.98	8.9	26.4
220	1.60	6.3	12.9				
240	1.65	6.5	14.1				
260	1.70	6.6	14.8		$w=15$ m/sec		
280	1.75	6.7	15.4	5	0.30	2.2	0.4
300	1.80	6.9	16.8	10	0.45	2.7	0.7
400	2.05	7.5	21.4	20	0.65	3.3	1.3
432	2.19	7.7	22.7	40	0.95	4.1	2.4
				60	1.10	4.6	3.4
	$w=13$ m/sec			80	1.25	5.0	4.3
5	0.25	2.1	0.4	100	1.40	5.4	5.4
10	0.40	2.6	0.8	120	1.55	5.7	6.3
20	0.60	3.2	1.5	140	1.65	6.0	7.3
40	0.80	3.9	2.8	160	1.75	6.2	8.1
60	0.95	4.4	3.9	180	1.85	6.5	9.2
80	1.10	4.8	5.0	200	1.95	6.7	10.1
100	1.20	5.1	6.0	220	2.00	6.9	11.0
120	1.30	5.4	7.1	240	2.10	7.1	11.9

x, km	$\bar{h}$, m	$\bar{\tau}$, sec	t, hours
260	2.15	7.2	12.4
280	2.25	7.4	13.4
300	2.30	7.5	14.0
400	2.65	8.2	18.0
500	2.90	8.8	22.3
600	3.15	9.4	26.1
675	3.42	9.6	28.4

$w = 16$ m/sec

x, km	$\bar{h}$, m	$\bar{\tau}$, sec	t, hours
5	0.30	2.2	0.4
10	0.45	2.8	0.7
20	0.70	3.3	1.2
40	1.00	4.2	2.3
60	1.20	4.8	3.3
80	1.35	5.2	4.2
100	1.50	5.6	5.3
120	1.65	5.9	6.2
140	1.75	6.1	6.8
160	1.85	6.4	7.8
180	1.95	6.6	8.5
200	2.05	6.8	9.2
220	2.15	7.0	10.1
240	2.25	7.2	11.0
260	2.35	7.4	11.9
280	2.45	7.6	12.9
300	2.50	7.7	13.4
400	2.80	8.4	16.8
500	3.10	9.0	21.0
600	3.45	9.6	25.3
700	3.70	10.0	28.5
768	3.90	10.3	30.1

$w = 17$ m/sec

x, km	$\bar{h}$, m	$\bar{\tau}$, sec	t, hours
5	0.30	2.3	0.4
10	0.50	2.9	0.7
20	0.75	3.5	1.2
40	1.05	4.3	2.2
60	1.30	4.9	3.2
80	1.45	5.3	4.0
100	1.60	5.7	5.0
120	1.75	6.0	5.8
140	1.90	6.3	6.6
160	1.95	6.5	7.3
180	2.10	6.8	8.3
200	2.20	7.0	9.0
220	2.30	7.2	9.8
240	2.40	7.4	10.6
260	2.50	7.6	11.3
280	2.60	7.8	12.4
300	2.65	7.9	12.8

x, km	$\bar{h}$, m	$\bar{\tau}$, sec	t, hours
400	3.00	8.6	16.4
500	3.30	9.2	20.1
600	3.65	9.8	23.9
700	3.90	10.2	27.0
800	4.10	10.6	30.1
867	4.39	10.9	32.1

$w = 18$ m/sec

x, km	$\bar{h}$, m	$\bar{\tau}$, sec	t, hours
5	0.35	2.4	0.4
10	0.50	2.9	0.6
20	0.75	3.6	1.1
40	1.15	4.4	2.1
60	1.35	5.0	3.1
80	1.55	5.4	3.8
100	1.70	5.8	4.6
120	1.90	6.2	5.7
140	2.05	6.5	6.5
160	2.15	6.7	7.1
180	2.30	7.0	8.1
200	2.40	7.2	8.8
220	2.50	7.4	9.5
240	2.60	7.6	10.3
260	2.70	7.8	11.1
280	2.75	7.9	11.5
300	2.85	8.1	12.4
400	3.20	8.8	15.8
500	3.50	9.4	19.1
600	3.90	10.0	22.8
700	4.15	10.4	25.5
800	4.40	10.8	28.5
900	4.70	11.2	31.7
972	4.92	11.5	34.0

$w = 19$ m/sec

x, km	$\bar{h}$, m	$\bar{\tau}$, sec	t, hours
5	0.35	2.4	0.3
10	0.55	3.0	0.6
20	0.80	3.7	1.1
40	1.20	4.5	2.0
60	1.45	5.1	2.9
80	1.65	5.6	3.7
100	1.85	6.0	4.5
120	2.00	6.3	5.4
140	2.15	6.6	6.2
160	2.30	6.9	7.0
180	2.40	7.1	7.6
200	2.50	7.3	8.2
220	2.65	7.6	9.3
240	2.75	7.8	10.0
260	2.80	7.9	10.4
280	2.90	8.1	11.1
300	2.95	8.2	11.5

x, km	$\bar{h}$, m	$\bar{\tau}$, sec	t, hours	x, km	$\bar{h}$, m	$\bar{\tau}$, sec	t, hours
400	3.40	9.0	15.2	200	2.80	7.6	7.7
500	3.75	9.6	18.3	220	2.95	7.9	8.6
600	4.10	10.2	21.8	240	3.05	8.1	9.2
700	4.35	10.6	24.4	260	3.15	8.2	9.5
800	4.65	11.1	27.8	280	3.25	8.4	10.2
900	4.90	11.5	30.9	300	3.35	8.6	11.0
1000	5.20	11.9	34.1	400	3.85	9.4	14.2
1083	5.48	12.2	36.1	500	4.15	10.0	17.0
				600	4.60	10.6	20.2
				700	4.90	11.1	23.0
	$w = 20$ m/sec			800	5.20	11.5	25.5
				900	5.55	12.0	28.9
5	0.35	2.5	0.3	1000	5.75	12.3	31.0
10	0.55	3.0	0.6	1100	6.00	12.7	34.0
20	0.85	3.7	1.0	1200	6.30	13.0	36.4
40	1.25	4.6	1.95	1300	6.60	13.3	38.9
60	1.55	5.2	2.8	1323	6.70	13.4	39.8
80	1.75	5.7	3.6				
100	1.95	6.1	4.4				
120	2.10	6.4	5.1				
140	2.25	6.7	5.8				
160	2.40	7.0	6.6				
180	2.55	7.3	7.4		$w = 22$ m/sec		
200	2.65	7.5	8.1				
220	2.80	7.7	8.8	5	0.40	2.5	0.3
240	2.95	8.0	9.8	10	0.60	3.2	0.5
260	3.00	8.1	9.9	20	0.90	3.9	1.0
280	3.05	8.2	10.4	40	1.35	4.8	1.8
300	3.15	8.4	11.3	60	1.70	5.4	2.6
400	3.65	9.2	14.8	80	1.95	5.9	3.4
500	4.00	9.8	17.6	100	2.15	6.3	4.1
600	4.35	10.4	20.9	120	2.30	6.6	4.7
700	4.60	10.8	23.3	140	2.50	7.0	5.5
800	4.95	11.3	26.6	160	2.70	7.3	6.2
900	5.20	11.7	29.4	180	2.80	7.5	6.8
1000	5.50	12.1	32.5	200	2.95	7.8	7.6
1100	5.75	12.4	34.8	220	3.10	8.0	8.2
1200	6.08	12.8	37.8	240	3.20	8.2	8.7
				260	3.30	8.4	9.4
				280	3.35	8.5	9.7
	$w = 21$ m/sec			300	3.50	8.6	10.4
				400	4.00	9.5	13.4
5	0.35	2.5	0.3	500	4.45	10.2	16.5
10	0.55	3.1	0.6	600	4.80	10.8	19.5
20	0.85	3.8	1.0	700	5.15	11.3	22.2
40	1.30	4.7	1.9	800	5.50	11.8	25.2
60	1.60	5.3	2.7	900	5.80	12.2	27.7
80	1.90	5.8	3.5	1000	6.00	12.5	29.8
100	2.10	6.2	4.2	1100	6.25	12.9	32.6
120	2.20	6.5	4.9	1200	6.60	13.3	35.6
140	2.35	6.8	5.7	1300	6.80	13.6	38.0
160	2.55	7.1	6.3	1400	7.05	13.9	40.5
180	2.70	7.4	7.1	1452	7.12	14.0	41.5

Left columns:

x, km	$\bar{h}$, m	$\bar{\tau}$, sec	t, hours
	$w=23$ m/sec		
5	0.40	2.6	0.3
10	0.60	3.2	0.5
20	0.95	4.0	1.0
40	1.40	4.9	1.8
60	1.80	5.5	2.5
80	2.05	6.0	3.3
100	2.25	6.4	3.9
120	2.40	6.8	4.6
140	2.65	7.1	5.3
160	2.80	7.4	6.0
180	2.95	7.7	6.7
200	3.10	7.9	7.2
220	3.25	8.2	8.0
240	3.40	8.4	8.6
260	3.50	8.6	9.2
280	3.55	8.7	9.5
300	3.70	8.9	10.2
400	4.20	9.7	13.1
500	4.70	10.4	16.0
600	5.05	11.0	18.9
700	5.40	11.5	21.5
800	5.80	12.0	24.3
900	6.05	12.4	26.7
1000	6.35	12.8	29.3
1100	6.60	13.1	31.3
1200	6.90	13.5	34.2
1300	7.10	13.8	36.4
1400	7.40	14.2	39.6
1500	7.70	14.5	42.1
1587	8.05	14.7	43.5
	$w=24$ m/sec		
0	0.40	2.7	0.3
15	0.65	3.3	0.5
20	1.00	4.0	0.9
40	1.45	5.0	1.7
60	1.90	5.6	2.4
80	2.15	6.1	3.2
100	2.35	6.5	3.8
120	2.55	6.9	4.5
140	2.75	7.2	5.1
160	2.90	7.5	5.7
180	3.10	7.8	6.4
200	3.20	8.0	7.0
220	3.40	8.3	7.7
240	3.50	8.5	8.2
260	3.65	8.7	8.8
280	3.70	8.8	9.1
300	3.90	9.1	10.0
400	4.40	9.9	12.8
500	4,80	10.5	15.2

Right columns:

x, km	$\bar{h}$, m	$\bar{\tau}$, sec	t, hours
600	5.30	11.2	18.3
700	5.65	11.7	20.8
800	6.05	12.2	24.0
900	6.35	12.6	25.8
1000	6.65	13.0	28.2
1100	6.90	13.3	30.2
1200	7.20	13.7	32.9
1300	7.50	14.1	35.8
1400	7.75	14.4	38.0
1500	7.90	14.7	40.3
1728	8.75	15.3	45.5
	$w=25$ m/sec		
5	0.45	2.7	0.3
10	0.65	3.3	0.5
20	1.00	4.1	0.9
40	1.55	5.0	1.6
60	2.00	5.7	2.4
80	2.20	6.2	3.1
100	2.45	6.6	3.7
120	2.70	7.0	4.3
140	2.85	7.3	4.9
160	3.00	7.6	5.5
180	3.20	7.9	6.1
200	3.40	8.2	6.8
220	3.55	8.4	7.4
240	3.65	8.6	7.9
260	3.85	8.9	8.7
280	3.90	9.0	9.0
300	4.05	9.2	9.6
400	4.60	10.0	12.2
500	5.05	10.7	14.9
600	5.60	11.4	17.4
700	5.95	11.9	20.2
800	6.35	12.4	22.8
900	6.65	12.8	25.0
1000	6.95	13.2	27.3
1100	7.25	13.6	29.1
1200	7.60	14.0	32.4
1300	7.85	14.3	34.5
1400	8.20	14.6	36.6
1500	8.40	15.0	39.6
1875	9.50	15.9	47.5
	$w=26$ m/sec		
5	0.45	2.7	0.3
10	0.70	3.4	0.5
20	1.05	4.2	0.9
40	1.55	5.1	1.6
60	2.00	5.8	2.3

TABLE 3.3.1 (Continued)

x, km	$\overline{h}$, m	$\overline{\tau}$, sec	t, hours	x, km	$\overline{h}$, m	$\overline{\tau}$, sec	t, hours
80	2.35	6.3	3.0	1200	8.20	14.4	30.4
100	2.55	6.7	3.6	1300	8.45	14.8	32.9
120	2.70	7.1	4.2	1400	8.80	15.1	35.7
140	3.00	7.4	4.8	1500	9.15	15.5	37.6
160	3.20	7.7	5.3	2000	10.30	16.8	47.5
180	3.35	8.0	5.4	2183	11.20	17.2	51.0
200	3.55	8.3	6.6				
220	3.75	8.6	7.2		$w = 28$ m/sec		
240	3.80	8.7	7.6				
260	4.00	9.0	8.3	5	0.45	2.8	0.2
280	4.05	9.1	8.6	10	0.70	3.5	0.5
300	4.25	9.4	9.4	20	1.10	4.3	0.8
400	4.80	10.2	12.0	40	1.65	5.3	1.5
500	5.30	10.9	14.5	60	2.15	6.0	2.2
600	5.80	11.5	17.0	80	2.55	6.5	2.8
700	6.25	12.1	19.7	100	2.75	6.9	3.4
800	6.65	12.6	22.1	120	3.00	7.3	4.0
900	6.95	13.0	24.2	140	3.25	7.7	4.6
1000	7.25	13.4	26.5	160	3.50	8.0	5.1
1100	7.55	13.8	28.8	180	3.65	8.3	5.7
1200	7.90	14.2	31.3	200	3.80	8.5	6.2
1300	8.15	14.5	33.3	220	4.00	8.8	6.8
1400	8.40	14.8	35.3	240	4.20	9.1	7.4
1500	8.75	15.2	38.2	260	4.30	9.2	7.8
2000	9.95	16.5	48.4	280	4.42	9.4	8.2
2028	10.03	16.6	49.3	300	4.55	9.6	8.8
				400	5.20	10.5	11.3
	$w = 27$ m/sec			500	5.75	11.2	13.5
				600	6.20	11.8	15.9
5	0.45	2.8	0.3	700	6.70	12.4	18.4
10	0.70	3.4	0.5	800	7.15	13.0	21.1
20	1.05	4.2	0.8	900	7.50	13.4	23.0
40	1.60	5.2	1.5	1000	7.85	13.8	25.1
60	2.05	5.9	2.3	1100	8.20	14.2	27.2
80	2.45	6.4	2.9	1200	8.55	14.6	29.5
100	2.65	6.8	3.5	1300	8.90	15.0	31.9
120	2.85	7.2	4.1	1400	9.15	15.3	33.9
140	3.15	7.6	4.7	1500	9.50	15.7	36.4
160	3.30	7.9	5.2	2000	10.80	17.0	45.8
180	3.50	8.1	5.8	2352	11.90	17.9	53.0
200	3.65	8.4	6.4				
220	3.85	8.7	7.0		$w = 29$ m/sec		
240	4.00	8.9	7.5				
260	4.10	9.1	8.0	5	0.50	2.9	0.2
280	4.25	9.3	8.4	10	0.75	3.5	0.5
300	4.40	9.5	9.1	20	1.15	4.4	0.8
400	5.00	10.4	11.8	40	1.70	5.4	1.4
500	5.55	11.1	14.3	60	2.20	6.1	2.2
600	6.00	11.7	16.6	80	2.60	6.6	2.7
700	6.45	12.3	19.2	100	2.95	7.1	3.3
800	6.85	12.8	21.6	120	3.15	7.4	3.8
900	7.20	13.2	23.6	140	3.40	7.8	4.5
1000	7.50	13.6	25.7	160	3.60	8.1	5.0
1100	7.85	14.0	28.0				

TABLE 3.3.1 (Continued)

x, km	$\bar{h}$, m	$\bar{\tau}$, sec	t, hours	x, km	$\bar{h}$, m	$\bar{\tau}$, sec	t, hours
180	3.80	8.4	5.6	40	1.75	5.4	1.4
200	4.00	8.7	6.1	60	2.25	6.2	2.1
220	4.20	9.0	6.7	80	2.65	6.7	2.7
240	4.35	9.2	7.2	100	3.05	7.2	3.3
260	4.50	9.4	7.6	120	3.25	7.5	3.7
280	4.60	9.5	8.0	140	3.55	7.9	4.4
300	4.80	9.8	8.6	160	3.70	8.2	4.8
400	5.45	10.7	11.0	180	3.95	8.5	5.4
500	6.00	11.4	13.1	200	4.15	8.8	6.0
600	6.55	12.1	15.6	220	4.35	9.1	6.6
700	7.00	12.6	18.0	240	4.50	9.3	7.0
800	7.50	13.2	20.6	260	4.65	9.5	7.4
900	7.85	13.6	22.5	280	4.80	9.7	7.9
1000	8.15	14.0	24.4	300	4.95	9.9	8.4
1100	8.55	14.4	26.5	400	5.65	10.8	10.8
1200	8.85	14.8	28.7	500	6.20	11.5	12.9
1300	9.25	15.2	31.1	600	6.80	12.2	15.4
1400	9.50	15.5	32.8	700	7.30	12.8	17.7
1500	9.90	15.9	35.4	800	7.70	13.3	19.7
2000	11.40	17.3	45.1	900	8.15	13.8	22.0
2523	12.80	18.5	55.0	1000	8.50	14.2	23.9
				1100	8.90	14.6	25.9
				1200	9.25	15.0	28.0
$w = 30$ m/sec				1300	9.60	15.4	30.2
				1400	9.90	15.7	31.9
5	0.50	2.9	0.2	1500	10.30	16.1	34.3
10	0.75	3.6	0.4	2000	11.90	17.5	43.8
20	1.15	4.4	0.8	2700	13.70	19.2	56.7

TABLE 3.3.2. A. Height and period of swell as a function of the distance over which it propagates

x_{sw}, km	$k_{sw}=\dfrac{h_{sw}}{h_0}$	τ_{sw}, sec	t_{sw}, hours	x_{sw}, km	$k_{sw}=\dfrac{h_{sw}}{h_0}$	τ_{sw}, sec	t_{sw}, hours
\multicolumn{4}{c}{$\tau_0 = 2.4$ sec}	$\tau_0 = 3.2$ sec			
40	0.23	2.7	6.2	40	0.35	3.5	4.6
80	0.15	3.0	11.8	80	0.31	3.8	9.0
120	0.11	3.1	16.8	120	0.23	4.0	13.0
160	0.08	3.3	21.8	160	0.19	4.2	16.8
				200	0.17	4.3	20.5
$\tau_0 = 2.8$ sec							
40	0.30	3.1	5.2	$\tau_0 = 3.6$ sec			
80	0.22	3.4	10.0				
120	0.17	3.6	14.8	40	0.41	3.9	4.2
160	0.14	3.8	19.0	80	0.34	4.2	8.0
200	0.11	3.9	21.0	120	0.29	4.5	11.6

TABLE 3.3.2 (Continued)

x_{sw}, km	$k_{sw} = \dfrac{h_{sw}}{h_0}$	τ_{sw}, sec	t_{sw}, hours
160	0.25	4.7	15.2
200	0.22	4.8	18.4
300	0.16	5.1	26.4
	$\tau_0 = 4.0$ sec		
40	0.45	4.4	3.8
80	0.39	4.6	7.2
120	0.34	4.9	10.6
160	0.30	5.1	13.8
200	0.27	5.2	16.8
300	0.21	5.7	24.2
	$\tau_0 = 4.4$ sec		
40	0.49	4.8	3.4
80	0.43	5.1	6.6
120	0.39	5.3	9.6
160	0.35	5.5	12.6
200	0.31	5.7	15.4
300	0.26	6.7	22.2
	$\tau_0 = 4.8$ sec		
40	0.58	5.2	3.2
80	0.48	5.5	6.2
120	0.43	5.7	9.0
160	0.39	6.0	11.6
200	0.36	6.1	14.2
300	0.30	6.5	20.6
	$\tau_0 = 5.2$ sec		
40	0.56	5.6	3.0
80	0.50	5.9	5.8
120	0.48	6.1	8.2
160	0.43	6.4	10.8
200	0.41	6.6	13.2
300	0.34	7.0	19.2
400	0.29	7.2	24.6
	$\tau_0 = 5.6$ sec		
40	0.59	5.9	2.6
80	0.55	6.3	5.2
120	0.50	6.5	7.8
160	0.47	6.8	10.0
200	0.44	7.0	12.4
300	0.38	7.4	17.8
400	0.33	7.8	23.0

x_{sw}, km	$k_{sw} = \dfrac{h_{sw}}{h_0}$	τ_{sw}, sec	t_{sw}, hours
	$\tau_0 = 6.0$ sec		
40	0.61	6.3	2.6
80	0.57	6.7	5.0
120	0.54	6.9	7.2
160	0.50	7.2	9.6
200	0.47	7.5	11.6
300	0.42	7.9	16.8
400	0.37	8.3	21.8
500	0.34	8.5	26.4
	$\tau_0 = 6.4$ sec		
40	0.63	6.7	2.4
80	0.59	7.1	4.6
120	0.56	7.4	6.8
160	0.53	7.7	9.0
200	0.50	7.9	11.0
300	0.45	8.3	17.8
400	0.41	8.6	20.4
500	0.37	8.9	25.0
	$\tau_0 = 6.8$ sec		
40	0.65	7.1	2.4
80	0.63	7.5	4.4
120	0.60	7.8	6.4
160	0.57	8.1	8.4
200	0.53	8.3	10.4
300	0.48	8.8	14.8
400	0.44	9.1	19.4
500	0.40	9.4	23.8
600	0.37	9.7	28.0
	$\tau_0 = 7.2$ sec		
40	0.67	7.5	2.2
80	0.64	7.9	4.2
120	0.61	8.2	6.2
160	0.59	8.4	8.0
200	0.56	8.7	9.8
300	0.51	9.2	14.2
400	0.47	9.6	18.4
500	0.43	10.0	22.6
600	0.40	10.2	26.4
	$\tau_0 = 7.6$ sec		
40	0.68	7.9	2.0
80	0.65	8.3	4.0
120	0.62	8.6	6.0
160	0.60	8.8	7.6
200	0.58	9.1	9.4

TABLE 3.3.2 (Continued)

x_{sw}, km	$k_{sw}=\dfrac{h_{sw}}{h_0}$	τ_{sw}, sec	t_{sw}, hours	x_{sw}, km	$k_{sw}=\dfrac{h_{sw}}{h_0}$	τ_{sw}, sec	t_{sw}, hours
300	0.53	9.7	13.6				
400	0.49	10.0	17.6				
500	0.46	10.4	21.4				
600	0.43	10.7	25.2				

$\tau_0 = 10.4$ sec

x_{sw}, km	k_{sw}	τ_{sw}, sec	t_{sw}, hours
40	0.76	10.7	1.4
80	0.74	11.1	3.0
120	0.73	11.4	4.2
160	0.71	11.9	5.8
200	0.70	12.0	7.0
300	0.67	12.7	10.2
400	0.64	13.1	12.8
500	0.61	13.5	16.2
600	0.59	14.0	19.2
700	0.56	14.4	22.0
800	0.54	14.5	24.6
900	0.52	14.8	27.4
1000	0.50	15.0	30.0

$\tau_0 = 8.0$ sec

x_{sw}, km	k_{sw}	τ_{sw}, sec	t_{sw}, hours
40	0.69	8.3	2.0
80	0.67	8.7	3.8
120	0.65	9.1	5.6
160	0.63	9.3	7.2
200	0.60	9.6	9.0
300	0.56	10.1	12.0
400	0.52	10.4	16.8
500	0.49	10.8	20.4
600	0.46	11.1	24.0
700	0.43	11.4	27.6
800	0.41	11.5	31.0

$\tau_0 = 11.2$ sec

x_{sw}, km	k_{sw}	τ_{sw}, sec	t_{sw}, hours
40	0.78	11.5	1.4
80	0.76	12.0	2.8
120	0.75	12.2	4.0
160	0.74	12.7	5.4
200	0.72	12.8	6.6
300	0.69	13.5	9.4
400	0.67	14.0	12.4
500	0.64	14.6	15.2
600	0.62	14.9	17.8
700	0.60	15.1	20.4
800	0.58	15.6	23.2
900	0.56	15.7	25.6
1000	0.55	15.9	28.2

$\tau_0 = 8.8$ sec

x_{sw}, km	k_{sw}	τ_{sw}, sec	t_{sw}, hours
40	0.72	9.1	1.8
80	0.70	9.5	3.4
120	0.68	9.8	5.0
160	0.66	10.1	6.6
200	0.64	10.3	8.2
300	0.60	11.0	12.0
400	0.56	11.4	15.4
500	0.53	11.8	19.0
600	0.50	12.1	22.2
700	0.48	12.2	25.2
800	0.45	12.6	28.6
900	0.44	12.8	31.6

$\tau_0 = 12.0$ sec

x_{sw}, km	k_{sw}	τ_{sw}, sec	t_{sw}, hours
40	0.80	12.2	1.2
80	0.77	12.7	2.6
120	0.76	13.1	3.8
160	0.75	13.4	5.0
200	0.74	13.7	6.2
300	0.71	14.4	9.0
400	0.69	14.9	11.6
500	0.67	15.4	14.2
600	0.65	15.7	16.8
700	0.63	16.1	19.4
800	0.61	16.6	20.0
900	0.59	16.8	24.0
1000	0.58	17.0	26.4
1200	0.55	17.4	31.0

$\tau_0 = 9.6$ sec

x_{sw}, km	k_{sw}	τ_{sw}, sec	t_{sw}, hours
40	0.75	9.9	1.6
80	0.73	10.3	3.2
120	0.70	10.6	4.6
160	0.69	10.8	6.2
200	0.67	11.2	7.6
300	0.63	11.8	11.0
400	0.60	12.3	14.2
500	0.58	12.7	17.4
600	0.54	13.1	21.0
700	0.52	13.3	23.6
800	0.50	13.5	26.0
900	0.48	13.8	29.4

TABLE 3.3.2 (Continued)

x_{sw}, km	$k_{sw}=\dfrac{h_{sw}}{h_0}$	τ_{sw}, sec	t_{sw}, hours	x_{sw}, km	$k_{sw}=\dfrac{h_{sw}}{h_0}$	τ_{sw}, sec	t_{sw}, hours
	$\tau_0 = 12.8$ sec			500	0.71	17.3	12.8
				600	0.69	17.7	15.0
40	0.81	13.0	1.2	700	0.68	17.8	17.2
80	0.80	13.5	2.4	800	0.67	18.2	19.4
120	0.78	13.9	3.6	900	0.64	18.7	21.8
160	0.77	14.2	4.6	1000	0.63	18.9	23.6
200	0.76	14.6	5.8	1200	0.61	19.3	29.8
300	0.74	15.2	8.4				
400	0.71	15.8	11.0				
500	0.69	16.2	13.4		$\tau_0 = 14.4$ sec		
600	0.67	16.6	15.8				
700	0.65	17.0	18.2	40	0.83	14.7	1.0
800	0.64	17.3	20.4	80	0.82	15.0	2.2
900	0.62	17.7	22.8	120	0.81	15.5	9.2
1000	0.61	17.8	25.0	160	0.80	15.7	4.0
1200	0.58	18.4	29.4	200	0.79	16.2	5.2
				300	0.77	16.8	7.8
				400	0.74	17.6	9.8
	$\tau_0 = 13.6$ sec			500	0.73	18.0	12.2
				600	0.71	18.4	14.2
40	0.82	13.9	1.0	700	0.70	18.7	16.4
80	0.81	14.2	2.2	800	0.69	19.2	18.4
120	0.80	14.7	3.4	900	0.67	19.4	20.4
160	0.79	14.9	4.2	1000	0.66	20.0	22.6
200	0.78	15.3	5.4	1200	0.63	20.4	26.4
300	0.75	15.9	8.0	1400	0.61	20.8	30.4
400	0.73	16.6	10.4				

B. Height and period of swell as a function of the duration of its propagation

t, hours	$k_{sw}=\dfrac{h_{sw}}{h_0}$	τ_{sw}, sec	x_{sw}, km	t, hours	$k_{sw}=\dfrac{h_{sw}}{h_0}$	τ_{sw}, sec	x_{sw}, km
	$\tau_0 = 2.4$ sec				$\tau_0 = 2.8$ sec		
2	0.30	2.5	12	2	0.36	2.9	14
4	0.26	2.6	26	4	0.32	3.1	30
6	0.23	2.7	39	6	0.28	3.2	45
8	0.20	2.8	53	8	0.25	3.3	62
10	0.17	2.9	68	10	0.22	3.4	79
12	0.15	3.0	82	12	0.20	3.5	96
14	0.13	3.1	98	14	0.18	3.6	114
16	0.12	3.1	114	16	0.16	3.6	133
18	0.10	3.2	130	18	0.14	3.7	152
20	0.10	3.2	146	20	0.13	3.8	170
22	0.08	3.3	162	22	0.12	3.9	189
24	0.07	3.3	179	24	0.11	3.9	209
26	0.07	3.4	196	26	0.10	3.9	228
28	0.06	3.4	213	28	0.09	4.0	248
30	0.05	3.5	231	30	0.08	4.0	269

t, hours	$k_{sw}=\dfrac{h_{sw}}{h_0}$	τ_{sw}, sec	x_{sw}, km	t, hours	$k_{sw}=\dfrac{h_{sw}}{h_0}$	τ_{sw}, sec	x_{sw}, km
	$\tau_0 = 3.2$ sec				$\tau_0 = 4.4$ sec		
2	0.41	3.3	16	2	0.52	4.6	22
4	0.36	3.5	35	4	0.48	4.8	48
6	0.33	3.6	52	6	0.44	5.0	71
8	0.30	3.7	71	8	0.41	5.2	98
10	0.27	3.9	90	10	0.38	5.4	124
12	0.24	4.0	110	12	0.36	5.5	151
14	0.21	4.1	130	14	0.33	5.6	179
16	0.20	4.2	152	16	0.31	5.7	208
18	0.18	4.3	173	18	0.29	5.8	238
20	0.17	4.3	195	20	0.28	5.9	268
22	0.15	4.4	216	22	0.26	6.0	297
24	0.14	4.4	239	24	0.24	6.1	328
26	0.13	4.5	261	26	0.23	6.2	359
28	0.12	4.5	284	28	0.21	6.2	390
30	0.11	4.6	308	30	0.20	6.3	423
	$\tau_0 = 3.6$ sec				$\tau_0 = 4.8$ sec		
2	0.45	3.7	18	2	0.55	5.0	24
4	0.41	3.9	39	4	0.51	5.2	52
6	0.37	4.1	58	6	0.48	5.5	77
8	0.34	4.2	80	8	0.44	5.6	107
10	0.31	4.4	102	10	0.41	5.8	135
12	0.28	4.5	124	12	0.39	6.0	165
14	0.26	4.6	147	14	0.36	6.1	196
16	0.24	4.7	171	16	0.34	6.2	228
18	0.22	4.8	195	18	0.32	6.4	260
20	0.21	4.8	219	20	0.30	6.4	292
22	0.19	5.0	243	22	0.29	6.6	324
24	0.18	5.0	268	24	0.27	6.7	358
26	0.16	5.1	294	26	0.26	6.8	392
28	0.15	5.1	319	28	0.24	6.8	425
30	0.14	5.2	346	30	0.23	6.9	461
	$\tau_0 = 4.0$ sec				$\tau_0 = 5.2$ sec		
2	0.49	4.2	20	2	0.57	5.4	27
4	0.45	4.4	43	4	0.54	5.7	56
6	0.41	4.6	64	6	0.50	5.3	84
8	0.38	4.7	89	8	0.48	6.1	116
10	0.35	4.9	113	10	0.44	6.3	147
12	0.32	5.0	137	12	0.42	6.5	178
14	0.30	5.1	163	14	0.39	6.7	212
16	0.28	5.2	190	16	0.37	6.8	246
18	0.26	5.3	216	18	0.35	6.9	281
20	0.24	5.4	243	20	0.34	7.0	316
22	0.22	5.4	270	22	0.31	7.1	351
24	0.21	5.6	298	24	0.30	7.2	388
26	0.20	5.6	326	26	0.28	7.3	424
28	0.18	5.7	355	28	0.27	7.4	461
30	0.17	5.8	384	30	0.26	7.5	500

t, hours	$k_{sw} = \dfrac{h_{sw}}{h_0}$	τ_{sw}, sec	x_{sw}, km	t, hours	$k_{sw} = \dfrac{h_{sw}}{h_0}$	τ_{sw}, sec	x_{sw}, km
	$\tau_0 = 5.6$ sec				$\tau_0 = 6.8$ sec		
2	0.60	5.8	29	2	0.65	7.1	35
4	0.56	6.1	60	4	0.62	7.4	73
6	0.53	6.4	90	6	0.59	7.7	110
8	0.50	6.6	124	8	0.56	8.0	151
10	0.47	6.8	158	10	0.54	8.3	192
12	0.44	7.0	192	12	0.52	8.5	233
14	0.42	7.2	228	14	0.49	8.7	277
16	0.40	7.3	266	16	0.47	8.8	322
18	0.38	7.4	303	18	0.45	9.0	368
20	0.36	7.5	340	20	0.44	9.1	413
22	0.34	7.7	378	22	0.41	9.4	459
24	0.33	7.8	418	24	0.40	9.5	507
26	0.31	7.9	457	26	0.38	9.6	555
28	0.30	8.0	496	28	0.37	9.7	603
30	0.28	8.1	538	30	0.35	9.8	654
	$\tau_0 = 6.0$ sec				$\tau_0 = 7.2$ sec		
2	0.62	6.2	31	2	0.67	7.5	37
4	0.59	6.5	65	4	0.64	7.8	78
6	0.55	6.8	97	6	0.61	8.2	116
8	0.52	7.0	138	8	0.59	8.4	160
10	0.50	7.3	169	10	0.56	8.8	203
12	0.47	7.5	206	12	0.53	9.0	247
14	0.45	7.7	244	14	0.51	9.2	293
16	0.43	7.8	284	16	0.49	9.4	341
18	0.41	8.0	325	18	0.47	9.6	390
20	0.39	8.0	365	20	0.45	9.7	438
22	0.37	8.3	405	22	0.44	10.0	486
24	0.35	8.3	448	24	0.42	10.0	537
26	0.34	8.5	490	26	0.40	10.1	588
28	0.32	8.5	532	28	0.39	10.2	639
30	0.30	8.6	576	30	0.38	10.4	692
	$\tau_0 = 6.4$ sec				$\tau_0 = 7.6$ sec		
2	0.64	6.7	33	2	0.68	7.9	39
4	0.60	7.0	69	4	0.65	8.3	82
6	0.57	7.3	103	6	0.62	8.7	122
8	0.55	7.5	142	8	0.60	8.9	169
10	0.52	7.8	180	10	0.57	9.3	214
12	0.49	8.0	220	12	0.55	9.5	261
14	0.47	8.2	260	14	0.53	9.7	309
16	0.45	8.3	304	16	0.51	9.9	360
18	0.43	8.5	346	18	0.49	10.1	411
20	0.41	8.6	389	20	0.48	10.2	462
22	0.39	8.8	432	22	0.45	10.5	513
24	0.38	8.9	478	24	0.44	10.6	567
26	0.36	9.0	522	26	0.42	10.7	620
28	0.35	9.1	568	28	0.41	10.8	674
30	0.33	9.2	615	30	0.39	10.9	730

TABLE 3.3.2 (Continued)

t, hours	$k_{sw}=\dfrac{h_{sw}}{h_0}$	τ_{sw}, sec	x_{sw}, km	t, hours	$k_{sw}=\dfrac{h_{sw}}{h_0}$	τ_{sw}, sec	x_{sw}, km
	$\tau_0=8.0$ sec				$\tau_0=10.4$ sec		
2	0.69	8.3	41	2	0.76	10.8	53
4	0.67	8.7	86	4	0.74	11.3	112
6	0.64	9.1	129	6	0.71	11.9	167
8	0.62	9.4	178	8	0.69	12.2	230
10	0.59	9.8	226	10	0.67	12.7	293
12	0.57	10.0	274	12	0.65	13.0	357
14	0.55	10.2	326	14	0.63	13.3	423
16	0.52	10.4	379	16	0.61	13.5	493
18	0.51	10.6	433	18	0.59	13.8	563
20	0.49	10.7	486	20	0.58	14.0	632
22	0.47	11.0	540	22	0.56	14.4	702
24	0.46	11.1	597	24	0.55	14.5	776
26	0.44	11.3	653	26	0.53	14.7	849
28	0.43	11.4	710	28	0.52	14.8	922
30	0.41	11.5	769	30	0.50	15.0	999
	$\tau_0=8.8$ sec				$\tau_0=11.2$ sec		
2	0.72	9.2	45	2	0.77	11.6	57
4	0.69	9.6	95	4	0.75	12.2	121
6	0.67	10.0	142	6	0.73	12.8	180
8	0.64	10.3	195	8	0.71	13.1	248
10	0.62	10.7	248	10	0.69	13.7	316
12	0.60	11.0	302	12	0.67	14.0	384
14	0.57	11.3	358	14	0.65	14.3	456
16	0.56	11.4	417	16	0.63	14.6	531
18	0.54	11.7	476	18	0.62	14.9	606
20	0.52	11.8	535	20	0.60	15.0	681
22	0.50	12.1	594	22	0.59	15.5	757
24	0.49	12.2	657	24	0.57	15.6	836
26	0.48	12.4	718	26	0.56	15.8	914
28	0.46	12.5	781	28	0.55	15.9	994
30	0.45	12.7	846	30	0.53	16.1	1076
	$\tau_0=9.6$ sec				$\tau_0=12.0$ sec		
2	0.74	10.0	49	2	0.79	12.5	61
4	0.71	10.5	104	4	0.77	13.1	130
6	0.69	10.9	155	6	0.74	13.7	193
8	0.67	11.2	213	8	0.72	14.0	266
10	0.64	11.7	271	10	0.70	14.6	338
12	0.62	12.0	329	12	0.69	15.0	412
14	0.60	12.3	391	14	0.67	15.4	489
16	0.59	12.5	455	16	0.65	15.6	569
18	0.57	12.8	519	18	0.63	16.0	650
20	0.55	12.9	584	20	0.62	16.1	729
22	0.54	13.2	646	22	0.60	16.6	810
24	0.52	13.3	716	24	0.59	16.7	895
26	0.50	13.5	783	26	0.58	16.9	980
28	0.49	13.6	852	28	0.57	17.0	1065
30	0.48	13.8	923	30	0.55	17.3	1153

TABLE 3.3.2 (Continued)

t, hours	$k_{sw}=\dfrac{h_{sw}}{h_0}$	τ_{sw}, sec	x_{sw}, km	t, hours	$k_{sw}=\dfrac{h_{sw}}{h_0}$	τ_{sw}, sec	x_{sw}, km
	$\tau_0 = 12.8$ sec			16	0.69	17.7	645
				18	0.67	18.1	736
2	0.80	13.3	65	20	0.66	18.2	826
4	0.78	14.0	138	22	0.64	18.8	918
6	0.76	14.6	206	24	0.63	18.9	1016
8	0.74	15.0	284	26	0.62	19.2	1110
10	0.72	15.6	361	28	0.61	19.3	1207
12	0.70	16.0	439	30	0.60	19.6	1307
14	0.69	16.4	521				
16	0.67	16.6	607		$\tau_0 = 14.4$ sec		
18	0.65	17.0	692				
20	0.64	17.2	778	2	0.82	15.0	73
22	0.63	17.7	864	4	0.80	15.7	156
24	0.62	17.8	955	6	0.78	16.4	232
26	0.60	18.0	1046	8	0.77	16.8	320
28	0.59	18.2	1136	10	0.74	17.6	406
30	0.57	18.4	1230	12	0.73	18.0	494
				14	0.71	18.4	586
	$\tau_0 = 13.6$ sec			16	0.70	18.7	683
				18	0.69	19.2	779
2	0.81	14.1	69	20	0.67	19.3	876
4	0.79	14.8	147	22	0.66	19.9	972
6	0.77	15.5	219	24	0.65	20.0	1076
8	0.75	15.9	302	26	0.63	20.3	1175
10	0.74	16.6	384	28	0.62	20.4	1278
12	0.71	17.0	467	30	0.61	20.8	1384
14	0.70	17.4	554				

1. Wind speed and direction unchanged

The variation in the mean wave period ($\bar{\tau}$) as a function of wind speed (w), its duration (t_w) and fetch (x) may have the form shown in Figure 3.3.1. The average wave period will be $\bar{\tau}_x$ at the end of fetch x if the duration of the wind (t_w) is greater than or equal to the duration of wave development (t) over a fetch x for given wind speed (w). Consequently,

$$t_w \geqslant t. \qquad (3.3.1)$$

If

$$t_w < t, \qquad (3.3.2)$$

for example, if

$$t_w = t_i, \qquad (3.3.3)$$

then (Figure 3.3.1) the wave period at the end of x will be smaller and equal to $\bar{\tau}_i$. This will also be the period over the fetch segment $x - x_i$ (Figure 3.3.1). It is required that the wind continue acting during the additional time interval

$$t - t_i, \qquad (3.3.4)$$

in order that the waves be fully developed at the end of the fetch. If
condition (3.3.1) is satisfied, then

$$\bar{\tau}_x = f_1(x,\ w). \qquad (3.3.5)$$

When condition (3.3.2) is satisfied

$$\bar{\tau}_t = f_2(t,\ w). \qquad (3.3.6)$$

Subscript x shows that the wave period depends, in addition to w, also
on x, while subscript t of $\bar{\tau}$ denotes the dependence of τ, in addition to w,
also on t. Consequently (Chapter 2, Section 7), the wind-wave elements should be computed either from (3.3.5) or (3.3.6), depending on whether conditions (3.3.1) or (3.3.2) apply. These considerations apply to calculation of any wind-wave element, not only to its period. Calculating the latter from (3.3.5) or (3.3.6), one selects the smaller of the two results. These calculations can be avoided by using the "bounding" curve (Chapter 2, Section 7), by means of which one determines from given values of x and t whether (3.3.5) or (3.3.6) should be used (Chapter 1, Section 8).

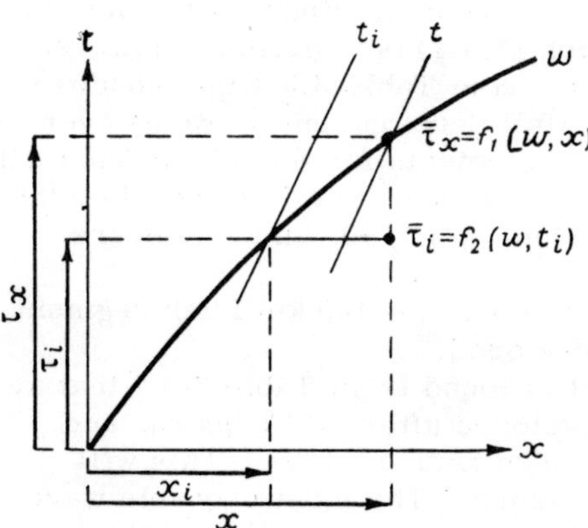

FIGURE 3.3.1.

Example 3.3.1. It is required to determine the wave dimensions
and the wave strength for $w = 16$ m/sec, $x = 200$ km and $t_w = 10$ hours.

Solution. We use, for example, relationship (3.3.5). Hence we find
from Table 3.3.1 for $w = 16$ m/sec and $x = 200$ km, that $\bar{h} = 2.05$ m, $\tau = 6.8$ sec
and $t = 9.2$ hours. In this case 9.2 hours < 10 hours, i.e., condition (3.3.1)
is satisfied. Using this value of $\bar{\tau}$ one finds from Table 1.6.2 that $\bar{\lambda} = 72$ m
and $\bar{\delta} = 1/36$. In determining the wave strength one must first find the
height of a wave with 3% cumulative probability in the given wave system.
It is found from Table 3.3.2 that

$$h_{3\%} = 2.1\bar{h} = 2.1 \cdot 2.05 = 4.3 \text{ m.}$$

It follows from the relevant table that this wave height corresponds to a wave
strength of force 6, which should be observed at the given point of the fetch
~9 hours after the start of the action of the 16 m/sec wind over the entire
200 km fetch.

Example 3.3.2. It is required to determine the wave dimensions and
wave strength for the same values of x and w as in Example 3.3.1, but for
a wind duration of only 4 hours.

Solution. Now (3.3.5) cannot be used (solution of Example 3.3.1),
since condition (3.3.1) is not satisfied. Hence relationship (3.3.6) is used
instead. For $w = 16$ m/sec and $t = 4$ hours one finds from Table 3.3.1

$\overline{h} \approx 1.35\,\text{m}$ $\overline{\tau} \approx 5.2$ sec. Waves of these dimensions will be observed over a fetch segment from 80 to 200 km. Proceeding now in the same manner as in Example 3.3.1, we find that $h_{3\%} = 2.8\,\text{m}$, $\lambda = 41\,\text{m}$, and $\overline{\delta} = 1/15$. The wave strength is force 5. Approximately five more hours of wind action is needed in order that the waves attain at the end of the fetch the dimensions calculated in Example 3.3.1.

Example 3.3.3. Wind at a speed of 16 m/sec has been acting for 2 hours over a distance of 300 km. It is required to determine: a) the wave dimensions after 4 hours of wind action over a distance of from 40 to 100 km from the start of the fetch; b) the time of full development of waves at the end of the fetch and the maximum wave height and its corresponding period and length.

Solution. a) for $x = 40$ km and $w = 16$ m/sec one finds from Table 3.3.1 that $t = 2.3$ hours. Consequently, condition (3.3.1) is satisfied. However, for $x = 100$ km and $w = 16$ m/sec it appears from Table 3.3.1 that condition (3.3.2) should be used, since the time for full development of waves over this fetch is 5.3 hours. Consequently the wave elements for the 40 km fetch will be: $\overline{h} = 1.0\,\text{m}$, $\overline{\tau} = 4.2$ sec, $\lambda = 27\,\text{m}$, $\overline{\delta} = 1/27$, and $t = 2.3$ hours. For the 100 km fetch $\overline{h} = 1.35$ m, $\overline{\tau} = 5.2$ sec, $\overline{\lambda} = 41\,\text{m}$, $\overline{\delta} = 1/30$, and $t \approx 4$ hours, i.e., in 4 hours + 2 hours = 6 hours.

Here the latter wave dimensions apply to the $80 - 100$ km fetch segment, over which the waves are not yet fully developed.

b) For $w = 16$ m/sec and $x = 300$ km it is found from Table 3.3.1 that at this point the waves will become fully developed after ~ 13.5 hours, and will then have $\overline{h} = 2.5\,\text{m}$, $\overline{\tau} = 7.7$ sec, $\overline{\lambda} = 92\,\text{m}$ and $\overline{\delta} = 1/37$. This will happen after 2 hours + 13.5 hours ≈ 15.5 hours. The most probable wave length for waves of any height is $0.8\,\overline{\lambda}$ (Chapter 2, Section 4), and the most probable period is the average period (Chapter 2, Section 4). It is permissible to take the height of a wave with $\sim 1\%$ cumulative probability as the maximum wave height (Table 2.4.2). Hence it is found that $h_{1\%} = 2.5 \times$ $\times 2.4 = 6.0\,\text{m}$, $\lambda = 92 \times 0.8 = 74\,\text{m}$, and $\overline{\tau} = 7.7$ sec.

2. Variable wind speed with invariable wind direction

For this situation a distinction is made between three cases: a) wind speed variable in time only; b) wind speed variable in space only; c) wind speed variable in both time and space.

2a. Wind speed variable in time only

Wind at speed w_1 acts for t_{w_1} hours over fetch x, which remains unchanged (Figure 3.2.2). Then the wind speed increases over the entire fetch. It starts acting at speed w_2 for a time of t_{w_2} hours. Here

$$w_2 > w_1. \tag{3.3.7}$$

In addition, condition (3.3.1) is satisfied:

$$t_{w_1} > t_1,$$
$$t_{w_2} > t_2,$$

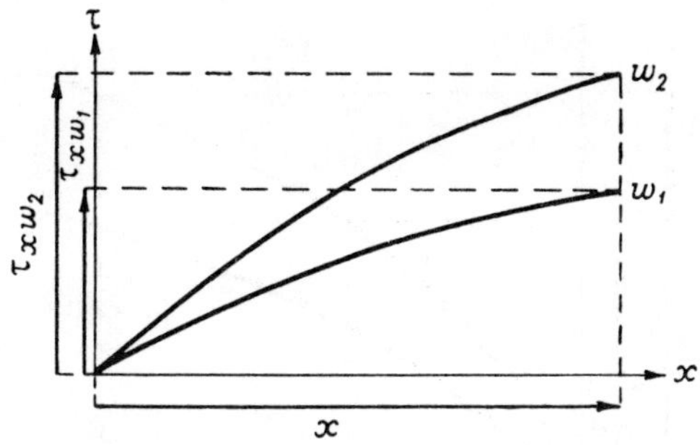

FIGURE 3.3.2.

where t_1 and t_2 are the wave development times over the given fetch due to winds at speeds w_1 and w_2. In order to satisfy condition (3.3.1), the wave dimensions at the end of the fetch for wind speed w_1 are found from (3.3.5):

$$\tau_{xw_1} = f_1(x,\ w_1),\qquad(3.3.8)$$

and for wind speed w_2:

$$\tau_{xw_2} = f_2(x,\ w_2).\qquad(3.3.9)$$

The wind, upon attaining its higher speed w_2, will be acting on waves generated over fetch x by a wind possessing speed w_1. These waves already contain some oscillatory energy transmitted to them by wind speed w_1. Hence the further wave growth due to wind with speed w_2 will be somewhat accelerated, and the more so, the larger w_2 as compared with w_1. On the other hand, wind at speed w_2 cannot generate waves at the end of the fetch larger than those which it is capable of generating over any fetch x, if condition (3.3.1) is satisfied for both the above wind speeds. The acceleration in the wave development is taken into account by a correction, denoted subsequently by t_{bi} (Figure 3.3.3), for the wave development duration calculated from (3.3.36), by solving this equation for t and substituting into it $\bar{\tau}_1$ and w_2:

$$t_{bi} = f_3(\bar{\tau}_1,\ w_2).\qquad(3.3.10)$$

Consequently, we determine the time which would have been required for a wind with speed w_2 to develop waves at a given fetch x, with a period corresponding to that of already existing waves (Figure 3.3.2). This time interval (t_{bi}) is subtracted from the time (t_2) needed for a wind with speed w_2 to develop waves with period $\bar{\tau}_2$. Thus, the required true time needed for wave development by a wind with increasing velocity (w_i) is calculated from the expression

$$t = t_2 - t_{bi}.\qquad(3.3.11)$$

Example 3.3.4. A wind with speed 16 m/sec (w_1) acts over a fetch (x) for 10 hours (t_{w_1}), after which the wind speed increases to 20 m/sec (w_2) and maintains this speed for 10 more hours (t_{w_2}). It is required to determine the wave dimensions at the terminal point of the fetch and the time of their appearance.

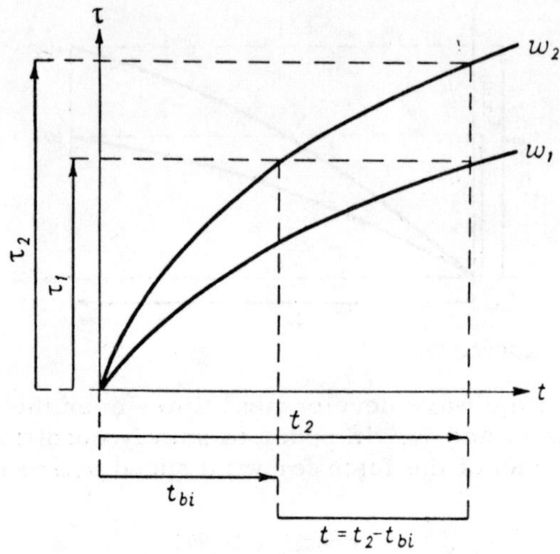

FIGURE 3.3.3.

Solution. In example 3.3.1 we found for $x = 200$ km, $w = 16$ m/sec and $t_w = 10$ hours that $\tau_x = 6.8$ sec, $\bar{h} = 2.05$ m and $t = 9.2$ hours. These waves appear at the end of the 200 km fetch as a result of the action of a wind with speed 16 m/sec. Then, for the same 200 km fetch, but for a wind speed of 20 m/sec, we find from Table 3.3.1 that now there should appear waves with $\bar{h} = 2.65$ m and $\bar{\tau} = 7.5$ sec after 8.1 hours (t_2). It must be remembered that these latter waves will appear earlier, since when wind with speed 20 m/sec started blowing the entire fetch was already covered by waves produced by the 16 m/sec wind. Hence according to (3.3.10) one finds from Table 3.3.1 for $w = 20$ m/sec and $\tau_x = 6.8$ sec that

$$-t_{bi} \cong -5.8 \text{ hours.}$$

Consequently, according to (3.3.11), the wave development time due to the 20 m/sec wind is

$$t_2 = 8.1 \text{ hours} - 5.8 \text{ hours} = 2.3 \text{ hours.}$$

Hence the waves calculated above for the 20 m/sec wind will appear at the terminal point of the 200 km fetch after

$$t = t_{w_1} + t_2 = 10 \text{ hours} + 2.3 \text{ hours} = 12.3 \text{ hours.}$$

The above computation can be represented graphically (Figure 3.3.4). The time in hours is plotted on the abscissa and the average wave height and period are plotted on the ordinate. Then the variation in wave height for a 16 m/sec wind are found from Table 3.3.1 for any time interval. It is also given for the time instant corresponding to the appearance of waves at the terminal point of the fetch, i.e., 9.2 hours. The variations in the wave period are found similarly. Then the values of $\bar{h}$ and $\bar{\tau}$ thus found are joined by appropriate curves. Then the wave heights corresponding to a period of 6.8 sec for a 20 m/sec wind are plotted on the ordinate axis for $t = 10$ hours (according to Example 3.3.4). This point is found to lie above that defining the wave height at the same instant by a 16 m/sec wind. After 10 hours the wave height with a 16 m/sec wind was 2.05 m, while that with the 20 m/sec wind at the same time instant should be 2.25 m, corresponding to the same period of 6.8 sec as with the 16 m/sec wind (Table 3.3.1). The wave heights determined by calculations with variable wind speed just increase abruptly, the degree of which is smaller, the lesser the difference between w_1 and w_2. If the starting element in the transition from wind speed w_1 to w_2 would have been the wave height instead of the period, as in the present example, then a similar abrupt change of the variation in wave period would have occurred. This discontinuity in the values of wave elements calculated for variable wind speed is a result of the form of relationship $h(\tau)$ (or of relationship $\beta(\delta)$ replacing it), which is the second principal equation for solving the wave energy balance equation (Chapter 2, Section 7). The same expression is also contained in phenomenological relationships between wave elements and wave-generating factors (Chapter 2, Section 5). When these relationships do not contain the wind speed, then there will be no discontinuity in the value of h (or τ) when it is calculated for one wind speed and then for the other (higher or lower). Whenever the wind speed is contained in the expression for $h(\tau)$ the discontinuity cannot be avoided. Hence, if we take as the starting element in transition from one wind speed to another the wave height rather than the period, then the results which are obtained are not unique. In addition, the presence of a discontinuity, for example, in the wave height, as is shown in Figure 3.3.4, makes it necessary to smooth it out by an approximate method (dashed curve in the variation of wave heights in Figure 3.3.4). As it is, since the calculations are approximate to a certain extent, this smoothing does not have a substantial effect on the final computational results. There is no consensus of opinion on the question as to which of the wave elements (height or period) should be used as the principal variable in calculations. However, all the considerations presented in this section on computational techniques for wind elements remain the same when the wave height rather than period is taken as the principal parameter. Below, in solving Example 3.3.9 we show a different technique for Example 3.3.4, which yields a more detailed description of the variation in wave elements.

Returning to the solution of Example 3.3.4 we note that condition (3.3.1) may not be satisfied for wind speed w_1. Then the wave elements are calculated as a function of w according to condition (3.3.2), i.e., in the same manner as was done in Example 3.3.2. If condition (3.3.1) is not satisfied with respect to w_2, then one must satisfy the condition

$$t_2 - t_{bi} = t_{w_2};\qquad (3.3.12)$$

t_{bi} is constant for the given w_1 and w_2, while t_{w_2} is given and hence (3.3.12) can be solved for t_2, which yields

$$t_2 = t_{w_2} + t_{bi}.\qquad (3.3.13)$$

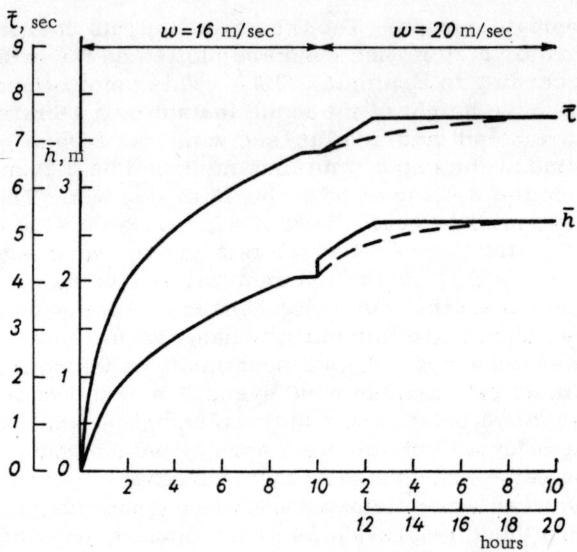

FIGURE 3.3.4. Pertaining to Example 3.3.4.

Now the wave dimensions at the end of the fetch, subject to satisfaction of condition (3.3.12), are determined from formula (3.3.6) upon substitution into it of t_2 obtained from (3.3.13):

$$\tau_t = f_4(w_2, \, t_2).$$

Example 3.3.5. The values $x = 200$ km, $t_{w_1} = 10$ hours, $w_1 = 16$ m/sec and $w_2 = 20$ m/sec are the same as in Example 3.3.4, but $t_{w_2} = 1$ hour. The wave dimensions are determined at the end of the fetch as a result of a wind with $w_2 = 20$ m/sec.

Solution. It was found earlier (Example 3.3.4) that $t_{bi} = 5.8$ hours. Hence, according to (3.3.13)

$$t_2 = 1 + 5.8 = 6.8 \text{ hours.}$$

For this value of t_2 and $w_2 = 20$ m/sec it is found from Table 3.3.1 that

$$\bar\tau = 7.1 \text{ sec,}$$

$$\bar h \cong 2.4 \text{ m,}$$

and the average wave length is ≈ 78 m. As compared with the results obtained in Example 3.3.4 the wave elements, naturally, were found to be somewhat smaller.

Sometimes the wind changes in such a manner that condition (3.3.7) is not satisfied, i.e., $w_2 < w_1$. In this case, for known relationships between x, t, w_1 and w_2, the elements of waves produced by a wind with speed w_2 may be found to be smaller than those due to a wind with speed w_1. In these cases the latter are treated as swell.

Example 3.3.6. A 16 m/sec wind (w_1) acts over a fetch of 200 km (x) for 10 hours (t_{w_1}), then it decreases to 10 m/sec (w_2) and continues for 10 more hours (t_{w_2}). It is required to determine the wave dimensions at the terminal point of the fetch due to the 10 m/sec wind.

Solution. According to calculations in Examples 3.3.1 and 3.3.4 it was found that at $w_1 = 16$ m/sec and $x = 200$ km, $\bar{\tau} = 6.8$ sec and $\bar{h} = 2.05$ m. We find from Table 3.3.1 for $w_2 = 10$ m/sec and $\bar{\tau} = 6.8$ sec that a 10 m/sec wind cannot produce waves with this period. Hence the latter should be treated as wind swell (Chapter 2, Section 10), which may propagate together with the wind waves produced by the 10 m/sec wind. This wind can produce over the 200 km fetch (Table 3.3.1) waves with height 1.25 m and period 5.7 sec. Waves of this size obviously already existed due to the 16 m/sec wind at the terminal point of the 200 km fetch. Now these waves will remain, being maintained by the 10 m/sec wind. The maximum wave period which can be produced by the 10 m/sec wind (Table 3.3.1) is 6.4 sec. Consequently, the waves produced by the 16 m/sec wind with periods of 6.4 sec and more will now become swell. Waves with periods from 5.7 to 6.4 sec will decay upon reaching the terminal point of the 200 km fetch, since the 10 m/sec wind cannot maintain them over this fetch, and these will be replaced by waves with period 5.7 sec and height 1.25 m. This variation in wave dimensions is calculated approximately, assuming that there exist at the terminal point wind waves with $\bar{h} = 1.25$ m and $\bar{\tau} = 5.7$ sec, and also swell which appears at this point of the fetch due to its propagation from those points of the fetch where it existed as wind waves, when the wind velocity decreases. The latter is calculated by the corresponding relationships for taking into account variations in the dimensions of swell (Chapter 2, Section 9). For example, one may use Table 3.3.2. Obviously, the use of relationships described in Chapter 2, Section 9 in the given case, for propagation of "wind swell," is approximate. One should first estimate at which points of the given fetch there exist waves with a period of 6.4 sec and more due to a 16 m/sec wind. For this it is found from Table 3.3.1 that the wind-wave periods at distances of 160, 180, and 200 km are, respectively, 6.4, 6.6 and 6.8 sec (Table 3.3.3). The distance which these waves should travel as swell (x_{sw}) until they reach the terminal point of the fetch is calculated from

$$x_{sw} = x - x_0,$$

where x is the given fetch; x_0 is the distance at which wind waves (which are not swell) existed due to reduction in wind speed. It is found from Table 3.3.2A that for the initial wind-wave period (τ_0) and for x_{sw}, waves which became swell at the 160 km point of the fetch will have a period of 6.7 sec at the terminal point of the fetch, and that their height will be given by the expression

$$h_{sw} = h_0 k_{sw} = 1.85 \cdot 0.63 = 1.16 \text{ m.}$$

Proceeding similarly with $h_0 = 1.95$ and $\bar{\tau}_0 = 6.6$, we find the respective values of elements of swell and the time of its appearance at the terminal point of the fetch (Table 3.3.2A). The simultaneous existence of swell and wind waves results in the formation of a mixed wave pattern (Chapter 2,

Section 1). The wave height for the mixed wave motion can be obtained
from

$$h_{mix} = \left[(h_w)^2 + (h_{sw})^2 \right]^{0.5}, \qquad (3.3.14)$$

which follows from the relationship between the wind wave and swell
energies, which are separately proportional to the square of the height of
each of them.

TABLE 3.3.3

x_0, km	$\bar{h}_0$, m	$\bar{\tau}_0$, sec	$x_{sw} = x - x_0$, km	k_{sw}	t_{sw}, hours	$\bar{h}_{sw}$, m	$\bar{\tau}_{sw}$, sec	$\bar{h}_{mix}$, m	t, hours
160	1.85	6.4	40	0.63	2.4	1.16	6.7	1.56	12.4
180	1.95	6.6	20	0.82	1.2	1.42	6.8	1.78	11.2
200	2.05	6.8	0	1.00	0	2.05	6.8	2.05	10.0

Using (3.3.14) it is possible to calculate the height of mixed waves; these
data are listed in Table 3.3.3. The results of computations can be
expressed in the manner in which they are shown in Figure 3.3.5. As can
be seen, at $t = 10$ hours (Figure 3.3.5) the wave height and period will
decrease at the terminal point of the fetch. Mixed wave motion with
gradually decreasing height exists during the first two hours. Here the
calculations yield discontinuities in wave height and period, the latter
being greater than the former. This circumstance is a result of the fact
that the calculations are approximate.

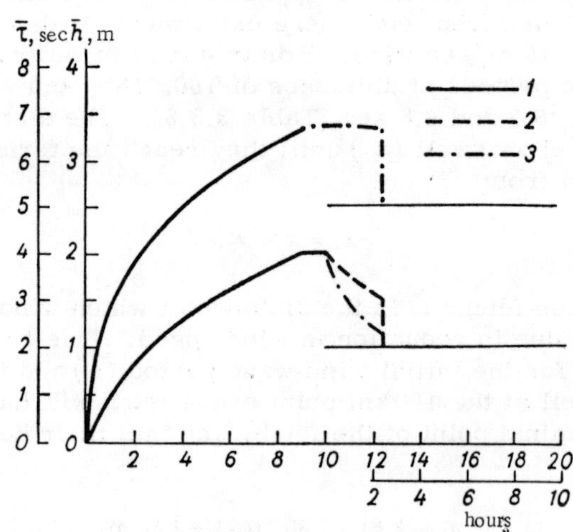

FIGURE 3.3.5. Pertaining to Example 3.3.6:

1 — wind waves; 2 — mixed wave motion; 3 — swell.

2b. Wind speed variable in space only

It is assumed that the fetch x (Figure 3.3.6) is divided into two segments. Over the first segment (x_1) the wind speed is w_1, and over the second (x_2) the wind speed is w_2. Here $w_2 > w_1$. The dependence of the wave period on the value of w and x has the form shown in Figure 3.3.6. Here it is assumed that the duration of wind with speed w_1 as well as with speed w_2 is much greater than the duration of wave developments, i.e., condition (3.3.1) is satisfied:

$$t_w > t.$$

In this case the wave elements should be determined from (3.3.5):

$$\tau_x = f_1(w, \ x).$$

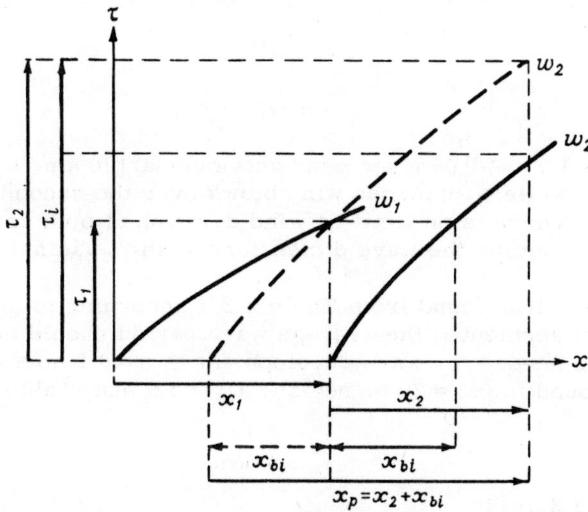

FIGURE 3.3.6.

However, the use of (3.3.5) is valid only for the first segment of the fetch. Waves produced over segment x_1 will appear at the start of segment x_2. This will accelerate the wave development over segment x_2, i.e., we have a case similar to that examined in Section 2a (Example 3.3.4). To take into account this acceleration of wave development over segment x_2 it is necessary to, on the one hand, decrease the duration of wave development over this segment and, on the other, to increase the fetch. The correction for the wave development time (t_{bi}) is calculated from (3.3.10), and the correction for the fetch (x_{bi}) is obtained from (3.3.5), which is solved for x upon substitution of τ_{x_1}. Consequently,

$$x_{bi} = f_1(\tau_{x_1}, \ w_2). \tag{3.3.15}$$

Now the wave dimensions at the end of the fetch are determined from

$$\tau = f_5(w_2,\ x_p),\qquad(3.3.16)$$

where

$$x_p = x_2 + x_{bi}.\qquad(3.3.17)$$

The time of wave development over the entire fetch (t_x) is calculated from the expression

$$t_x = t_1 + t_2 - t_{bi},\qquad(3.3.18)$$

where t_1 and t_2 are the times of wave development over segments x_1 and x_2, which are determined for each segment from (3.3.6) by solving it for t and by substituting respectively τ_{x_1} and τ_{x_2} (Figure 3.3.3):

$$t_1 = f_6(\tau_{x_1},\ w_1),\qquad(3.3.19)$$

$$t_2 = f_7(\tau_{x_2},\ w_2).\qquad(3.3.20)$$

Example 3.3.7. A 16 m/sec wind acts over a 120 km (x_1) segment of a 240 km fetch, while a 20 m/sec wind blows over the second 120 km segment (x_2). There is no limit on wind duration at both speeds. It is required to determine the wave dimensions at the terminal point of the entire fetch.

Solution. It is found from Table 3.3.1 according to (3.3.5) that at the termination of segment x_1 the average wave period should be $\tau_1 = 5.9$ sec, while the time needed for wave development is $t_1 = 6.2$ hours. According to (3.3.7) it is found for $w_2 = 20$ m/sec and $\tau_{xw_1} = 5.9$ sec (Table 3.3.1) that

$$x_{bi} = 90\ \text{km},$$

and t_{bi} from (3.3.10) is

$$t_{bi} = 4\ \text{hours}.$$

Now, according to (3.3.17)

$$x_p = 120\,\text{km} + 90\,\text{km} = 210\ \text{km};$$

from Table 3.3.1 for these values of x_p and $w = 20$ m/sec, it is found according to (3.3.16) that the wave period at the end of the entire 240 km fetch is

$$\tau = 7.6\ \text{sec},$$

while

$$h \cong 2.7\ \text{m}.$$

The time of development of these waves is determined from (3.3.9). First, according to (3.3.11) one finds from Table 3.3.1 for $\tau = 7.6$ sec and $w = 20$ m/sec, that $t_2 = 8.3$ hours and $t_1 = 6.2$ hours, which was already determined before. Hence according to (3.3.18)

$$t_x = 6.2 \text{ hours} + 8.4 \text{ hours} - 4 \text{ hours} = 10.6 \text{ hours}.$$

Example 3.3.8. For the same values of wave-generating factors as in Example 3.3.7, it is required to determine the wave dimensions and the time of their development in 40 km intervals.

TABLE 3.3.4. Computation of wave elements for Example 3.3.8

	w, m/sec					
	16			20		
	x, km					
	40	80	120	160 (40)	200 (80)	240 (120)
$\bar{\tau}$, sec	4.2	4.2	4.2	5.4	6.2	6.8
$\bar{h}$, m	1.0	1.0	1.0	1.6	2.0	2.3
t, hours	2.3	2.3	2.3	3.2	4.7	6.2
t_{bi}, hours				—1.5	—1.5	—1.5
t_1, hours				2.3	2.3	2.3
t_x, hours				4.0	5.5	7.0
x_{bi}, km			+30			
t_{bi}, hours			—1.5			
x_p, km				70	110	150
$\bar{\tau}$, sec	4.2	5.2	5.2	6.1	6.7	7,3
$\bar{h}$, m	1.0	1.4	1,4	1.9	2.2	2.5
t, hours	2.3	4,2	4.2	4.4	5.8	7.4
t_{bi}, hours				—2.8	—2.8	—2.8
t_1, hours				4.2	4.2	4.2
t_x, hours				5.8	7.2	8.8
x_{bi}, km			+60			
t_{bi}, hours			—2.8			
x_p, km				100	140	180
$\bar{\tau}$, sec	4.2	5.2	5.9	6.6	7.2	7.6
$\bar{h}$, m	1.0	1.4	1.6	2.2	2.5	2.7
t, hours	2.3	4.2	6.2	5.4	7.0	8,4
t_{bi}, hours				—4.2	—4.2	—4.0
t_1, hours				6.2	6.2	6.2
t_x, hours				7,6	9.2	10.6
x_{bi}, km			+90			
t_{bi}, hours			—4.0			
x_p, km				140	170	210

Solution. It is best to carry out the solution by the same techniques as in Example 3.3.7, but in the form of a table (Table 3.3.4). The entire computational procedure is clear from this table. First the wave

213

dimensions are determined over the first segment of the fetch, for a wind of 16 m/sec over a distance of 40 km. These wave dimensions will exist also at the 80 and 120 km points of the same segment 2.3 hours after the start of the action of the 16 m/sec wind. These waves, with a period of 4.2 sec and a height of 1.0 m, will come under the 20 m/sec wind. The values of t_{bi} and x_{bi} are determined from (3.3.10) and (3.3.15). The value of x_p is calculated from (3.3.17) and the wave period is obtained from (3.3.16). All the computational results are listed in Table 3.3.4. Then one determines elements of waves which form over the first 80 km of the first segment of the fetch. Calculations are carried out to determine the wave elements in this case at the second segment of the fetch, and so on. The calculations are terminated when the dimensions of waves are determined, with consideration of waves which are fully developed at the end of the first fetch segment, for each point of the entire fetch. The results of calculations presented in Table 3.3.4 can be depicted graphically. Such a graph can have any form, i.e., it may express variations in wave elements for any point of the fetch in time or for any time instant over the fetch. For·example, it is possible to estimate the variation in wave elements at the 200 km point of the fetch (Figure 3.3.7). First one finds from Table 3.3.1 the values of wave elements for a fetch of 80 km (200 km) due to a 20 m/sec wind. The

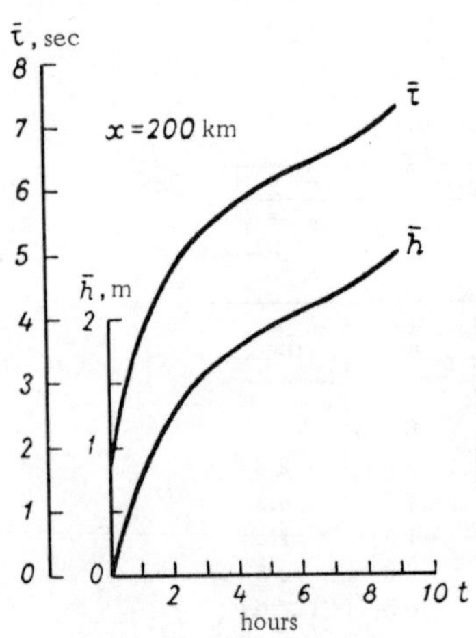

FIGURE 3.3.7. Pertaining to Example 3.3.8.

values of wind elements for $t = 5.5$, 7.2 and 9 hours are found from Table 3.3.4. The values thus obtained are joined by a smooth curve. It is seen from Figure 3.3.7 that after approximately 5 hours of wind action over a fetch of 200 km the wave height and period increase, due to the fact that at the start of the segment subjected to the 20 m/sec wind there will appear waves produced by the 16 m/sec wind.

The distribution of w over the fetch can be such that $w_2 < w_1$, but condition (3.3.1) is satisfied, as before, i.e., for velocity w_1 as well as w_2. In this case waves produced by the wind with speed w_1, upon moving into the region of action of wind with speed w_2, smaller than w_1, can become swell under certain conditions.

Example 3.3.9. A 16 m/sec wind acts over the first 120 km of a 200 km fetch, and an 8 m/sec wind prevails over the remaining 80 km. It is required to determine the wave dimensions at the terminal point of the entire fetch.

S o l u t i o n. It is best to carry out the solution as was done for the previous example, in the form of a table (Table 3.3.5). In calculations for the 40 km point of the fetch, the waves which appear after 2.3 hours at the start of the second segment will continue developing due to the 8 m/sec wind. However, the waves appearing at the start of this segment after 4.2 hours will already become "wind swell." The elements of these waves are calculated from Table 3.3.2A. These calculations, as in the preceding case, are best carried out in the form shown in Table 3.3.6. The variations in wave elements at the terminal point of the fetch (200 km) are shown in Figure 3.3.8.

TABLE 3.3.5. Pertaining to Example 3.3.9

	w, m/sec					
	16				8	
	x, km					
	20	40	80	120	160 (40)	200 (80)
$\bar{\tau}$, sec	3.3	3.3	3.3	3.3	3.9	4.5
$\bar{h}$, m	0.70	0.70	0.70	0.70	0.65	0.80
t, hours	1.2	1.2	1.2	1.2	6.9	10.5
t_{bi}, hours					—3.9	—3.9
t_1, hours					1.3	1.3
t_x, hours					4.2	7.8
x_{bi}, km				40		
t_{bi}, hours				—3.9		
x_p, km					80	120
$\bar{\tau}$, sec	3.3	4.2	4.2	4.2	4.7	5.0
$\bar{h}$, m	0.70	1.0	1.0	1.0	0.85	0.95
t, hours	1.25	2.3	2.3	2.3	11.9	14.3
t_{bi}, hours					—8.6	—8.6
t_1, hours					2.3	2.3
t_x, hours					5.6	8.0
x_{bi}, km				100		
t_{bi}, hours				—8.6		
x_p, km					140	180
$\bar{\tau}$, sec	3.3	4.2	5.2	5.2		
$\bar{h}$, m	0.70	1.0	1.4	1.4	Swell	Swell
t, hours	1.25	2.3	4.2	4.2		
$\bar{\tau}$, sec	3.3	4.2	5.2	5.9		
$\bar{h}$, m	0.70	1.0	1.35	1.65	Swell	Swell
t, hours	1.25	2.3	4.2	6.2		

In calculating wave elements produced by a wind with speed w_2 in Example 3.3.4, we used a technique for taking into account the wave motion previously produced by a wind with speed w_1, smaller than w_2. Now this problem can be solved by the same computational technique applied to Example 3.3.7. According to the statement of Example 3.3.4, a 16 m/sec (w_1) wind acts over a 200 km fetch (x) for 10 hours, and then the wind speed increases to 20 m/sec (w_2) and is maintained for another 10 hours. The wave elements at the terminal point of the fetch are now calculated by analogous relationships (3.3.15)—(3.3.18). The 16 m/sec wind (w_1), according to Table 3.3.1, at different points of the fetch (column 1 of Table 3.3.7) will produce waves, the elements of which are listed in columns 2—4 of Table 3.3.7.

TABLE 3.3.6. Computation of swell elements for Example 3.3.9

x_0, km	$\bar{h}_0$, m	$\bar{\tau}_0$, sec	t, hours	$x_{sw} = x - x_0$, km	R_{sw}	τ_{sw}, sec	$\bar{h}_{sw}$, m	t_{sw}, hours	t, hours
0	1.35	5.2	4.2	40	0.56	5.6	0.76	3.0	7.2
				80	0.50	5.9	0.68	5.8	10.0
0	1.65	5.9	6.2	40	0.61	6.3	1.0	2.6	8.8
				80	0.57	6.7	0.94	5.0	11.2

These waves are then subjected to the 20 m/sec wind (w_2). The variations in their size can be found from Table 3.3.1. However, it is first necessary to calculate x_{bi}, i.e., the correction for the fetch, discussed in examining the solution of Example 3.3.7. This correction is now calculated from the expression

$$x_{bi} = f(w_2, \tau_1). \qquad (3.3.21)$$

The quantities calculated from this expression are listed in column 5 of Table 3.3.7. Then it is necessary to calculate x_p (column 7 of Table 3.3.7), i.e., the fetch which will determine the dimensions of waves produced by the 20 m/sec wind. This latter quantity is found from the expression

$$x_p = x_b + x_{bi}, \qquad (3.3.22)$$

in which

$$x_b = x - x_0, \qquad (3.3.23)$$

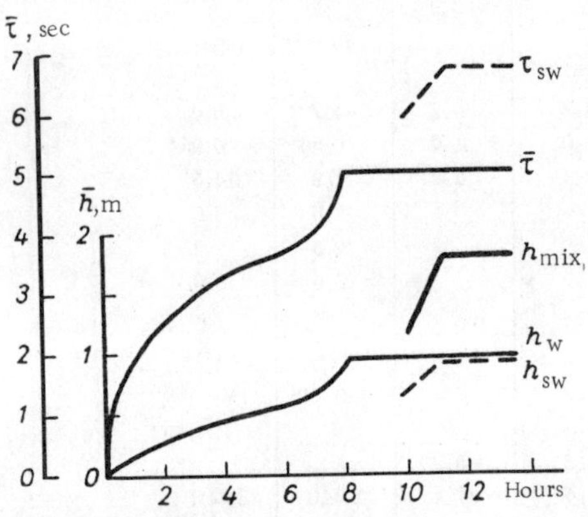

FIGURE 3.3.8. Pertaining to Example 3.3.9.

where x is the entire specified fetch, in the given case 200 km, while x_0 is the distance over which waves developed due to the wind with speed w_1 at individual points of the entire fetch (column 1 of Table 3.3.7). The value of x_b is listed in column 6 of Table 3.3.7. Finally, t_{bi} (column 8 of Table 3.3.7) is calculated from (3.3.10). Then, from Table 3.3.1 for the given x_p and w_2 one finds the wave dimensions at the terminal point of the fetch. These values are listed in columns 9—11 of Table 3.3.7. The true time of growth of these waves is determined from (3.3.10). The values of t are listed in column 12 of the above table. Finally column 13 shows the time of appearance of the calculated wave elements, measured from the start of action of the 16 m/sec wind (w_1) given by the expression

$$t_x = t_w + t. \qquad (3.3.24)$$

The computational results are shown by a dash-dot curve in Figure 3.3.4. As can be seen from comparing these latter data with those obtained in Example 3.3.4, the wave elements grow more slowly, but their limiting

possible values at the terminal point of the fetch remain as before. The
above computational method is applicable for a more detailed description of
changes with time of wave elements with variable wind speed over the entire
fetch.

TABLE 3.3.7. Computation of wave elements pertaining to Example 3.3.4
using a different technique

x_0, km	$w_1 = 16$ m/sec, $x = 200$ km			$w_2 = 20$ m/sec, $x = 200$ km								
	h_1, m	τ_1, sec	t_1, hours	x_{bi}, km	x_b, km	x_p, km	t_{bi}, hours	h_2, m	τ_2, sec	t_2, hours	t, hours	t_x, hours
1	2	3	4	5	6	7	8	9	10	11	12	13
0	0	0	0	0	0	200	0	2.65	7.5	8.1	8.1	18.1
40	1.00	4.2	2.3	30	160	190	1.5	2.60	7.4	7.7	6.2	16.2
80	1.35	5.2	4.2	60	120	180	2.8	2.55	7.3	7.4	4.6	14.6
120	1.65	5.9	6.2	90	80	170	4.0	2.48	7.15	7.0	3.0	13.0
160	1.85	6.4	7.8	120	40	160	5.1	2.40	7.0	6.6	1.5	11.5
200	2.05	6.8	9.2	0	0	0	0	2.05	6.8	0	10.0	10.0

2c. Wind speed variable in both time and space

Computation of wave elements in these cases reduces to a combination of
techniques presented in Sections 2a and 2b.

Example 3.3.10. A 15 m/sec wind acts over a 300 km fetch for 15 hours.
Then the wind strength increases to 20 m/sec at which speed it blows for
10 more hours over a 140 km segment of the fetch, while over the remaining segment, from 140 to 300 km, it now blows at 25 m/sec. It is
required to determine the wave dimensions at the terminal point of the fetch.

Solution. First the wave dimensions are calculated for the 15 m/sec
wind. We use (3.3.5) subject to condition (3.3.1), which is satisfied in the
case at hand, since for $x = 300$ km and $w = 15$ m/sec we have (Table 3.3.1)

$$t_w > t.$$

We find from Table 3.3.1 for $x = 300$ km and $w = 15$ m/sec that

$$\overline{h} = 2.3 \, \text{m}; \quad \overline{\tau} = 7.5 \text{ sec and } \quad t = 14 \text{ hours.}$$

These will be the dimensions of waves at the terminal point of the fetch
after 14 hours of a 15 m/sec wind action. Now we determine the wave
dimensions at the same point for the 25 m/sec wind. First one finds the
wave dimensions at the end of the 140 km segment using (3.3.8). It is found
from Table 3.3.1 for $x = 140$ km and $w = 15$ m/sec that

$$\overline{h} = 1.65 \, \text{m}, \quad \overline{\tau} = 6.0 \text{ sec and } t = 7.3 \text{ hours.}$$

217

According to (3.3.9), we again find from Table 3.3.1, but now for $x = 140$ km and $w = 20$ m/sec, that

$$\bar{h} = 2.25 \text{ m}, \quad \bar{\tau} = 6.7 \text{ sec} \quad \text{and} \quad t_2 = 5.8 \text{ hours}.$$

Using (3.3.10) one now finds the correction for the wave development time with the 20 m/sec wind. For $\bar{\tau}_1 = 6.0$ sec and $w_2 = 20$ m/sec Table 3.3.1 indicates that

$$t_{bi} = 4.2 \text{ hours}.$$

According to (3.3.11),

$$t = t_2 - t_{bi} = 5.8 \text{ hours} - 4.2 \text{ hours} = 1.6 \text{ hours}.$$

Consequently, the waves over the 140 km segment generated by the 20 m/sec wind appear 1.6 hours after wind with this speed starts blowing. Moving further over the fetch they are subjected to the 25 m/sec wind (w_2) over the fetch segment of $300 - 140 = 160$ km.

Further calculations are carried out with (3.3.15), and the correction for the fetch, i.e., x_{bi}, is calculated. For this purpose we find from Table 3.3.1 for $\bar{\tau}_{x1} = 6.7$ sec and $w_2 = 25$ m/sec that

$$x_{bi} \cong 100 \text{ km}.$$

Now, according to (3.3.17),

$$x_p = 160 \text{ km} + 100 \text{ km} = 260 \text{ km}.$$

For $x_p = 260$ km and $w = 25$ m/sec one finds from Table 3.3.1 according to (3.3.16) that the wave dimensions at the terminal point of the 300 km fetch are

$$\bar{h} = 3.85 \text{ m}, \quad h_{1\%} = 3.85 \cdot 2.42 = 9.3 \text{ m}, \quad \bar{\lambda} = 124 \text{ m}, \quad \bar{\tau}_{x_2} = 8.9 \text{ sec},$$
$$t_2 = 8.7 \text{ hours}.$$

To determine the time when these waves appear at the end of the fetch it is first necessary to calculate t_{bi} from (3.3.10). For this one finds from Table 3.3.1 that, for $\bar{\tau} = 6.7$ sec, $w_2 = 25$ m/sec. It is found from this table that $t_{bi} = 3.9$ hours.

Now, using (3.3.18) we have

$$t_{x = 300 \text{ km}} = 1.6 \text{ hours} + 8.7 \text{ hours} - 3.9 \text{ hours} = 6.7 \text{ hours}.$$

Since, according to the statement of the problem, the wind initially blew for 15 hours, waves of this size will appear at the end of the 300 km fetch after a time

$$15 \text{ hours} + 6.7 \text{ hours} = 21.7 \text{ hours}$$

and then will no longer change as long as the wind blows at 25 m/sec.

In the case under study the dimensions of waves at the end of the 140 km fetch segment due to the 20 m/sec wind were estimated by a technique

presented in solving Example 3.3.4. Let us now calculate the wave
dimensions by the technique used in composing Table 3.3.7.

This calculation is shown in Table 3.3.8, which is similar in form to
Table 3.3.7. At the terminal point of the 140 km fetch there thus appear
waves, generated by the 20 m/sec wind, with $h_2 = 2.25$ m and $\tau_2 = 6.7$ sec,
after the wind with the above speed acted for 5.8 hours, and not after
1.6 hours as was calculated above. After 1.6 hours waves generated by the
20 m/sec wind will not attain their limiting dimensions (Table 3.3.8). They
will be somewhat smaller, their height will be ~2.0 m with period ~6.3 sec.
Now, however, due to the slower development of the limiting-size waves due
to the 20 m/sec wind, waves of limiting size will appear at the end of the
140 km fetch segment not after 2.6 hours, but only after 5.8 hours. Hence,
according to (3.3.19),

$$t_{x=300\,\mathrm{km}} = 5.8 \text{ hours} + 8.7 \text{ hours} - 3.9 \text{ hours} = 10.6 \text{ hours,}$$

and consequently, for the duration of wind action of our example

$$t_x = 15 \text{ hours} + 10.6 \text{ hours} = 25.6 \text{ hours,}$$

which is 0.6 hours more than the duration of action of the 25 m/sec wind.
The calculations thus show that the wave dimensions produced by the 25 m/sec
wind at the end of the 300 km fetch should be smaller. Further calculations,
using the duration of wind action, should be carried out in the manner shown
in Example 3.3.5.

TABLE 3.3.8. Computation of wave elements for Example 3.3.10

x_0, km	$w = 15$ m/sec, $x = 140$ km			$w = 20$ m/sec, $x = 140$ km								
	h_1, m	τ_1, sec	t_1, hours	x_{bi}, km	x_b, km	x_p, km	t_{bi}, hours	h_2, m	τ_2, sec	t_2, hours	t, hours	t_{x}, hours
0	0	0	0	0	140	140	0	2.25	6.7	5.8	5.8	20.8
40	0.95	4.1	2.4	30	100	130	1.5	2.18	6.6	5.4	3.9	18.9
80	1.25	5.0	4.3	55	60	115	2.6	2.00	6.3	5.0	2.4	17.4
120	1.55	5.7	6.3	80	20	100	3.6	1.95	6.1	4.4	0.8	15.8
140	1.65	6.0	7.3	100	0	0	0	1.65	6.0	15.0	0	15.0

3. The wind speed is constant, but its direction varies by more than 45°

The wave elements are calculated with allowance for changes in the
fetch. Here the wind waves produced by the wind over a previous fetch are
treated as swell. These are calculated by a method examined in Examples
3.3.6 and 3.3.9.

Example 3.3.11. A northerly (N) wind with speed 20 m/sec acts for
12 hours over a 300 km fetch. Then it becomes easterly (E), and the fetch
becomes 100 km. This wind also acts for 12 hours. It is required to
determine the wave elements at the terminal point, common to both fetches.

TABLE 3.3.9. Computation of swell for example 3.3.11

x_0,km	$\bar{h}_0$,m	$\bar{\tau}_0$,sec	$x_{sw}=x-x_0$, km	k_{sw}	τ_{sw},sec	$\bar{h}_{sw}$,m	τ_{sw}, hours	t_0, hours	t,hours
100	1.95	6.1	200	0.47	7.5	0.91	11.6	12.0	23.6
200	2.65	7.5	100	0.64	8.5	1.70	6.0	12.0	18.0
260	3.00	8.1	40	0.70	8.5	2.10	1.9	12.0	13.9

Solution. The change in fetches produces at this point an appearance of swell from the north and simultaneous development of wind waves from the east. The wind waves are calculated from Table 3.3.1, first for the 20 m/sec wind over a 300 km fetch and then for a wind at the same speed over the 100 km fetch. The swell is calculated in the same manner as in Examples 3.3.6 and 3.3.9. The results of computations for the swell are listed in Table 3.3.9.

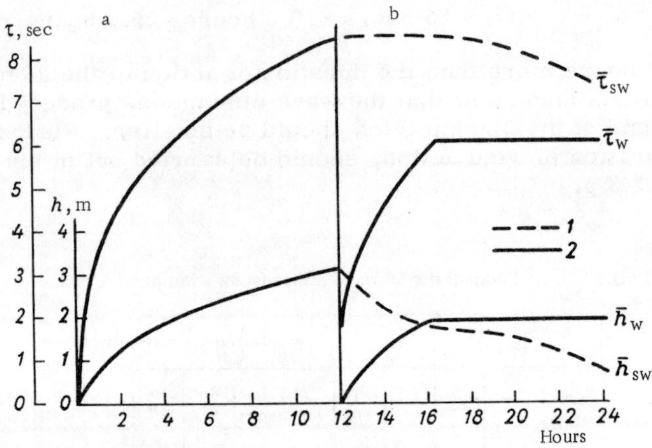

FIGURE 3.3.9. Pertaining to Example 3.3.11:

1 — swell; 2 — wind waves; a — wind from the north, 20 m/sec; b — wind from the east, 20 m/sec, swell from the north.

Figure 3.3.9 shows the results of calculations for changes in wave elements at the terminal points of the fetches. Had the wind speed changed in addition to the change in the fetch, then the calculations would be made for these changes also. For example, had the wind acting in this example over the 100 km fetch possessed a speed of, say, 24 m/sec rather than 20 m/sec, then calculations for this fetch would have been carried out from Table 3.3.1 for this new speed.

A special distinction should be made for the case when the change in wind direction does not exceed 45°, but is greater than 25°. Then one must also use new fetches. However, waves generated by winds over a previous fetch are now regarded as wind waves. Their effect on the wave development over the new fetch is taken into account by techniques presented in Examples 3.3.4 and 3.3.7.

Cases may be encountered in practical wave element calculations when wind speeds and fetches vary in a large number of combinations. Calculations in such cases are carried out by a combination of the above methods. However, since the number of such combinations is large it is always expedient to simplify the computation by disregarding changes of ± 2 m/sec in wind speed and of $\pm 25°$ in direction and using average values. All these simplifications become habitual with practice. It is sometimes necessary to estimate the extent over which swell propagates during a given time. In these cases one may use appropriate tables (for example, similar to Table 3.3.2B) or nomograms, which simplify these calculations.

Example 3.3.12. A wind wave produced by a 25 m/sec wind has an average height of 3.0 m and an average period of 7.6 sec. The wind became calm and the wave became swell. It is required to determine the distance that the swell will travel in 10 hours and its dimensions. From Table 3.3.2B for $\tau_0 = 7.6$ sec and $t_{sw} = 10$ hours one finds that the wave will travel 214 km during this time and its average height will be $3 \cdot 0.57 = 1.71$ m, while the average period will be 9.3 sec.

4. Computation of wind-wave elements in shallow water and shore shoals

These calculations are made in solving many applied problems, including those pertaining to engineering needs, for example, in designing all kinds of hydraulic structures, for which data are needed on the size of waves which may act on these structures. The large variety of the latter results also in differences in the requirements put to the calculated data. There exist various recommendations and specifications which govern these calculations (TU [Tekhnicheskie Usloviya] *, 1960).

Considering only the most elementary of these calculations we note that calculations of wind-wave elements for shallow bodies of water such as water storage lakes, as well as natural lakes and more or less closed sea gulfs, are usually made (Selyuk, 1961) without consideration of wind duration, i.e., one calculates waves of limiting size for given w and x (Chapter 2, Sections 2 and 3). Hence the calculations are rather simple (Chapter 3, Section 3) and are carried out with formulas which allow for the effect of the bottom on the wave development (Chapter 2, Sections 6 and 9).

If it is necessary to calculate wave elements for the shore shoals of a sea, then one first calculates elements of wind waves developing in adjoining deep-water regions using techniques presented in Section 3 of this chapter. If the wave elements in deep water are calculated from their climatic-regime values, then one uses techniques presented below (Chapter 3, Section 5). For wave elements in deep-water regions calculated by one of the available methods, it is necessary to estimate changes in their dimensions when arriving in the shore shoal zone. This zone, as mentioned above, is subdivided into segments (Figure 2.6.2) depending on the relative depth, expressed by $\dfrac{H}{\lambda_d}$ (Chapter 2, Section 6).

Limiting ourselves to the presentation of the simplest types of calculations for deformation of waves reaching the shore, we can illustrate these by the following examples.

* [Technical specifications.]

Example 3.4.1. It was found by computing elements of wind waves produced by a storm that waves approaching the shore shoals normal to the shore line will be 11 m high (1% cumulative probability of wave height) with average length $\bar{\lambda} = 132$ m and average period $\bar{\tau} = 9.2$ sec. It is required to determine the variation in elements of these waves when entering the shore shoals.

Solution. Since $\alpha° = 0$, we may use Figure 2.6.7. The steepness of the starting wave is $\delta_d = 1/12$. It follows that waves of this steepness break at $\dfrac{H}{\lambda_d} = 0.1$, i. e., in this case at a depth of $H_{cr} \approx 13$ m. Their height at $\dfrac{H}{\lambda_d} = 0.1$ is $0.93 \cdot 11$ m $= 10$ m according to Table 2.6.4, and the steepness will be $\sim 1/9$. After the first break the wave height will be reduced to ~ 8.5 m. The first breakers will appear on this wave at $\dfrac{H}{\lambda_d} \approx 0.15$, i. e., in a region of depth ~ 20 m when the wave height is ~ 9.9 m and its steepness is $\sim 1/11$. Consequently, a heavy surf will be observed in the sea strip between depths of 20 and 13 m. According to Table 2.6.2, at the time of breaking the length of this wave will be close to 132 m $\cdot 0.71 = 94$ m, and at the instant of breaking, according to the same table, its length will be ~ 108 m. The period remains unchanged.

Example 3.4.2. Wind waves of the same dimensions as in Example 3.4.1 approach the shore line at angle $\alpha = 45°$. It is required to estimate the changes in elements of these waves in a region where the depth is 15 m.

Solution. Since $\alpha = 45°$, we use Table 2.6.4. Hence for $\dfrac{H}{\lambda_d} = 15/132 \approx 0.11$ we find that $\dfrac{h_{sh}}{h_d} = 0.83$. Consequently, $h_{sh} = 11$ m $\cdot 0.83 = 9.2$ m. The direction of wave propagation for the given depth will, according to Table 2.5.3, make an angle of $\sim 34°$ with the 15 m isobath. The average wave length, according to Table 2.6.2, is about 94 m. The wave does not break upon reaching the 15 m depth, since its critical depth (H_{cr}) is $1.3\, h_{sh} \approx 12$ m. The wave period remains the same as before.

Example 3.4.3. Swell of wave length 100 m and wave height 1 m approaches the shore shoals normal to the shore line. It is required to determine the elements of these waves at the 3 m depth.

Solution. In the case under study $\dfrac{H}{\lambda_d} = 0.03$, while $\delta_d = 1/100$. It follows from Figure 2.6.7 that the steepness of this wave at the specified depth is $\sim 1/30$. Its height, according to Table 2.6.4, increases to 1.14 m, while the length (Table 2.6.2) decreases to 41 m.

Example 3.4.4. It is required to determine wind-wave elements for a shallow body of water when $w = 20$ m/sec, $x = 100$ km and $\overline{H} = 15$ m.

Solution. This problem can be solved, for example, using Table 2.6.1. It is hence found that when $w = 20$ m/sec and $\overline{H} = 15$ m, $\bar{h} = 1.1$ m and $\bar{\tau} = 4.4$ sec for $x = 36$ km and $t = 1.5$ hours. These waves will exist at fetches of $\geqslant 36$ km, i. e., in the given case also at $x = 100$ km and after $\geqslant 1.5$ hours, without change in their elements. It is found from Table 2.4.2 that for a height with 1% cumulative probability

$$h_{1\%} = 2.42 \cdot 1.1 = 2.66 \text{ m}$$

and $\bar{\lambda} = 30$ m, according to Table 1.6.2.

Had it been required in this case to determine the wave dimensions over a fetch, of for example, 20 km, i.e., smaller than 36 m, then this could have been done from Table 3.3.1.

Example 3.4.5. Figure 3.4.1 is a schematic of a shore shoal, at individual points of which (denoted in Figure 3.4.1 by numbers 1 through 4) it is required to calculate the wave elements for a 10 m depth. A 25 m/sec wind blows over this region, as well as over the open sea, for 15 hours.

Solution. Table 3.4.1 presents the computational results and lists the sequence of operations and the tables and graphs used. Below are presented some explanations on computing waves at point 2.

TABLE 3.4.1. Computation of wind-wave elements in shore shoals, pertaining to Example 3.4.5 (Figure 3.4.1).

Serial number of points	w, m/sec	x, km	t_w, hours	$\bar h_d$, m	$h_{d\,1\%}$, m	$\bar\tau_d$, sec	$\bar\lambda_d$, m	δ_d	t, hours	α_d°
1	2	3	4	5	6	7	8	9	10	11
1, 2	25	60	15	2.00	4.7	5.7	51	$^1/_9$	2.4	0
A	25	400	15	4.6	11.2	10.0	156	$^1/_{14}$	12.2	75
2 (B)	25	50*	15	—	6.4	10.3	166	$^1/_{26}$	1.3 (13.5)	40
3	25	450	15	4.8	11.7	10.3	166	$^1/_{14}$	13.5	0
4	25	60	15	1.45**	3.5	4.6**	33	$^1/_9$	1.3	0
Figure 3.4.1	Original data	Original data	Original data	Table 3.3.1	Table 2.3.2	Table 3.3.1	Table 1.6.2	Column 8/Column 6	Table 3.3.1	Measured

Serial number of points	H_d, m	$\dfrac{H_{10}}{\bar\lambda_d}$	α_{sh}°	$\dfrac{h_{sh}}{h_d}$	h_{sh}, m	$\dfrac{\lambda_{sh}}{\bar\lambda_d}$	λ_{sh}, m	δ_{sh}	H_δ, m	H_{cr}, m	h_s, m
1	12	13	14	15	16	17	18	19	20	21	22
1, 2	$\geqslant 15$	0.19	0	0.92	4.3	0.89	45	$^1/_{10}$	~ 9	~ 6.0	~ 3.7
A	$\geqslant 47$	0.06	32	0.54	6.1	0.58	91	$^1/_{15}$	~ 12	~ 8	~ 5.2
2 (B)	$\geqslant 50$	0.06	19	0.96	6.1	0.58	96	$^1/_{16}$	~ 12	~ 8	~ 5.2
3	$\geqslant 50$	0.06	0	1.0	11.7	0.58	96	$^1/_8$	~ 23	~ 15	~ 10
4	$\geqslant 10$	0.3	0	1.0	3.5	1.0	33	$^1/_9$	~ 7	~ 4.5	~ 3
Figure 3.4.1	$0.3\,\bar\lambda_d$	$\dfrac{10}{\bar\lambda_d}$	Table 2.10.3	Table 2.11.5	Columns 15 and 6	Table 2.11.3	Columns 8 and 17	Column 18/Column 16	$2h_{sh}$	$1.3\,h_{sh}$	$0.85\,h_{sh}$

* Explanation in the text.

** Values of $\bar h_d$ and $\bar\tau_d$ were calculated using Table 2.11.1.

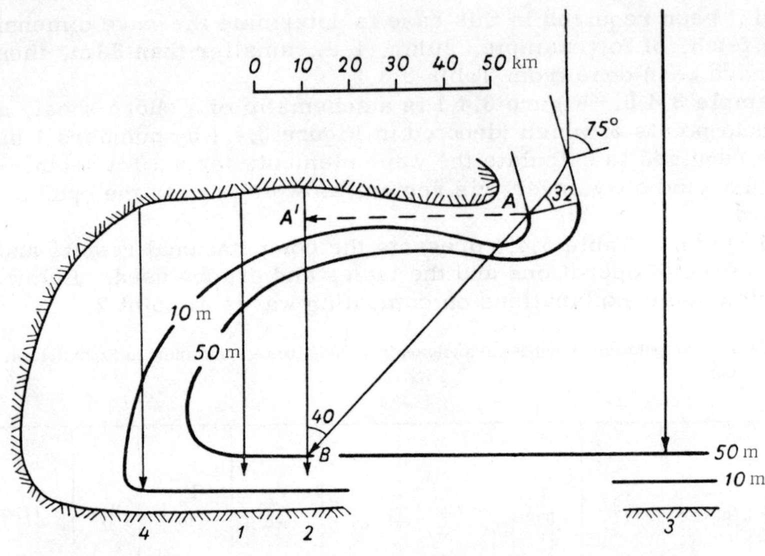

FIGURE 3.4.1. Schematic of a shore shoal of a sea (pertaining to Example 3.4.5). The solid arrows indicate the direction of wave propagation. The isobaths are in meters.

To compute wave elements at this point one must estimate not only the dimensions of waves which were developed by the northerly wind over a fetch of 60 km (Table 3.4.1), but also of those which appear from the NE. To calculate elements of these waves one first carries out computations at point A (Figure 3.4.1, Table 3.4.1). These waves, propagating to the SW, flow into a region of depth below 10 m. It is assumed that the depths in the region of wave propagation are not smaller than 8 m, i.e., are greater than the critical depth of these waves (Table 3.4.1). Then these waves again move into deeper water. The effect of wind on these waves at the segment between points A and B is taken into account using Table 3.3.1. The fetch of 50 km is determined by measuring segment A'B, which is the projection of the wave travel between points A and B onto the wind direction. This computational technique is necessitated by the fact that the wave propagates over this segment in a direction differing from that of the wind by more than 25°.

It is found from Table 3.3.1 that the wave height for $w = 25$ m/sec and $\bar{\tau} = 10$ sec is 4.6 m. The wave height for a fetch of 400 km + 50 km = 450 km will be 4.8 m with a period of 10.3 sec. Consequently, the wave increases by $(4.8 - 4.6)/4.6 \simeq 5\%$. Increasing by this percentage the wave height of 6.1 m, one finds the wave height at point 2 (B); it is ~ 6.4 m with a period of 10.3 sec (Table 3.4.1, columns 6 and 7).

Table 3.4.1 shows that after ~ 1.3 hours waves 4.7 m high appear at point 2, and attain a height of 6.4 m after a further 13.5 hours. It is obviously possible to carry out a detailed calculation of variations in wave elements at point 2 for the entire storm by computing wave elements at points A and B by methods presented in Chapter 3, Section 3 (Examples 3.3.2 and 3.3.3).

5. Computation of climatic-regime characteristics of wind waves

The probability characteristics of waves of different dimensions in individual regions of seas and oceans over an extended period of time, i.e., over individual months, seasons, years, or over a long-time period, are called regime or climatic-regime wave-motion characteristics (Chapter 3, Section 1). The latter is used in many applications.

Such characteristics can be determined by several methods. One of these consists in utilizing typical wind-field maps (Chapter 3, Section 2) for calculating wind-wave elements by techniques described above (Chapter 3, Section 3). If reference is had to the frequency of those types of weather for which maps of the wind field are in effect for a month, season or year, then the required probability of wind-wave elements is identified with the probability of corresponding types of weather, and this yields the probability of wind-wave elements for the entire period under study. This technique was used by Zubova. Subsequently a similar approach was used for obtaining information on the regime characteristics of wave motion in many seas of the USSR (Handbook Data, 1962). This method is difficult to implement when determining regime characteristics of waves in oceans. Typical maps of the wind field for ocean conditions are more difficult to compose than for the limited extents of seas. These maps for oceans are schematic, and the frequency of weather types for which these maps are composed is not always exact. Hence this method of determining regime characteristics of wave motion is not extensively used for ocean conditions. For these conditions one may calculate wind-wave elements from current, daily, synoptic maps. Selecting a system of fixed points on the ocean surface and determining the corresponding wind speed and fetch from synoptic maps, one computes the wave elements by relevant techniques (Chapter 2, Sections 5 and 8). If such systematic calculations are carried out for a longer period of time, for example, a year or more, then this information on wave dimensions can be employed for obtaining some idea on the regime characteristics of waves (Rzheplinskii, 1961). This method naturally requires a large volume of computations and the availability of daily synoptic maps, which cannot always be obtained with the required detail for oceans. Hence other techniques were developed for calculating the climatic-regime wave characteristics, enabling one to determine them with less computational time and to avoid difficulties encountered in using raw data such as synoptic maps of wind fields (Davidan, 1961, 1967; Rzheplinskii, 1965).

The climatic-regime characteristics of wave elements can be estimated by the method examined in Chapter 2, Section 6. For a given natural wind speed frequency, for example, that listed in Table 3.5.3, one can employ Tables 3.5.1 and 3.5.2 (Chapter 2, Section 6) for calculating the climatic-regime characteristic of wave heights and periods for the region of weather ship J. For this one writes the wind speed and its frequency in the first two columns of Table 3.5.4. Then one enters into the following columns, pertaining to the corresponding wave heights, the values of $F(k_x)$ from Table 3.5.1 (first column). The value of $f\%$ for each wind speed is multiplied by the corresponding value of $F(k_x)$ of each column. These results are listed in the second column. Summation, over each vertical column, of

TABLE 3.5.1. Cumulative probability of relative fetches $F(k_x)$

w, m/sec	Height of waves with 3% cumulative probability, m														
	$\geqslant 0.5$	$\geqslant 1$	$\geqslant 2$	$\geqslant 3$	$\geqslant 4$	$\geqslant 5$	$\geqslant 6$	$\geqslant 7$	$\geqslant 8$	$\geqslant 9$	$\geqslant 10$	$\geqslant 11$	$\geqslant 12$	$\geqslant 13$	$\geqslant 14$	$\geqslant 15$
4	0.905														
5	0.942														
6	0.961	0.844													
7	0.970	0.887													
8	0.980	0.914	0.677												
9	0.985	0.946	0.748												
10	0.990	0.951	0.795	0.566											
11	0.990	0.961	0.818	0.613											
12	0.991	0.966	0.835	0.644	0.440										
13	0.992	0.970	0.852	0.677	0.482	0.301									
14	0.993	0.970	0.861	0.705	0.522	0.343	0.202								
15	0.993	0.970	0.878	0.733	0.554	0.375	0.235	0.131							
16	0.994	0.970	0.887	0.748	0.583	0.411	0.265	0.156	0.080						
17	0.995	0.980	0.896	0.771	0.601	0.445	0.298	0.183	0.099	0.050					
18	0.995	0.980	0.905	0.787	0.631	0.472	0.326	0.208	0.121	0.063	0.030				
19	0.995	0.980	0.914	0.803	0.657	0.502	0.354	0.235	0.141	0.086	0.040	0.018			
20	0.996	0.980	0.923	0.811	0.677	0.522	0.383	0.259	0.162	0.097	0.051	0.025	0.011		
21	0.996	0.980	0.923	0.827	0.698	0.548	0.407	0.284	0.185	0.110	0.062	0.031	0.015	0.0067	0.0034
22	0.996	0.980	0.932	0.835	0.712	0.566	0.432	0.307	0.196	0.128	0.073	0.038	0.019	0.0091	0.0045	0.0015
23	0.997	0.980	0.932	0.844	0.726	0.589	0.458	0.333	0.228	0.145	0.087	0.048	0.025	0.012	0.0067	0.0025
24	0.997	0.990	0.932	0.852	0.741	0.607	0.477	0.354	0.247	0.162	0.100	0.057	0.031	0.015	0.0091	0.0034
25	0.997	0.990	0.942	0.861	0.756	0.625	0.502	0.375	0.270	0.179	0.115	0.068	0.038	0.020	0.0111	0.0045
26	0.998	0.990	0.942	0.869	0.763	0.644	0.512	0.395	0.289	0.196	0.130	0.080	0.046	0.025	0.016	0.0061
27	0.998	0.990	0.951	0.869	0.779	0.657	0.538	0.415	0.307	0.214	0.144	0.090	0.053	0.030	0.017	0.0082
28	0.998	0.990	0.951	0.878	0.787	0.670	0.554	0.436	0.323	0.223	0.157	0.101	0.062	0.036	0.023	0.010
29	0.999	0.990	0.951	0.887	0.795	0.684	0.571	0.454	0.343	0.249	0.174	0.113	0.071	0.042	0.027	0.014
30	0.999	0.990	0.951	0.887	0.803	0.698	0.587	0.492	0.364	0.267	0.190	0.128	0.081	0.050	0.033	0.017

TABLE 3.5.2. Cumulative probability of relative fetches $F(k_x)$

w, m/sec	Average wave period, sec													
	$\geqslant 2$	$\geqslant 3$	$\geqslant 4$	$\geqslant 5$	$\geqslant 6$	$\geqslant 7$	$\geqslant 8$	$\geqslant 9$	$\geqslant 10$	$\geqslant 11$	$\geqslant 12$	$\geqslant 13$	$\geqslant 14$	$\geqslant 15$
4	0.961													
5	0.970	0.887												
6	0.975	0.905												
7	0.979	0.923	0.811											
8	0.983	0.932	0.841	0.691										
9	0.985	0.942	0.861	0.733										
10	0.987	0.952	0.878	0.763	0.607									
11	0.988	0.953	0.883	0.769	0.613	0.445								
12	0.988	0.954	0.884	0.775	0.625	0.458								
13	0.988	0.955	0.888	0.779	0.631	0.463	0.301							
14	0.988	0.956	0.891	0.784	0.638	0.472	0.316							
15	0.989	0.957	0.893	0.788	0.644	0.482	0.320	0.186						
16	0.989	0.958	0.895	0.792	0.651	0.487	0.326	0.192	0.095					
17	0.989	0.959	0.896	0.796	0.657	0.497	0.336	0.198	0.100					
18	0.989	0.960	0.896	0.799	0.664	0.502	0.343	0.204	0.105	0.045				
19	0.990	0.961	0.904	0.803	0.667	0.507	0.350	0.210	0.110	0.047	0.017			
20	0.990	0.962	0.905	0.806	0.672	0.512	0.357	0.217	0.113	0.050	0.018			
21	0.990	0.962	0.905	0.808	0.676	0.517	0.361	0.221	0.117	0.052	0.019	0.006		
22	0.990	0.963	0.905	0.813	0.678	0.522	0.364	0.225	0.120	0.054	0.021	0.006	0.001	
23	0.990	0.963	0.906	0.815	0.685	0.527	0.371	0.230	0.125	0.057	0.022	0.007	0.002	
24	0.990	0.964	0.908	0.816	0.688	0.533	0.375	0.235	0.129	0.060	0.023	0.007	0.002	0.0003
25	0.990	0.964	0.909	0.818	0.691	0.538	0.383	0.239	0.133	0.062	0.024	0.008	0.002	0.0004
26	0.991	0.965	0.910	0.820	0.696	0.543	0.387	0.247	0.137	0.065	0.026	0.008	0.002	0.0005
27	0.991	0.965	0.911	0.823	0.697	0.549	0.391	0.249	0.140	0.067	0.027	0.009	0.002	0.0005
28	0.991	0.966	0.912	0.823	0.702	0.553	0.395	0.254	0.142	0.069	0.028	0.009	0.003	0.0006
29	0.991	0.966	0.913	0.824	0.705	0.555	0.399	0.257	0.145	0.071	0.029	0.010	0.003	0.0006
30	0.991	0.967	0.914	0.826	0.711	0.560	0.403	0.264	0.150	0.073	0.030	0.011	0.003	0.0007

the results of multiplication, corresponding to each value $\geqslant h_i$, yields the required climatic-regime characteristic of the wave height distribution, with a 3% cumulative probability in the given case. The distribution thus obtained describes the wind wave distribution. To obtain the actual, natural distribution of mixed waves, one must make a correction for so-called "background" swell.

TABLE 3.5.3. Frequency of wind and cumulative probability of wave heights from observations of weather ship J (56°N, 20°W) in winter (January — March)
(Climatological and Oceanographic Atlas, 1959)

$\overline{w}$, m/sec	$f\,(w)$, %	h, m*	$F\,(h)$, %
3	4.7	0.6	94
4	7.0	1.2	92
7	16.3	1.8	80
10	19.3	2.4	68
13	25.3	3.0	53
16	16.0	3.6	36
19	8.0	4.2	22
23	2.3	4.8	18
27	0.2	5.4	13
30	0.03	6.0	9

* The cumulative probability of the wave height is $\sim 3{-}5\%$.

It is seen from Figure 2.6.3 that the distribution of $k_{h_{50\%}}$ deviates from its natural distribution at the $\sim 70\%$ cumulative probability level and above. Similar differences were detected by Rzheplinskii (1965) in analyzing the natural climatic-regime distribution of the heights of significant waves for individual regions in the North Atlantic (Climatological and Oceanographic Atlas, 1959). Rzheplinskii quite correctly attributes this difference to the effect of swell, which distorts the distribution of wave heights. The strong wind wave motion which corresponds to low cumulative probabilities is not affected by swell. Hence distortions of the climatic-regime distributions are found for [waves with] high cumulative probabilities. This distortion is estimated by Rzheplinskii by means of (3.3.14), which he expresses in the form

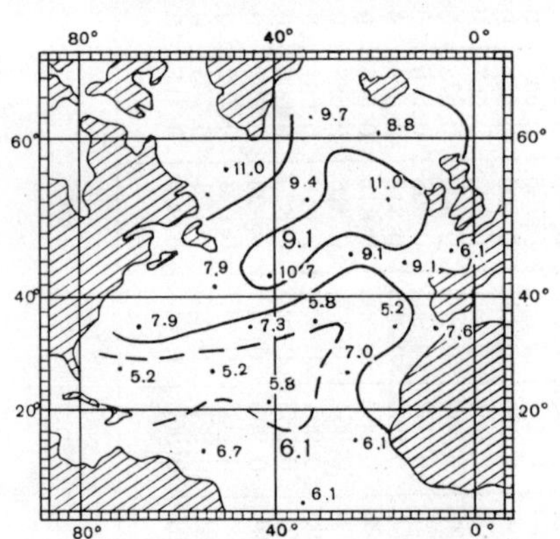

Figure 3.5.1. h_{sw} (background swell) (in decimeters) for the North Atlantic, averaged over a year (after Rzheplinskii).

$$h_{sw} = \left(h_{mix}^2 - h_w^2\right)^{\frac{1}{2}}. \qquad (3.5.1)$$

Quantity h_{sw} is termed the background swell. The latter was calculated (Rzheplinskii, 1965) for 23 regions of the North Atlantic for individual seasons of the year and on the average for a year (Figure 3.5.1).

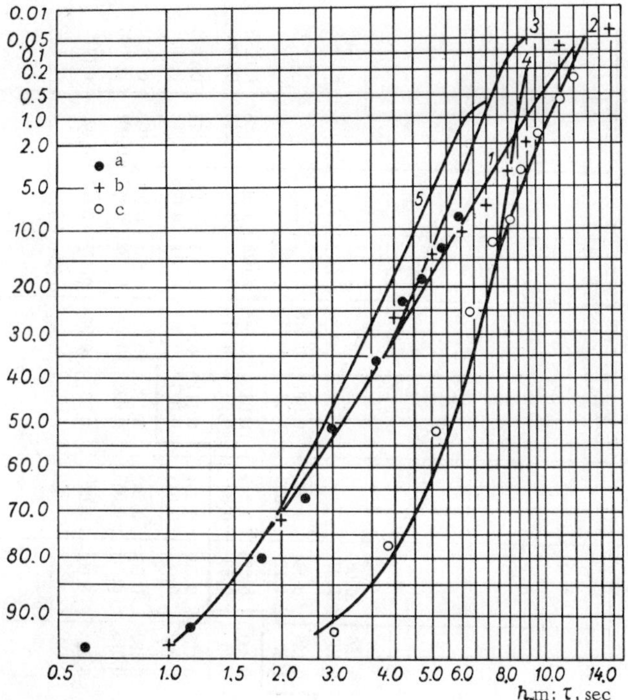

h, m; τ, sec

FIGURE 3.5.2:

1 — from Table 3.5.5 [$\varphi(h_{3\%})$]; 2 — from Table 3.5.5 [$\varphi(\overline{\tau})$]; 3 — from Table 3.5.4 [$\varphi(h_{3\%})$ for $x = 300$ km]; 4 — $\varphi(\overline{\tau})$ for $x = 300$ km; 5 — from Table 3.5.8 ($x = 300$ km, $H = 10$ m); a — data of Table 3.5.3 for the wave height; b — wave height; c — wave period according to Hogben (1967).

Background swell varies in different regions. For the region (Figure 3.5.1) situated to the north of ~40°N it is close to 0.9 m, and to the south of this latitude it is ~0.6 m. The known analogy between Figures 3.5.1 and 2.9.2 is found by comparing these two values. The regions with high background swell in Figure 3.5.1 coincide with the region of high frequency of swell 3 m and higher in Figure 2.9.2. If background swell is estimated on the basis of Figure 2.6.3, then it is found, using (3.5.1), that for a cumulative probability > 70% we obtain $h_{sw} \approx 0.35$ m. Assuming that the cumulative probability of significant waves is close to 3%, then using Table 2.3.2, we derive the background swell as equal to

$$h_{sw} = 0.35 \cdot 2.1 \approx 0.75 \text{ m.} \tag{3.5.2}$$

This value is approximately an average between the previously mentioned $h_{sw} = 0.6$ m and $h_{sw} = 0.9$ m (Figure 3.5.1). Apparently this is a result of the fact that the natural distribution of $k_{h_{50\%}}$ is based on observations in ocean regions lying between 11 and 62°N (Davidan, 1967). Turning to Figure 2.6.2 we see that in this case the difference between the

TABLE 3.5.4. Example of computing the climatic-regime characteristics of waves with 3% cumulative probability for the region of weather ship J

w, m/sec	f, %	Height of waves with 3% cumulative probability, m															
		$\geqslant 0.5$	$\geqslant 1.0$	$\geqslant 2.0$	$\geqslant 3.0$	$\geqslant 4.0$	$\geqslant 5.0$	$\geqslant 6.0$	$\geqslant 7.0$	$\geqslant 8.0$	$\geqslant 9.0$	$\geqslant 10.0$	$\geqslant 11.0$	$\geqslant 12.0$	$\geqslant 13.0$	$\geqslant 14.0$	$\geqslant 15.0$
1	2	3	4	5	6	7	8	9	10	11	12	13	14	15	16	17	18
4	7.0	0.905 6.33															
7	16.3	0.970 15.8	0.887 14.4														
10	19.3	0.990 19.1	0.951 18.3	0.795 15.4	0.566 10.9												
13	25.3	0.992 25.1	0.970 24.3	0.852 21.5	0.677 17.3	0.482 12.4	0.301 7.6										
16	16.0	0.994 15.9	0.970 15.5	0.887 14.2	0.748 11.9	0.583 9.3	0.411 6.6	0.265 4.25	0.156 2.50	0.080 1.27							
19	8.0	0.995 7.95	0.980 7.80	0.914 7.30	0.803 6.40	0.657 5.3	0.502 4.0	0.354 2.84	0.235 1.87	0.141 1.12	0.086 0.69	0.040 0.32	0.018 0.14				
23	2.3	0.997 2.29	0.980 2.25	0.932 2.14	0.844 1.92	0.726 1.67	0.589 1.35	0.458 1.06	0.333 0.76	0.228 0.53	0.145 0.33	0.087 0.20	0.048 0.11	0.025 0.06	0.012 0.03	0.0067 0.015	0.0025 0.006
27	0.20	0.998 0.19	0.990 0.19	0.951 0.19	0.869 0.17	0.779 0.16	0.657 0.13	0.538 0.11	0.415 0.082	0.307 0.061	0.214 0.043	0.144 0.029	0.090 0.018	0.053 0.011	0.030 0.006	0.017 0.003	0.0082 0.002
30	0.03	0.999 0.029	0.990 0.029	0.951 0.029	0.887 0.027	0.803 0.024	0.698 0.020	0.587 0.020	0.492 0.015	0.364 0.011	0.267 0.007	0.190 0.006	0.128 0.004	0.081 0.002	0.050 0.001	0.033 0.001	0.017 0.0006
$\varphi(h)$		92.69	82.77	60.76	48.62	28.85	19.70	8.28	5.22	3.00	1.07	0.56	0.27	0.07	0.04	0.02	0.01
Computation of heights of mixed waves																	
		$(0.5)^2$ (0.25)	$(1.0)^2$ 1.0	$(2.0)^2$ 4.0	$(3.0)^2$ 9.0	$(4.0)^2$ 16.0											
		$(1.1)^2$ 1.21	$(1.1)^2$ 1.21	$(1.1)^2$ 1.21	$(1.1)^2$ 1.21	$(1.1)^2$ 1.21											
		$(1.46)^{1/2}$ 1.21	$(2.21)^{1/2}$ 1.48	$(5.21)^{1/2}$ 2.28	$(10.21)^{1/2}$ 3.2	$(17.21)^{1/2}$ 4.1											
$\varphi(h)$ when $x \leqslant 300$ km		92.69	82.77	60.76	48.62	28.85	12.10	4.03	0.85	0.07	0.05	0.006					

TABLE 3.5.5. Computed average long-time climatic-regime characteristics $\varphi(h, \tau)$ of wind waves for the region of weather ship J (56°N, 20°W) for winter (January — March)

	Height of waves with 3% cumulative probability, m															
	≥1.2	≥1.5	≥2.3	≥3.2	≥4.1	≥5.0	≥6.0	≥7.0	≥8.0	≥9.0	≥10.0	≥11.0	≥12.0	≥13.0	≥14.0	≥15.0
Height of waves with 1% cumulative probability, m	1.4	1.6	2.6	3.7	4.7	5.8	6.9	8.0	9.2	10.3	11.5	12.6	13.8	14.9	16.0	17.3
$\varphi(h)$, %	92.7	82.8	60.8	48.6	28.9	19.7	8.3	5.2	3.0	1.07	0.56	0.27	0.07	0.04	0.02	~0.01
Number of seasons during which it is observed once											<1	1	4	7	14	~28
Number of days per season	83	75	55	44	25	18	7	5	3	1	<1					

	Average wave period, sec													
	2.4	3.3	4	5	6	7	8	9	10	11	12	13	14	15
$\varphi(\tau)$, %	92.9	83.0	78.3	55.1	44.8	24.8	8.1	5.5	1.2	0.52	0.2	0.02	0.01	<0.01
Number of seasons during which it is observed once										<1	1	14	28	>28
Number of days per season	84	75	70	50	40	22	7	5	1	<1				

distribution curves of relative wave periods is also a result of the existence of swell. Consequently, it is necessary to increase the values of the average wave periods corresponding to 90% by a factor of 1.2, those corresponding to 85% should be multiplied by 1.1, those corresponding to 80% should be multiplied by 1.07, and those corresponding to 75% should be multiplied by 1.03, in order to obtain the true climatic-regime distribution of average wave periods.

In this case instead of the value $h_{sw} = 0.7$ m one can use, according to (3.5.2), the data of Figure 3.5.1 for the region of weather ship J. At this point (Figure 3.5.1) $h_{sw} = 1.1$ m. Hence, using (3.3.14), one obtains values of h_{mix}, which are listed in Table 3.5.4. For $h_{3\%} > 4$ m background swell need not be considered.

The climatic-regime characteristics of wave periods are calculated in quite a similar manner. All the computational results are shown in Figure 3.5.2 and Table 3.5.5. This table lists data on the number of winter seasons when wave heights of 1 and 3% cumulative probability may be encountered once, as well as the number of days per season with waves of a given height with given cumulative probability in the given wave system. This also applies to the average wave period.

This information was obtained from (3.2.6), which is solved for n. One first must estimate the value of m, i.e., the duration of wind action which determines the wave development. It follows from Table 2.4.2 that if the wind remains constant within the limits of 2 m/sec, and its direction does not change by more than 2 compass points, then this wind acts for ~ 6 hours. Hence m should be estimated at approximately 6 hours. This wind action duration can be sufficient for development of heavy wave motion if it is considered that the wind may be acting not on a smooth sea surface, but on already existing wave motion, produced by the previously acting wind (Chapter 3, Section 3, Example 3 of Section 4). These conditions are apparently typical for waves developing in oceans. Observations are usually made each 6 hours; hence $N = 4$, and consequently, $t = 6$ hours. The number of days per season is 90.

Examination of Table 3.5.5 shows that if one takes as the maximum height of waves (1% cumulative probability) that height which is encountered once in 28 seasons, then each such wave height should be 17 m and more, and the period of waves of this probability should be close to 14 sec and more. It can also be noted that, for example, in winter in the region of weather ship J one may observe for ~ 50 days (four observations per 24 hours) per winter, waves ~ 3.7 m high (1% cumulative probability in the given wave system), with a period of ~ 5 sec.

Figure 3.5.2 shows observations correlated in American editions (Climatological and Oceanographic Atlas, 1959) and in the British Atlas (Hogben, 1967) corresponding to the region of weather ship J. It follows from Figure 3.5.2 that all these data are in satisfactory agreement with the calculated distributions of wave heights and periods. In making these comparisons one must estimate the cumulative probability of wave elements listed in Western publications. Lately reference is had to so-called significant waves. The cumulative probability of the height of the latter was determined by Davidan (Handbook Data, 1965):

Observed wave height, m	0.5	1	2	3	4	5	6	7	8	9	10
Cumulative probability, %	24	19	11	10	9	8	7	6	5	5	5

Observed wave height, m	11	12	13	14	15	16	17
Cumulative probability, %	4	4	4	4	3	3	3

According to British sources (Hogben, 1967), the height of significant waves is related to the maximum wave height by the expression

$$h_{max} \cong 1.6 h_{1/3},$$
(3.5.3)

where $h_{1/3}$ is the height of significant waves.

In addition,

$$h_{1/3} \cong 1.06 h_{obs}.$$
(3.5.4)

If it is assumed that (Chapter 2, Section 3)

$$h_{max} \cong 3\bar{h},$$
(3.5.5)

then the observed waves can be estimated as waves with a height having a cumulative probability close to 5%:

$$h_{obs} \approx 1.8\bar{h}.$$
(3.5.6)

The average wave period is related to its observed values (Hogben, 1967) by the following approximate relationship:

$$\tau_{max} \cong 0.73 \tau_{obs},$$
(3.5.7)

where $\tau_{max} \approx 1.2\bar{\tau}$ (Chapter 2, Section 10).

Consequently,

$$\bar{\tau} \cong 0.61 \tau_{obs}.$$
(3.5.8)

The above method of computing climatic-regime wave characteristics can be also used for ocean regions with a limited open extent.

This is possible if it is assumed that the law governing the distribution of wind fluxes (2.4.2), which is identified with the distribution of fetches (Chapter 2, Section 4), is valid also when baric systems form not only over the water surface, but also over the land. This may occur near shores, in gulfs, etc. In this case calculations of the form shown in Table 3.5.4 should be restricted to wave heights which are possible over the given limiting fetches. For example, if in the given computation (Table 3.5.4) the open water extent in all directions is limited to 300 km, then one may proceed as follows.

Table 3.3.1 is used to determine that limiting height of waves with 3% cumulative probability which is possible at wind speeds shown in column 1 of Table 3.5.4. According to data of Table 3.3.1, the limiting possible waves are shown in Table 3.5.6.

TABLE 3.5.6. Limiting wave dimensions for $x = 300$ km (according to Table 3.3.1)

w, m/sec	$h_{3\%}$, m	$\bar{\tau}$, sec	w, m/sec	$h_{3\%}$, m	$\bar{\tau}$, sec
4	0.5	2.5	19	6.1	8.2
7	1.6	4.5	23	7.8	8.9
10	3.2	6.4	27	9.3	9.5
13	4.2	7.1	30	10.5	9.9
16	5.3	7.7			

Now calculation of $\varphi(h)$ in the form of Table 3.5.4 should be limited to levels of corresponding wave heights. These limits are delineated in Table 3.5.4 by lines. The obtained values of $\varphi(h)$ are listed in the lower lines of Table 3.5.4. The same is also carried out for $\varphi(\tau)$. Figure 3.5.2 shows the distribution of $\varphi(h)$ and $\varphi(\tau)$, corresponding to a limiting fetch of 300 km. One notes that in this latter case the cumulative probability of high waves and large wave periods has decreased markedly.

The above method is suitable for calculating climatic-regime characteristics of wind waves also when the fetches are limited only in some directions. It is then necessary to determine for these directions the probability of the wind and to find $\varphi(h)$ and $\varphi(\tau)$ for each wind direction separately, and then to sum the probability distribution of h and τ thus obtained over all the directions. It is also possible to take into account the effect of shallow waters and the effect of this factor on climatic-regime distribution of wind-wave elements. The solution of the latter problem is best examined by means of a specific example.

It is assumed that in the previously considered case of calculating climatic-regime characteristics of wave elements for a limited fetch of 300 km, the sea depth at the point for which the climatic-regime characteristic is calculated is 10 m. Above we found the limiting possible wave heights with 3% cumulative probability for conditions pertaining to the limited 300 km fetch (Table 3.5.6). Now it is necessary to determine the length of this wave, which is easily done on the basis of the values of periods (Table 3.5.7). The required values of $\bar{\lambda}$ are found from Table 1.6.2, and quantities H/λ_d, where H is equal to 10 m according to the problem's formulation, are calculated. Values of H/λ_d thus found are used for determining h_{sh}/h_d from Table 2.11.5 when $\alpha = 0$. Now it is easy to find the limiting heights of wind waves which arrived at the 10 m depth. It is only necessary to first estimate the possible height of waves which will break at the assumed 10 m depth. According to (2.11.21) the wave height cannot exceed

$$h_p = H_{cr} : 1.3 = 7.7 \text{ m.}$$

Now one should restrict the computation of the climatic-regime characteristics to corresponding wave-height levels for each wind speed. All the aforementioned computational results are listed in Table 3.5.7.

TABLE 3.5.7. Limiting wave dimensions for $x = 300$ km with consideration of limited depth at the point of computation ($H = 10$ m)

w, m/sec	$h_{3\%}$, m	$\bar{\tau}_d$, sec	λ_d, m	$\dfrac{H}{\lambda_d}$	$\dfrac{h_{sh}}{h_d}$	h_{sh}, m	Wave-height level
4	0.5	2.5	10	1.0	1.0	0.5	0.5
7	1.6	4.5	32	0.31	0.95	1.5	$\geqslant 1$
10	3.2	6.4	64	0.16	0.90	2.9	$\geqslant 2$
13	4.2	7.1	78	0.13	0.91	3.8	$\geqslant 3$
16	5.3	7.7	92	0.11	0.93	5.0	$\geqslant 4$
19	6.1	8.2	105	0.1	0.94	5.7	$\geqslant 5$
23	7.8	8.9	124	0.08	0.96	7.5	$\geqslant 7$
27	9.3	9.5	141	0.07	0.97	9.0	Broken
30	10.5	9.9	153	0.06	1.0	10.5	Broken

Consequently, computation of $\varphi(h)$ in the form of Table 3.5.4 should be limited to those wave-height levels which are given in the last column of Table 3.5.7. This is denoted by a dashed line in Table 3.5.4.

The climatic-regime characteristics of wave heights in the case under study are given in Table 3.5.8.

TABLE 3.5.8. Climatic-regime characteristics $\varphi(h)$ of wave heights at a point with a limited 300 km fetch and a sea depth of 10 m

$h_{3\%}$, m	$\geqslant 1.2$	$\geqslant 1.5$	$\geqslant 2.3$	$\geqslant 3.2$	$\geqslant 4.1$	$\geqslant 5.0$	$\geqslant 6.0$	$\geqslant 7.0$
$\varphi(h)$, %	92.69	82.77	60.76	37.7	16.4	5.5	1.19	0.85

It is seen by comparing these results with climatic-regime wave-height characteristics calculated previously for the same point, but for a deep sea (lower line in Table 3.5.4), that the probability of waves with height 4 m and more has decreased markedly. To determine the climatic-regime characteristics of wave periods for the given depth it is necessary to eliminate wave periods of 9 sec and more from calculations according to Table 3.5.7.

It is frequently necessary in engineering practice to determine the maximum dimensions of wind waves which appear as a result of individual storms, i. e., to estimate the probability of their appearance. These calculations are carried out as follows (TU, 1960). On the basis of data on the probability of winds of a given limiting strength, allowance is made for the frequency of the highest observed wind speed. Using the applicable formulas, one calculates wave elements, which are then assigned the same cumulative probability of appearance as that assumed in calculating the wind speed. For example, the probability of winds of different speeds has the frequency listed in Table 3.5.9. The free sea surface (measured from the shore), above which a given wind speed may be observed, is approximately equal (for the two cases of computation) to 300 and 500 km. The sea is deep. Using, for example, Table 3.3.1, one finds the wave dimensions for $w = 30$ m/sec and $x = 300$ km and 500 km (Table 3.5.10). Now, identifying the frequency of computed wave elements with that of the design wind at a speed of 30 m/sec (Table 3.5.9), it is assumed that the waves given in Table 3.5.10 will be observed once in 20 years.

TABLE 3.5.9. Frequency of winds of different velocities

w, m/sec	Observed once	Frequency, %
22	in 1 year	0.272
24	7 years	0.039
26	10 years	0.027
28	15 years	0.018
30	20 years	0.014

The same calculation can be carried out using Table 3.5.1. First one assumes that the wind accompanying the storm, for example, at 30 m/sec, acts over the entire fetch (300 or 500 km) and its speed and direction do not change. This assumption is obviously approximate, since the wind speed in a storm region is not uniform, even if it may be assumed that the speed is constant. However, this assumption makes it possible to simplify the calculations using Table 3.5.1, since then the climatic-regime character-istics of the waves can be calculated only for the 30 m/sec wind and for $h_{3\%} = 10$ and $h_{3\%} = 13$ m (Table 3.5.10). Then it is easily found that for wave heights of 10 m $F(k_x) = 0.190$ from Table 3.5.1, and then $\varphi(h) = 0.190 \cdot 0.014 = 0.003\%$, or once in 92 years; and for wave heights of 13 m, similarly, $\varphi(h) = 0.050 \cdot 0.014 = 0.0007\%$, or once in 272 years.

TABLE 3.5.10. Highest possible wind waves according to computation

w, m/sec	x, km	$\bar{h}$, m	$h_{3\%}$, m	$h_{1\%}$, m	$\bar{\tau}$, sec	$\bar{\lambda}$, m	δ	t, hours
30	300	4.95	10.4	12.0	9.9	153	$^1/_{13}$	8.4
30	500	6.20	13.0	15.0	11.5	206	$^1/_{14}$	12.9

Consequently, the previous calculation exceeds the probability of appearance of the largest waves for a 500 km fetch by approximately a factor of 15. This is also understandable, since the probability of appearance of waves depends not only on the probability of the wind, but also on the proba-bility of fetches. It is more sensible to carry out the calculations for this problem employing Table 3.5.1, but taking into account (under the conditions of heavy storms being considered) all the wind speeds listed in Table 3.5.9. The calculations are carried out in the same form as those for the climatic-regime characteristics given in Table 3.5.4.

Entering the starting data into Table 3.5.11 and carrying out all the necessary, rather simple calculations, which were discussed previously in setting up Table 3.5.4, we derive the information of interest.

If one retains the assumption that each storm is accompanied by a wind of the same speed, i. e., by a wind of speed 22 or 23 m/sec only, and so on, then the maximum possible wave height is determined from Table 3.5.11. It is easy to see that a storm with 22 m/sec winds over a 300 km fetch generates waves with a limiting height (3% cumulative probability) of 7 m, which will be observed once in 3 years; for a 24 m/sec wind the limiting height will be 8 m with a probability of appearance once in 31 years, etc.

TABLE 3.5.11. Computation of the probability of wave heights

w, m/sec	$f(w)$, %	Height of waves with 3% cumulative probability, m						
		$\geqslant 7$	$\geqslant 8$	$\geqslant 9$	$\geqslant 10$	$\geqslant 11$	$\geqslant 12$	$\geqslant 13$
22	0.272	0.307 0.084 3.3	0.196 0.059	0.128 0.0035 7.7				
24	0.039	0.354 0.014	0.247 0.009 31	0.162 0.006	0.100 0.004 69			
26	0.027	0.395 0.011	0.289 0.008	0.196 0.005 55	0.130 0.004	0.080 0.002 137		
28	0.018	0.436 0.008	0.323 0.006	0.223 0.004 68	0.157 0.003	0.101 0.002	0.062 0.001 273	
30	0.014	0.492 0.007	0.364 0.005	0.267 0.004	0.190 0.003 92	0.128 0.002	0.081 0,001	0.050 0.0007 406
$x = 500$ km Number of years over which it appears once		0.124 2.3	0.087 3.4	0.054 5.1	0.014 19.5	0.006 45	0.002 137	0.001 272
$x = 300$ km Number of years over which it appears once		0.124 2.3	0.028 9.8	0.013 21	0.003 92			

The results of these calculations of limiting wave heights for the 300 km
fetch are plotted in Figure 3.5.3 (curve 1). It is easy to see that in order
for maximum wave heights to appear once in 20 years this limiting wave
height will be close to 8 m for the 300 km fetch and 9.5 m for the 500 km fetch
(curve 2 in Figure 3.5.3). Here the most dangerous storms will be those
with winds speeds of $\sim 22 - 24$ m/sec for the first fetch and of $22 - 23$ m/sec
for the second.

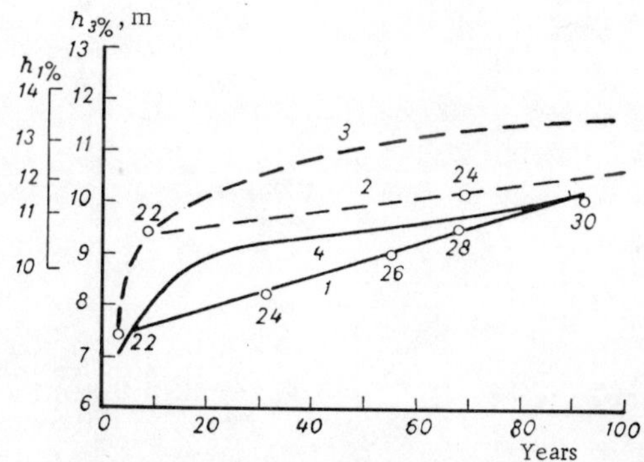

FIGURE 3.5.3. Numerals at the points denote wind speeds
in m/sec. Solid curves correspond to a 300 km fetch,
while dashed curves correspond to a 500 km fetch.

Consequently, the strongest storm may not generate the most dangerous
waves in the region under study. The decisive factor here is the frequency
of the given wind speed accompanying the storm. It is important to note

that, in the case under study, increasing the fetch from 300 to 500 km has only a slight effect on the height of maximum-size waves. As it is, the results of calculations depend extensively on the combination of a number of factors. The period, length and time of growth of the waves discussed above are determined directly from Table 3.3.1. For example, for a wind speed of ~23 m/sec it is found (by interpolation along curve 1 in Figure 3.5.3) that for the 300 km fetch waves with an average height of 3.7 m ($h_{3\%} \approx 8.0$ m) have a period of 8.9 sec and a length of 124 m, and the development time of these waves is ~10.2 hours. Consequently, the calculated duration of a storm with winds blowing at 23 m/sec should at least be ~10 hours. It is more expedient to assume that under the given wind conditions, winds in storms alternating with a different probability have speeds of 22 m/sec and more. Winds at speeds of 22 and 23 m/sec, etc., are encountered during a storm with a 24 m/sec wind speed. For example, during a storm with a wind speed of 30 m/sec there act winds at speeds of from 22 to 30 m/sec. Then these computations should be carried out for all wind speed graduations (lowermost two lines in Table 3.5.11). In this case one may observe once in 20 years (solid curve 4 in Figure 3.5.3) 9 m high waves over a 300 km fetch (curve 4) and waves ~10 m high over a 500 km fetch (curve 3). This computational technique yields somewhat larger wave heights than those presented before and apparently is more sensible, since these wind conditions are close to those encountered in nature.

Bibliography

Publications in Russian

Anapol'skaya,L.E. and L.S.Gandin. Rezhim bol'shikh skorostei
    vetra na territorii SSSR dlya ucheta vetrovykh nagruzok na
    sooruzheniya (The Regime of High Wind Speeds over the USSR for
    Taking into Account Wind Loads on Structures). — In: Sbornik
    "Voprosy prikladnoi klimatologii."Leningrad, Gidrometeoizdat. 1960.
Andreyanov,V.G. Vetrovaya volna ozerovidnykh vodoemov (The Wind
    Wave of Lakelike Bodies of Water). — Izvestiya Nauchno-Issledo-
    vatel'skogo Instituta Gidrotekhniki, Nos. 24 and 25. 1939.
Bonchkovskaya,T.V. Raspredelenie davleniya nad profilem volny
    (Pressure Distribution above a Wave Profile). — Trudy MGI AN SSSR,
    No. 6. 1955.
Braslavskii,A.P. Raschet vetrovykh voln (Wind-Wave Computation). —
    Trudy GGI, No. 35 (89). 1952.
Brovikov,I.S. Vetrovoe volnenie v melkovodnom more (Wind Waves in
    a Shallow Sea). — Trudy GOIN, No. 26 (38). 1954.
Davidan,I.N. Zakonomernosti mnogoletnogo raspredeleniya morskikh
    voln i ikh svyaz' so skorost'yu vetra (Regularities in the Long-Time
    Distribution of Sea Waves and their Relationship with Wind Speed). —
    Okeanologiya, No. 2. 1961.
Davidan,I.N. Nekotorye rezul'taty issledovaniya morskogo volneniya,
    razvivayushchegosya v usloviyakh, blizkikh k "ideal'nym" (Some
    Results of a Study of Sea Waves Developing under Nearly "Ideal"
    Conditions). — In: Sbornik "Morskie volneniya, neregulyarnaya kachka
    i ee uchet v razlichnykh oblastyakh morskoi tekhniki. " 1964.
Davidan,I.N. Zavisimost' veroyatnostnykh kharakteristik voln ot vetra
    (The Dependence of the Probability Characteristics of Waves on
    Wind). — In: Sbornik "Teoreticheskie i prakticheskie voprosy
    morekhodnosti sudov." Izdatel'stvo "Transport." 1967.
Davidan,I.N. Primenenie veroyatnostnykh metodov dlya analiza rezhima
    volneniya (Application of Probability Methods to the Analysis of Wave
    Regimes). — Trudy GOIN, No. 91. 1967.
Davidan,I.N. and A.V.Smirnova. Raschet paramctrov vetrovykh voln
    (Computation of Wind-Wave Parameters). — Trudy koordinatsionnogo
    soveshchaniya po gidrotekhnike, No. 50. VNII (Vsesoyuznyi Nauchno-
    Issledovatel'skii Institut) im. Vedeneeva. Izdatel'stvo "Energiya". 1969
Dobroklonskii,S.V. Turbulentnaya vyazkost' v poverkhnostnom sloe
    morya i volnenie (Eddy Viscosity in the Surface Layer of the Sea and
    Wave Motion). — DAN SSSR, Vol. 58, No. 7. 1947.
Duvanin,A.I. Volnovoe dvizhenie v more (Wave Motion in the Sea). —
    Leningrad, Gidrometeoizdat. 1968.

Efimov, V.V. Nekotorye rezul'taty eksperimental'nogo issledovaniya peredachi energiii vetra morskim volnam (Some Results of an Experimental Study of Wind Energy Transmission to Sea Waves). — Okeanologiya, Vol. 4, No. 6. 1964.

Glukhovskii, B. Kh. Issledovanie morskogo vetrovogo volneniya (A Study of Wave Motion in the Sea). — Leningrad, Gidrometeoizdat. 1966.

Grushevskii, M. S. O postanovke i putyakh resheniya zadach pri raschete morskikh vetrovykh voln (The Formulation and Techniques for Solving Problems in the Computation of Wind Waves in the Sea). — Trudy Okeanograficheskoi Komissii AN SSSR, Vol. 9. 1960.

Ivanov, A.A. K voprosu o predvychislenii elementov vetrovykh voln (Transformation of Wind-Wave Elements). — Trudy MGI AN SSSR, Vol. 5. 1955.

Kao Wen-Hsiu and Wang Hsiao-Nan. Raschet vetrovykh voln na more proizvol'noi glubiny (Computation of Wind Waves in a Sea of Arbitrary Depth). — Izvestiya AN SSSR, Seriya Geofizicheskaya, No. 10. 1961.

Kapitsa, P.L. K voprosu ob obrazovanii vetrom morskikh voln (Sea-Wave Generation by Winds). — DAN SSSR, Vol. 64, No. 4. 1949.

Khvan, A.P. Pole vetrovykh voln na melkovod'e (The Wind-Wave Field in Shallow Waters). — Izvestiya AN SSSR, Seriya Geofizicheskaya, No. 10. 1958.

Khvan, A.P. Raschet elementov vetrovykh voln na melkovod'e (Calculation of Wind-Wave Elements in Shallow Waters). — Izvestiya AN SSSR, Seriya Geofizicheskaya, No. 1. 1961.

Kitaigorodskii, S.A. K teorii turbulentnogo peremeshivaniya v more v svyazi s raschetom tolshchiny verkhnego izotermicheskogo sloya (Theory of Turbulent Mixing in the Sea in Conjunction with Calculating the Thickness of the Upper Isothermal Layer). — Trudy IOAN SSSR, Vol. 52. 1961.

Kochin, N.E., I.A. Kibel', and N.V. Roze. Teoreticheskaya gidromekhanika (Theoretical Hydromechanics). — OGIZ, Gostekhizdat. 1948.

Kondrat'ev, N.E. Raschet vetrovogo volneniya i transformatsii beregov vodokhranilishch (Calculation of Wind Waves and Transformation of Shores of Water-Storage Lakes). — Gidrometeoizdat. 1953.

Kononkova, G.E. Zarozhdenie vetrovykh voln na poverkhnosti vody (Generation of Wind Waves on the Surface of Water). — Trudy MGI AN SSSR, Vol. 3. 1953.

Korneva, L.A. and V.P. Liverdi. Statisticheskie kharakteristiki volneniya v glubokom more i okeane (Statistical Characteristics of Wave Motion in Deep Seas and in Oceans). — In: Sbornik "Okeanograficheskie issledovaniya." Izdatel'stvo "Naukova Dumka." 1964.

Krylov, Yu. M. Teoriya i raschet vetrovykh voln glubokogo morya (Theory and Computation of Wind Waves in a Deep Sea). — Trudy GOIN, No. 26 (38). 1954.

Krylov, Yu. M. Statisticheskaya teoriya i raschet morskikh vetrovykh voln. Ch. I (Statistical Theory and Computation of Wind Waves in the Sea. Part I). — Trudy GOIN, No. 33 (45). 1956.

Krylov, Yu. M. Statisticheskaya teoriya vetrovykh voln. Ch. II (Statistical Theory of Wind Waves. Part II). — Trudy GOIN, No. 42. 1958.

Krylov, Yu. M. Spektral'nye metody issledovaniya i rascheta vetrovykh voln
    (Spectral Methods of the Study and Calculation of Wind Waves). —
    Gidrometeoizdat. 1966.
Kuznetsov, D. S. Gidromekhanika (Hydromechanics). — Gidrometeoizdat.
    1951.
Labzovskii, N. A. Opredelenie elementov voln v zavisimosti ot skorosti
    vetra (Determination of Wave Elements as a Function of Wind
    Speed). — In: Sbornik trudov LONITOVT. 1952.
Lappo, D. D. K voprosu opredeleniya glubiny razrusheniya voln na
    melkovod'e (Determination of the Depth of Shallow Waters at which
    Waves Break). — Trudy Dal'nevostochnogo Politekhnicheskogo
    Instituta, No. 45. 1956.
Levich, V. G. O vliyanii turbulentnosti na vozbuzhdenie i zatukhanie
    zhidkosti (The Effect of Turbulence on Excitation and Decay in a
    Fluid). — DAN SSSR, Vol. 101, No. 4. 1955.
Makkaveev, V. M. O protsessakh vozrastaniya i zatukhaniya voln maloi
    dliny i o zavisimosti ikh vysoty ot rasstoyaniya po navetrennomu
    napravleniyu (Growth and Decay of Short Waves, and the Dependence
    of their Height on Distance in the Windward Direction). — Trudy GGI,
    No. 5. 1937.
Makkaveev, V. M. Uchet vetrovogo faktora, sherekhovatosti dna v
    dinamike voln i perenosnykh techenii (Consideration of the Wave
    Factors and of Bottom Roughness in the Dynamics of Waves and
    Transport Currents). — Trudy GGI, No. 28 (82). 1951.
Matushevskii, G. V. Issledovanie polei vetrovykh voln glubokogo morya
    vblizi ostrovov i v prolivakh (Study of Wind-Wave Fields in Deep
    Seas in the Vicinity of Islands and in Straits). — Trudy GOIN, No. 75.
    1964.
Matushevskii, G. V. and S. S. Strekalov. Raschet dvukhmernogo
    energeticheskogo spektra po dannym stereofotos"emki voln (Calcula-
    tion of the Two-Dimensional Energy Spectrum from Stereophotographs
    of Waves). — Okeanologiya, Vol. 3, No. 5. 1963.
Morozov, A. P. Issledovanie izmenchivosti elementov vetrovykh voln
    (Study of the Variability of Wind-Wave Elements). — Trudy GOIN,
    No. 23 (35). 1953.
Petrov, L. S. Nekotorye statisticheskie kharakteristiki izmenchivosti
    vetra k zapadu ot Britanskikh ostrovov (Some Statistical Character-
    istics of the Variability of Winds to the West of the British Isles). —
    Trudy GOIN, No. 91. 1967.
Rozhkov, V. A. O vyborochnoi izmenchivosti spektral'nykh kharakteristik
    morskogo volneniya (The Selective Variability of Spectral Character-
    istics of Sea Waves). — In: Sbornik "Teoreticheskie i prakticheskie
    voprosy morekhodnykh kachestv sudov." Izdatel'stvo "Transport."
    1967.
Rukovodstvo po raschetu morskogo volneniya i vetra nad morem (Instructions
    on Calculating Wave Motion in the Sea and the Wind Blowing over
    Seas). — Gidrometeoizdat. 1960.
Rzheplinskii, G. V. Opyt rascheta elementov voln v okeane (Experience
    in Calculating Elements of Ocean Waves). — Trudy Okeanograficheskoi
    Komissii AN SSSR, No. 11. 1961.

Rzheplinskii, G. V. Metod rascheta rezhimno-klimaticheskikh
    kharakteristik volneniya okeanov i ego obosnovanie (A Method of
    Calculating the Climatic-Regime Characteristics of Wave Motion in
    Oceans and its Substantiation). — Trudy GOIN, No. 84. 1965.
Selyuk, E. M. Metody issledovaniya volnovogo rezhima ozerovidnykh
    vodoemov (Methods of Studying the Wave Regime of Lakelike Bodies
    of Water). — Trudy GGI, No. 22. 1950.
Selyuk, E. M. Issledovaniya, raschety i prognozy volneniya na vodokhrani-
    lishchakh (Studies, Computation and Forecasting of Wave Motion in
    Water-Storage Lakes). — Gidrometeoizdat. 1961.
Shishov, N. A. K voprosu o raschete elementov vetrovykh voln na ograni-
    chennoi glubine (Calculation of Elements of Wind Waves in Limited
    Depths). — Meteorologiya i Gidrologiya, No. 1. 1949.
Shuleikin, V. V. Refraktsiya voln na materikovoi otmeli (Refraction of
    Waves on the Continental Shelf). — Izvestiya AN SSSR, No. 10. 1935.
Shuleikin, V. V. Teoriya morskikh voln (A Theory of Sea Waves). —
    Trudy MGI, Vol. 9. 1956.
Shuleikin, V. V. Edinaya kharakteristika turbulentnoi vyazkosti dlya
    morskikh voln i techenii (A Single Characteristic of Eddy Viscosity
    for Sea Waves and Currents). — DAN SSSR, Vol. 144, No. 4. 1962.
Shuleikin, V. V. Utochnennyi raschet vetrovykh voln dannoi obespechen-
    nosti (Refined Calculation of Wind Waves of a Given Cumulative
    Probability). — Izvestiya AN SSSR, Seriya Geofizicheskaya, No. 1. 1963.
Sidorova, L. G. and G. F. Krasnozhen. Raschet parametrov vetrovykh
    voln pri opredelenii volnovykh nagruzok na gidrotekhnicheskie
    sooruzheniya (Calculation of Wind-Wave Parameters in Determining
    Wave Loads on Hydraulic Structures). — Trudy Okeanograficheskoi
    Komissii AN SSSR, Vol. 9. 1960.
Sirotov, K. M. Zyb' pri vetrovom volnenii. Rezul'taty issledovanii po
    probleme MGG (Swell During Wind-Wave Motion. Results of Studies
    Pertaining to the IGY Program). — Okeanologicheskie Issledovaniya,
    No. 8. 1963.
Solntseva, N. O. and M. M. Zubova. Raschet elementov vetrovykh voln
    dlya severnoi chasti Atlanticheskogo okeana (Calculation of Wind-
    Wave Elements for the North Atlantic). — Trudy Okeanograficheskoi
    Komissii AN SSSR, Vol. 9. 1960.
Sorkina, A. I. Postroenie kart vetrovykh polei dlya morei i okeanov
    (Construction of Wind Field Maps for Seas and Oceans). — Trudy
    GOIN, No. 44. 1958.
Spravochnye dannye po rezhimu vetrov i volneniya na moryakh, omyva-
    yushchikh berega SSSR (Handbook Data on Wind and Wave Regimes
    in Seas Washing the Shores of the USSR). — Izdatel'stvo "Moskoi
    Transport." 1962.
Spravochnye dannye po rezhimu vetrov i volneniya v okeanakh (Handbook
    Data on Wind and Wave Regimes in Oceans). — Izdatel'stvo "Trans-
    port." 1965.
Sretenskii, L. N. Teoriya volnovykh dvizhenii zhidkosti (Theory of
    Wave Motions of Fluids). — ONTI. 1936.
Sretkalov, S. S. K opredeleniyu analiticheskogo vida energeticheskogo
    spektra razvitogo volneniya (Determination of the Analytic Form of
    the Energy Spectrum of a Fully Arisen Sea). — Okeanologiya, Vol. 1,
    No. 3. 1961.

Tekhnicheskie usloviya opredeleniya volnovykh vozdeistvii na morskie i
    rechnye sooruzheniya i berega (Technical Specifications for Deter-
    mining the Wave Action on Marine and Riverine Structures and on
    Shores) (TU) (SN 92-60). 1960.
Titov, L. F. Raschet stepeni volneniya po empiricheskim formulam
    (Calculation of Wave Strength by Empirical Formulas). — Trudy
    NIU GUGMS, Series 5, No. 12. 1946.
Titov, L. F. Instruktsiya po sostavleniyu prognozov volneniya (Instructions
    on Compiling Wave Forecasts). — Leningrad, Gidrometeoizdat. 1951.
Titov, L. F. Novye shkaly stepeni volneniya i sostoyaniya morya (New
    Scales of Wave Strength and State of the Sea). — Meteorologiya i
    Gidrologiya, No. 5. 1952.
Titov, L. F. Vetrovye volny na okeanakh i moryakh (Wind Waves in
    Oceans and Seas). — Leningrad, Gidrometeoizdat. 1955.
Titov, L. F. Sootnosheniya mezhdu elementami vetrovykh voln i vetrom
    (Relationships between Wind-Wave Elements and Wind). — In:
    Sbornik "Issledovaniya severnoi chasti Atlanticheskogo okeana,
    No. 4." Izdatel'stvo LGMI. 1965.
Titov, L. F. and M. M. Zubova. Statisticheskie zakonomernosti
    raspredeleniya volnoobrazuyushchikh faktorov (Statistical Regularities
    in the Distribution of Wave-Generating Factors). — Okeanologiya,
    Vol. 7, No. 3. 1967.
Ukazaniya po volnovym raschetam gidrotekhnicheskikh sooruzhenii (Direc-
    tives on Wave Design of Hydraulic Structures) (UVRGS-67). Moskva.
    1968.
Uspenskii, P. N. O zarozhdenii i razvitii voln na poverkhnosti vody pod
    deistviem vetra (Generation and Development of Waves on the Water
    Surface Due to Winds). — Izvestiya AN SSSR, Seriya Geograficheskaya
    i Geofizicheskaya, No. 3. 1937.
Venttsel', E. S. Teoriya veroyatnosti (Probability Theory). — Fizmatgiz.
    1962.
Vilenskii, Ya. G. and B. Kh. Glukhovskii. Nekotorye zakonomernosti
    vetrovogo volneniya (Some Regularities of Wind-Wave Motion). —
    Trudy GOIN, No. 29 (41). 1955.
Vilenskii, Ya. G. and B. Kh. Glukhovskii. Eksperimental'nye issledo-
    vaniya protsessov morskogo vetrovogo volneniya (Experimental
    Studies of Wind-Wave Motion in Seas). — Trudy GOIN, No. 36. 1957.
Vilenskii, Ya. G. and B. Kh. Glukhovskii. Vetrovoe volnenie v okeane
    (Wind Waves in Oceans). — Trudy GOIN, No. 62. 1961.
Zhukovets, A. M. Opredelenie poter' energii voln zybi, vyzvannykh
    deistviem turbulentnoi i kinematicheskoi vyazkosti (Determination of
    Energy Losses of Swell Due to Eddy and Kinematic Viscosities). —
    Okeanologiya, Vol. 3, No. 2. 1963.
Zubova, M. M. Razvitie vetrovykh voln po materialam nablyudenii GMI
    (Development of Wind Waves from GMI Observational Data). —
    In: Sbornik "Issledovaniya po probleme okean-atmosfera," No. 1.
    1967.
Zubova, M. M. Metod rascheta rezhimnykh kharakteristik vetrovykh voln
    (A Method for Calculating Regime Characteristics of Wind Waves). —
    Trudy koordinatsionnogo soveshchaniya po gidrotekhnike, No. 50.
    VNII im. Vedeneeva. Izdatel'stvo "Energiya." 1969.

Publications in Other Languages

Arthur, R.S. Variability in Direction of Wave Travel. Ocean Surface
        Waves. — Ann. N.Y. Acad. Sci., **51** (3): 511—522. 1949.
Burling, R. The Spectrum of Waves at Short Fetches. — Deutsche
        Hydrogr. Zeitschr. Vol. 12, Nos. 2 and 3. 1959.
Climatological and Oceanographic Atlas for Mariners. North Atlantic
        Ocean. 1959.
Hogben, N. and F. Lumb. Ocean Wave Statistics. London. 1967.
Jeffreys, H. On the Formation of Water Waves by Wind. — Proc. Roy.
        Soc., Series A. 1924.
Longuet-Higgins, M.S. The Statistical Analysis of a Random Moving
        Surface. — Phil. Trans. Roy. Soc. A 249 (966): 321—387. 1957.
Miles, L.W. On the Generation of Surface Waves by Turbulent Shear
        Flows. — J. Fluid Mech. Vol. 7, No. 3. 1960.
Munk, W.H. and M.A. Traylor. Refraction of Ocean Waves; a Process
        Linking Underwater Topography to Beach Erosion. — J. Geol., **55**,
        1—26, Chicago University Press. 1947.
Sverdrup, H.U. and W.H. Munk. Wind, Sea and Swell: Theory of
        Relations for Forecasting. — U.S. Navy Hydrographic Office Pub.
        No. 601. 1947.